THE ENVIRONMENTAL APOCALYPSE

This volume brings together scholars working in diverse traditions of the humanities in order to offer a comprehensive analysis of the environmental catastrophe as the modern-day apocalypse. Drawing on philosophy, theology, history, literature, art history, psychoanalysis, as well as queer and decolonial theories, the authors included in this book expound the meaning of the climate apocalypse, reveal its presence in our everyday experiences, and examine its impact on our intellectual, imaginative, and moral practices.

Importantly, the chapters show that eco-apocalypticism can inform progressively transformative discourses about climate change. In so doing, they demonstrate the fruitfulness of understanding the environmental catastrophe from within an apocalyptic framework, carving a much-needed path between two unsatisfactory approaches to the climate disaster: first, the conservative impulse to preserve the status quo responsible for today's crisis, and second, the reckless acceptance of the destructive effects of climate change.

This book will be an invaluable resource for students and scholars interested in the contributions of both apocalypticism and the humanities to contemporary ecological debates.

Jakub Kowalewski is an Associate Lecturer in the Department of Philosophy, Religions and Liberal Arts at the University of Winchester, UK.

ROUTLEDGE ENVIRONMENTAL HUMANITIES

The *Routledge Environmental Humanities* series is an original and inspiring venture recognising that today's world agricultural and water crises, ocean pollution and resource depletion, global warming from greenhouse gases, urban sprawl, overpopulation, food insecurity and environmental justice are all *crises of culture*.

The reality of understanding and finding adaptive solutions to our present and future environmental challenges has shifted the epicenter of environmental studies away from an exclusively scientific and technological framework to one that depends on the human-focused disciplines and ideas of the humanities and allied social sciences.

We thus welcome book proposals from all humanities and social sciences disciplines for an inclusive and interdisciplinary series. We favour manuscripts aimed at an international readership and written in a lively and accessible style. The readership comprises scholars and students from the humanities and social sciences and thoughtful readers concerned about the human dimensions of environmental change.

For more information about this series, please visit: www.routledge.com/Routledge-Environmental-Humanities/book-series/REH

THE ENVIRONMENTAL APOCALYPSE

Interdisciplinary Reflections on the Climate Crisis

Edited by Jakub Kowalewski

Routledge
Taylor & Francis Group

LONDON AND NEW YORK

from Routledge

Cover image: AerialPerspective Works @ iStock

First published 2023
by Routledge
4 Park Square, Milton Park, Abingdon, Oxon OX14 4RN

and by Routledge
605 Third Avenue, New York, NY 10158

Routledge is an imprint of the Taylor & Francis Group, an informa business

British Library Cataloguing-in-Publication Data
A catalogue record for this book is available from the British Library

Library of Congress Cataloging-in-Publication Data
A catalog record has been requested for this book

ISBN: 978-1-032-03821-6 (hbk)
ISBN: 978-1-032-03806-3 (pbk)
ISBN: 978-1-003-18919-0 (ebk)

DOI: 10.4324/9781003189190

Typeset in Bembo
by Taylor & Francis Books

CONTENTS

CONTRIBUTORS

Lindsay Atnip is a Tutor at St John's College in Santa Fe. She wrote her chapter while a postdoctoral scholar at the Humanities and Social Change Center at UC Santa Barbara.

Agata Bielik-Robson is Professor of Jewish Studies at the University of Nottingham and at the Institute of Philosophy and Sociology at the Polish Academy of Sciences in Warsaw.

Jonathon Catlin is a PhD candidate in the Department of History and the Interdisciplinary Doctoral Program in the Humanities at Princeton University. His dissertation is a history of the concept of catastrophe in twentieth-century thought with a focus on German-Jewish intellectuals including the members of the Frankfurt School of Critical Theory.

Francesca Laura Cavallo is a curator, art historian, and Honorary Research Fellow at the Centre for Indigenous and Settler Colonial Studies at the University of Kent, UK.

Vinita Damodaran is Professor in South Asian History and Director of the Centre for World Environmental History at the University of Sussex, UK.

Omar Rafael Regalado Fernandez is a Guest Researcher in the Palaeontological Collection of the Eberhard Karls Universität Tübingen, Germany.

Sarah France is a PhD candidate in the School of English Literature, Language and Linguistics at Newcastle University, UK. Her current research examines twenty-first-century pre-extinction narratives in relation to anxieties surrounding climate catastrophe, mapping the complexities in considering, narrating, and mourning extinction.

Marita Furehaug is an Associate Lecturer in the Faculty of Theology at the University of Oslo, Norway.

Jakub Kowalewski is an Associate Lecturer in the Department of Philosophy, Religions and Liberal Arts at the University of Winchester, UK.

Andrew Patrizio is Professor of Scottish Visual Culture in the History of Art department at Edinburgh College of Art, the University of Edinburgh, UK.

Elizabeth Pyne is an Assistant Professor of Religious Studies at Mercyhurst University in Erie, Pennsylvania, USA.

Timothy Secret is Head of the Department of Philosophy, Religions and Liberal Arts at the University of Winchester, UK.

Robert G. Seymour is a translator and independent researcher.

Stefan Skrimshire is an Associate Professor in the School of Philosophy, Religion and History of Science at the University of Leeds, UK.

Simon Thornton is a Teaching Associate in the Department of Philosophy at the University of Sheffield. He conducted the research for his chapter while on a scholarship at the Humanities and Social Change Center at UC Santa Barbara.

ACKNOWLEDGMENTS

I am extremely grateful to the contributors – thank you for your insights, advice, time, effort, and patience. Working with you has been exceptional.

I would like to thank Ben Randolph for his helpful comments on the early stages of this volume.

Darshan Cowles, in addition to being an incredible friend, has been an extraordinary reviewer throughout this project – thank you!

I benefited hugely from the support and advice of my colleagues Matt Burch, Luca Di Gregorio, Tommy Lynch, Elizabeth Mackintosh, Neil Messer, Thomas Nørgaard, Erica Peters, Marika Rose, Megan Reeves, Hannah M. Strømmen, Tim Watson, Hannah Whiting, and Adam Willows.

My special thanks to Timothy Secret for his ongoing help throughout the rocky post-doctoral years.

I am lucky enough to have had the best PhD supervisor imaginable – Irene McMullin. Her guidance during and after my doctorate has been absolutely invaluable.

I am thankful to the editors at Routledge – Grace Harrison, Rosie Anderson, and Matthew Shobbrook – for their assistance on all stages of this project.

Finally, I would like to thank my family – especially my parents Beata and Jacek – for their emotional and culinary backing; my friends in the UK and Poland for helping me forget the unbearable weight of being; Josh Boyd for his amazing company before, during, and after lockdowns; and Kimia Gashtili for everything.

INTRODUCTION

Jakub Kowalewski

Apocalypticism has become part and parcel of contemporary rhetoric surrounding the environmental crisis. A quick online search for "environmental apocalypse" (or its cognates "climate apocalypse" and "eco-apocalypse") results in page after page – including a whole Wikipedia article – on the topic. However, the framing of the climate disaster as a modern-day apocalypse can be challenged in two ways. First, because nobody seems to know if there is any determinate meaning to the notion of eco-apocalypse, the term can signify *anything* related to crises, disasters, and catastrophes, the end of history and the end of the world, or revelation, disclosure, and unveiling. In consequence, the idea of climate apocalypse seems unhelpful for rigorous theoretical discourses which operate with meaningful concepts. Second, even if we agree on the meaning of "environmental apocalypse," it is far from clear that eco-apocalypticism can inform a responsible debate about the climate crisis, or that it can generate productive attitudes. As critics point out, alarmist representations of climate change as an apocalypse may result in anxiety, hopelessness, and fatalism, which, in turn, could prevent effective responses to the ecological disaster. Consequently, when dealing with the environmental crisis, wouldn't it be better to abandon the apocalyptic tone altogether?

This volume responds to the above challenges by demonstrating the fruitfulness of understanding the environmental catastrophe from within an apocalyptic framework.

First, this book clarifies the notion of "environmental apocalypse." Drawing on philosophy, theology, history, literature, art history, psychoanalysis, as well as queer and decolonial theories, the authors included in this book expound the meaning of the climate apocalypse, reveal its presence in our everyday experiences, and examine its impact on our intellectual, imaginative, and moral practices.

Second, the authors share the belief that climate apocalypticism produces subjective experiences – importantly, however, they recognise that even though "eco-apocalyptic experience" might have an ambivalent valence, it is nonetheless underlined by complex normative intuitions which can productively inform both ethics and politics. Consequently, as the contributors make clear, understanding the subjective effects of eco-apocalypticism is necessary if we are to grasp ethical and political subjectivity in times of environmental crisis.

Third, the authors remain hopeful in the face of the catastrophe by demonstrating that eco-apocalypticism can motivate progressively transformative discourses about climate change. Here we can encounter two strategies: most contributions stay within an apocalyptic framework, demonstrating the benefits of eco-apocalypticism as a theoretical approach; others use the concept of climate apocalypse as a starting point for reflections which eventually lead beyond apocalypticism. Both perspectives, however, aim to carve a much-needed path between two unsatisfactory approaches to the climate disaster: the conservative impulse to preserve the status quo responsible for today's crisis, and the reckless acceptance of the destructive effects of climate change.

Overall, this volume aims to reorient ecological debates in a way which helps us to tackle recalcitrant problems, while attesting to the pertinence of both apocalypticism and the humanities for the contemporary climate crisis.

★★★

Part 1 explores the multiple ways in which the environmental apocalypse can be framed conceptually.

The opening chapter by Omar Rafael Regalado Fernandez demonstrates, somewhat counter-intuitively, that climate apocalypse could be understood as a scientific category. On Fernandez's reading, climate apocalypse is a concept situated within a long history of catastrophism – a theoretical position which emerges during the Renaissance and the Enlightenment to explain geological changes. Fernandez's chapter then explores the relation between scientific (neo) catastrophism and eco-political debates in the Global North and the Global South, respectively.

The second chapter, which I wrote, responds to a decolonial critique of climate apocalypticism – specifically, the charge that eco-apocalyptic discourses operate with a Western-centric model of time inherited from Christian eschatology. I turn to medieval and contemporary apocalyptic texts in order to show that eco-eschatological time can take different "shapes," and that, consequently, eco-apocalypticism doesn't necessarily commit one to a Western-centric temporality. In fact, as I suggest, climate apocalypticism can be useful for a decolonial project.

Elizabeth Pyne's chapter explores the presence of apocalyptic queer negativity in ecological debates. Her argument is powerfully centred on the figure of the child as it ambivalently functions in modern environmental discourses. Such a focus allows Pyne to identify the forms of negation traversing our representations of the future symbolised by the figure of the child, and to question the theoretical commitments of various eco-social critiques concerned with futurity. Throughout the chapter, Pyne defends the importance of apocalyptic queer negativity for both ecological thought and activism.

In the following chapter, Jonathon Catlin advances the notion of "slow catastrophe," which, as he argues, enables us to avoid both naïve optimism and unnecessary fatalism in the face of the climate disaster. Drawing on the writings of various critical theorists – most notably Walter Benjamin and Theodor Adorno – Catlin demonstrates how a "slow catastrophe" can highlight the imperceptible and prolonged effects of climate change while also motivating solutions to punctual disasters generated by the structural crisis. Catlin's chapter, therefore, makes an important conceptual contribution to critical theory, which, by avoiding the extremes of ungrounded confidence and pessimism, has the potential to inform effective environmental politics.

Marita Furehaug concludes Part 1 with a chapter on environmentalism and Islamic eschatology. Furehaug shows how Qur'anic representations of the apocalypse can help us rethink notions such as free will, human agency, and accountability. In addition to her theoretical analysis, Furehaug also examines the ways in which apocalypticism finds its expression in Muslim ecological activism. Her chapter, therefore, sketches a theology, an ethics, and a practice of Islamic eco-eschatology.

Part 2 focuses on the representations of the environmental apocalypse.

Andrew Patrizio's chapter examines the role apocalypticism has played in visual art since the 1960s. Patrizio offers a topology of apocalyptic art according to four categories of the apocalypse: extinction, mineral, marine, and nuclear. Such a mapping enables him to argue that the visualisation of the climate disaster has an overall disappointing character. Importantly, on Patrizio's reading the disappointment inherent in apocalyptic art can stimulate a productive engagement with our contemporary condition.

In the following chapter, Francesca Laura Cavallo offers a study of the socio-political functions of disaster how-to guides. She traces the history of the survival manuals from the Cold War up to the modern day. Cavallo's historical analysis allows her to explore the role and significance of coping with global catastrophes through action plans and straightforward images and to examine the different types of political strategies which emerge from the guidebooks for how to avoid the apocalypse.

Sarah France's chapter focuses on the works of fiction set before extinction events. These texts contain secular apocalyptic narratives which dispose of any promise of anticipated survival or redemption evocative of religious apocalypticism. For France, the impossibility of narrative closure found in pre-extinction fiction provides a means to critique "techno-utopian" approaches to climate disaster. Simultaneously, France offers a compelling account of how the representations of non-survivable extinction events can produce a type of planetary grief capable of informing productive responses to the eco-apocalypse.

The final chapter of Part 2, by Lindsay Atnip, offers a fascinating reading of Herman Melville's *Moby-Dick*, Cormac McCarthy's *Blood Meridian*, and the latter's unpublished screenplay "Whales and Men." Atnip's exegesis leads her to suggest that the environmental apocalypse is bound up with the transgression of normative and material human limits. Importantly for Atnip, the works of Melville and McCarthy suggest a possibility of a new form of consciousness responsive to the magnitude of our environment and aware of our limitations.

Part 3 examines the ethical consequences of the environmental apocalypse.

In his chapter, Simon Thornton engages with the works of William T. Vollmann and Søren Kierkegaard in order to identify three kinds of climate guilt: ethical guilt, which follows one's responsibility for the environmental crisis; tragic guilt, which involves a paradoxical experience of being and not being guilty; and apocalyptic guilt, which is generated by a possibility that one's life might be erroneous and yet which one might never be able to grasp. Overall, Thornton's chapter offers a captivating account of the normative role affectivity has been playing during the climate disaster.

Stefan Skrimshire asks how our moral responsibilities to present and future generations are affected by the possibility of human extinction. In his answer, Skrimshire successfully brings together two kinds of approaches: on the one hand, he explores the philosophical issues related to the ethics of apocalyptic scenarios; on the other hand, he draws on Jewish and Christian apocalyptic texts in order to investigate the possibility of an alternative morality for the end of the world.

Robert G. Seymour concludes Part 3 by critically reconstructing the environmental ethics of Hans Jonas. As Seymour makes clear, for Jonas technological developments effect concomitant changes in human agency. This, in turn, demands of us a rethinking of our moral systems in light of the new possibilities opened by technology – most importantly, the possibility of a future destruction of life. Seymour's chapter offers an insightful analysis of Jonas' attempt to identify forms of normativity compatible with our modern condition.

Part 4 aims to think beyond the environmental apocalypse.

Timothy Secret's chapter questions the extent to which psychoanalysis can provide a means to effectively engage with the climate crisis. For Secret, the failure of psychoanalysis lies in its commitment to a metaphysical and counterproductive concept of "proper death" found on a larger scale in eco-apocalypticism. In order to propose an alternative model of the end capable of a constructive influence on environmentalism, Secret turns to the works of Xavier Bichat. The latter's vitalism can supplement the psychoanalytic approach by offering a non-metaphysical and thus productive notion of death – neither "proper" nor apocalyptic.

In the following chapter, inspired by the messianic gesture of wiping away the tears of Esau found in the Zohar (Esau symbolising here a superseded yet consoled natural world), Agata Bielik-Robson suggests a novel possibility of theorising humanity's reconciliation with nature. To this end, she interprets Theodor Adorno and Emmanuel Levinas as neo-humanists for whom doing justice to the natural world involves maintaining the anthropological difference between humanity and its "other".

Vinita Damodaran concludes Part 4 – and the volume – by contrasting the environmental history of colonial powers with that of India's tribal communities. Damodaran argues that colonial responses to the environmental crisis – in analogy with contemporary top-down solutions to climate change – have failed to tackle the root cause of the problem, namely, capitalism. The alternative, for Damodaran, lies in the bottom-up environmental strategies found in the history of India's tribal communities, which can inform eco-apocalyptic arguments by pointing towards emancipatory futures.

★★★

An attentive reader may notice that the concept of environmental apocalypse operative in this volume is not fixed. On the contrary, each author employs their own notion of eco-apocalypse, which serves their respective purposes. If the attentive reader was also critically inclined, they may accuse this volume of failing to respond to the above-mentioned challenge. The present book, they may argue, instead of clarifying the term "environmental apocalypse", as it purports to do, contributes to its frustrating polysemy.

As a response to the attentive yet critical reader, I could suggest two possible exegetical tactics. The first one – which we may call an "either/or tactic" – would involve confronting the various meanings of "climate apocalypse" found across the chapters and choosing one. Thus, the reader could decide on the correct (or most useful) meaning of "environmental apocalypse" present in this volume, and discard the remaining ones as incorrect or counter-productive. That way, the reader would end up with a single, fixed category – a *simple* environmental apocalypse – effectively differentiated from its rejected competitors.

However – and here we are presented with the second exegetical tactic – it is also possible that the polysemy of the term "climate apocalypse" provides the only adequate way of grasping the complexity of the eco-apocalyptic situation. Rather than thinking of the unstable meaning of "climate apocalypse" as a failure on the part of our theory, the reader could think of it as reflective of its unstable and polysemic condition. In other words, if the environmental apocalypse is itself *complex*, then eco-apocalypticism must draw on multiple (and possibly incompatible) meanings in order to theorise its object. On this reading, the present volume clarifies the term "environmental apocalypse" because it demonstrates that it is *complex* and *overdetermined*.

The choice between these two types of readings – and thus between the concepts of *simple* and *complex* eco-apocalypses – is itself a question of the two tactics highlighted above: we can decide on one of them or, more creatively, attempt to combine both.

The attentive and critical reader will certainly know which tactic, if any, to choose when engaging with this book.

PART 1

Conceptualising the Environmental Apocalypse

1

ON THE APOCALYPTIC THEME IN MODERN SCIENTIFIC DISCOURSE

Omar Rafael Regalado Fernandez

Introduction: Apocalypse in the Scientific Imagination

The mass extinction at the end of the Cretaceous, 66 million years ago, that led to the extinction of many animals like dinosaurs, pterosaurs, and marine reptiles has become a symbol of what an extinction-level event is. Towards the last years of the Cold War, the only event on Earth powerful enough to cause human extinction was a nuclear fallout. In the 1980s, it was discovered that a meteorite triggered an ecological collapse, and the collision of a meteorite was likened to a nuclear winter, consolidating them as symbols of the end of times.[1]

The environmental degradation of the second half of the 20th century was seen not as a threat to human existence but as a threat to the human condition, and most of the efforts to prevent the collapse of human societies were focused on preventing a nuclear war. However, towards the end of the 20th century, scientists gathered enough evidence that human activities were causing a rapid climate change which could accelerate or worsen any previous environmental degradation. In this review, I outline the use of the term 'apocalypse' as a communication device to convey the urgency to mitigate climate change as a response to the rapid development of the Earth Sciences in the second half of the 20th century.

Since the Enlightenment, apocalypticism has been part of the scientific imagination, as the Bible and Christian literature were the foundation for many natural sciences. Catastrophism was part of understanding and explaining the natural landscape when looking for evidence of biblical catastrophes, such as the Universal Deluge, and its apocalypticist roots led to it being sidelined by the gradualist approach. The concept of extinction played a minor role in understanding evolution as a gradual and cumulative sequence of changes – gradualism – as it was seen as an accident or a consequence of natural selection. After the Second World War, the consequences of the rapid emission of carbon dioxide (CO_2) into the atmosphere gained attention as the

DOI: 10.4324/9781003189190-2

understanding of Earth as a complex system was formulated, and neocatastrophism as a natural philosophy was adopted. Notwithstanding this development, neocatastrophism brought to the fore the apocalypticist elements and discourses so prominent in the catastrophist natural philosophy.

The narrative on climate change is shaped to accommodate the 'reality' of capitalism as the only plausible alternative. The apocalypticist roots of neocatastrophism have made the discourse more popular and compatible with the central political discourse in the Global North, whereas anti-capitalist and decolonial discourses from the Global South are seen as marginal, unrealistic, or utopic.

Before the Climate Crisis: *Doomsday* in Science

Many scientists in Europe were interested in Christian eschatology, studying the 'end of the world', or the Apocalypse. It was frequent that through the Renaissance and the Enlightenment physicists, astronomers, and mathematicians would venture to make a prediction about the date of the 'end of times' based as much as possible on empirical evidence. The first mathematician to use the term 'exponent' and an early version of what we now call 'logarithms', Michael Stifel (1487–1567), calculated that the Apocalypse would come on October 19, 1533 at 8:00 am, and the failed prediction became the origin of the expression "to calculate a Stiefel" (Oettinger 2003). The German physician and astrologer Helisaeus Roeslin (1545–1616) predicted that the world would end in 1654 based on the observation in 1572 of Tycho's Supernova (SN 1572) (Granada 2011). More famously, Scottish mathematician and astronomer John Napier (1550–1617) used the Book of Revelation to calculate the end of times, predicting the world would end between 1688 and 1700 (Clouse 1974). Napier used his calculations to warn King James VI and urge him to ensure "that justice be done against the enemies of God's church", one of them being the Pope, whom Napier considered the Anti-Christ.

The examples above of eschatological scientists refer to the Christian beliefs that they upheld, and since they considered the Bible to be the starting point of their sciences, finding empirical evidence of the end of times made sense. Scientists started to pay attention to cosmic causes for the Apocalypse into the Enlightenment. Isaac Newton (1643–1727) and his student William Whiston (1667–1752) also made predictions for the end of times. Whiston articulated creationism and flood geology by studying the Universal Deluge evidence in the rocks (Whiston 1696). Whiston attributed the origin of the flood to the collision of a comet, and he predicted that in the same way the world would end on October 16, 1736 (Hegel 2014). As a form of natural philosophy, catastrophism was developed to understand the documental evidence of catastrophes in the Bible with the evidence of catastrophes in the natural world (Baker 1998).

Except for these and a few other examples, in the 18th and 19th centuries scientists started to depart from theological explanations for the natural world. Although catastrophism as a form of natural philosophy did not disappear, eschatological studies done by scientists started to fade, and it was astrologers, clergy members, and preachers who mostly made predictions for the end of times. Human extinction, or the extinction of any species, was not part of any scientific area of study.

The study of human extinction is a relatively new concept in the history of science. Before the 19th century, when evolution became the unifying theory of biology, the study of the natural world centred on the description and classification of living beings and their processes within what we refer to as a 'fixist' framework. Fixism has been recently defined as "the doctrine that the most important properties of the members of any species, those properties allowing for their classification as members of this species, cannot vary beyond definite limits" (Barberousse and Samedi 2010). In this framework, species are entities that always remain recognisable in time and space, and if a creature cannot be assigned to a species then that creature has always existed as it is and a new species has been discovered.

Traditionally, scholars consider that the definite abandonment of fixism occurred with the publication of *On the Origin of Species* by Charles Darwin in 1859, and creationists often use this historical benchmark to define "evolution". However, the scientific framework around evolution has drastically changed since it was first proposed; one of the most radical changes in our understanding of evolution is the concept of extinction (Benton 1995). The last drastic change in our understanding of evolution came with the ideas of continental drift and plate tectonics. The discovery of plate tectonics and the confirmation that continental drift occurs is a relatively young scientific advancement that was made around the 1960s, and was complemented by the idea that climate changes are the major drivers of extinction on a global scale (Pearson et al. 2014).

A gradualist approach considered that, although the environment changed, the ecosystems were slowly changing and adapting to the gradual changes they were exposed to. This paradigm is best illustrated in a paragraph that Carl Sagan wrote in 1983:

> The global ecosystem can be considered an intricately woven fabric composed of threads contributed by the millions of separate species that inhabit the planet [...] The system has developed considerable resiliency so that pulling a single thread is unlikely to unravel the entire fabric. Thus, most ordinary assaults on the biosphere are unlikely to have catastrophic consequences. For example, because of small natural changes in stratospheric ozone abundance, organisms have probably experienced, in the fairly recent geologic past, ten per cent fluctuations in the solar near-ultraviolet flux.
>
> *(Sagan 1983: 273)*

In the 19th century and most of the 20th century, the only changes geologists had any evidence of were sea-level changes. The transgression and regression of the seas were conceived as the only force of change in the environment. It used to be thought that organisms colonised islands through land bridges that disappeared as the shoreline moved inland. The second source of change in the environment, slower than the sea-level changes but more constant, was erosion. Gradualist evolution started to encounter problems as a theoretical framework when explaining the distribution of organisms, for in doing so it recurred to a dispersalist philosophy rooted in the biblical accounts of the Garden of Eden and the Universal Deluge (Morrone and Crisci 1995).

The difference between the catastrophism outlined by French naturalist Georges Cuvier and the gradualism proposed by Jean-Baptiste Lamarck was the nature of extinction. For Cuvier, extinction was a definitive state that forced life to restart again, whereas for Lamarck extinction was just apparent, as the organisms left behind were a transitional form to the more complex and perfect form we see today. Some species seemed unchanged in the fossil record, and the present time was considered evidence of the few survivors of the catastrophe that has helped to repopulate the Earth. Likewise, the species seen unchanged in the fossil record were considered to have achieved a state of equilibrium with the changing environment and constant creation through spontaneous generation.

This gradualism developed by the French naturalists was consolidated into Darwin's more concrete theory of evolution. Unlike Etienne Geoffroy and Lamarck, Darwin proposed a mechanism by which the organisms gradually change from one species to another: the better-adapted variants reproduce faster than the not-so-fitting ones, eventually replacing them in the habitat. Rapid changes in the environment, like earthquakes, deluges, or volcanic eruptions, are so quick that they may not allow species to reproduce, and extinction ensues. Sometimes, none of the variants in a species can respond to environmental change, thus becoming extinct. In other instances, what is preserved as a fossil is one of the transitional stages in this response to change. Natural selection is somewhat a misnomer, as Darwin tried to explain that evolution was a continuous physical process in line with the gradualist ideas developed first by Lamarck. *Archaeopteryx*, a fossil from Germany that had a mixture of bird and reptile features that was discovered a few years after he published his book *On the Origin of Species*, was seen as irrefutable evidence of gradualism through transitional forms.

Before the formulation of plate tectonics as a theoretical framework, it was believed that static land masses connected environments in the past that are currently separated by oceans. Catastrophism was incorporated into the gradualist framework to account for the sudden disappearance of these land masses, and extinction could be a consequence of the drastic changes in the environments that used to be connected. Land bridges and sunken land masses started to become the explanation for the distribution of species under the gradualist approach (e.g. Schuchert 1932). Extinctions were always explained by gradual changes that led to some species surviving and others thriving. The extinction of dinosaurs was intriguing, as it was clear that dinosaurs dominated the Earth for a significant amount of time until they suddenly disappeared from the fossil record.

In 1921, the American palaeontologist William Diller Matthew expanded on climate-driven extinctions, and thought that dinosaurs became extinct due to a geological uplifting that changed the humid lowlands to elevated terrain climates to which mammals were better suited. More explanations started to arise in the early

1920s that hinted at a climatic change as the cause of the extinction of dinosaurs. Catastrophist explanations became the most common ways to account for the extinction of dinosaurs, as the idea of a gradual change failed to account for all the caveats. The climate changes were also evident in the geological record, as the understanding of the relationship between rocks and the climates they were formed in started to consolidate. It became more evident that the Cretaceous period, in which dinosaurs lived, was warmer than the Palaeogene period, during which mammal evolution had started. Furthermore, whereas dinosaurs were the dominant elements of the terrestrial environments and mammals were restricted to smaller niches in the Cretaceous, the latter became the dominant elements in the Palaeogene. The quick turnover of fauna was better explained by climate change, which led to a rapid change in environments (e.g. McLean 1978).

These new catastrophists differ from Cuvier and the 19th-century catastrophists. They accept the gradualist model of evolution but acknowledge that catastrophes could trigger a collapse of the world's ecosystems, a mass extinction at points coinciding with certain worldwide events (Benton 1990). While gradualists struggled to explain the distribution of animals over the land using dispersalist hypotheses and land bridges, some geologists suggested that the similarities in fauna between regions separated by seas could be better explained if the seas separated continents that used to be together.

Formulating the New *Doomsday*: Fossil Fuel-Driven Climate Catastrophe

In 1861, English chemist John Tyndall established that water, carbon dioxide, and methane emit infrared radiation, and that this fact could explain the "mutations of climate which the researches of geologist reveal" (Tyndall 1861). In 1896, Swedish chemist Svante Arrhenius noted that towards the end of an ice age the CO_2 and the water vapour in the atmosphere would increase, whereas a halving of the concentration of CO_2 in the atmosphere would lead to the decrease in temperature necessary for an ice age to start, and he thereafter published the first model for climate change. In 1938, English engineer Guy Stewart Callendar published the first evidence that the climate was warming and that CO_2 levels had increased. The greenhouse effect was considered one of the few rapid changes that the Earth could experience, but these rapid changes were used to explain global changes in fixed continents such as ice ages.

The formulation of Plate Tectonics Theory produced a scientific revolution that forced a rapid change in the conceptualisation of the Earth as a complex system, and pushed scientists to analyse large amounts of data collected from all over the world (McKenna 1972). Indeed, the Plate Tectonics Revolution can be considered the Data Science Revolution, which was based on the use of large data sets and modelling to make predictions. Before the 1960s, the paradigm was a slowly changing Earth where extinction resulted from several adverse circumstances like an accident. Through the 1930s, 1940s, and 1950s, there was a boom in apocalyptic and post-apocalyptic fiction and non-fiction in Western culture (Booker and Thomas 2009). The threat of nuclear warfare and economic collapse inspired writers to imagine the possibilities in which the Apocalypse may be brought upon without divine intervention, and this topic was recurrent in the 1950s science-fiction genre. Furthermore, the fears that characterised the Cold War led to the creation of the survivalist movement in the 1960s, and mainly conservative and libertarian thinkers encouraged citizens to be prepared. The Plate Tectonics Revolution occurred during this time as well, and the acknowledgement that a global ecological collapse could be a real possibility was combined with the survivalist movement of the Cold War.

On the one hand, plate tectonics allowed scientists to understand that the position of continents was not fixed, which through geological time led to the different climate changes that preceded turnovers of the life preserved in the fossil record. On the other hand, it required scientists to start collecting data over extended periods worldwide to produce computer-based models to make predictions and deductions about the past. It was becoming clear that anthropogenic global warming would accelerate a process that in the fossil record spanned thousands of years, and that such changes were behind the mass extinction events.

After the Second World War, many scientists pondered the possibility of human extinction, as nuclear weapons were considered a threat (Teller 1947: 36): "It is not even impossible to imagine that the effects of an atomic war fought with greatly perfected weapons and pushed by the utmost determination will endanger the survival of man". At the dawn of the Cold War, in 1947, several former members of the Manhattan Project formalised the possibility of human extinction with the Doomsday Clock analogy in the *Bulletin of Atomic Scientists* (Rabinovitch 1947). It was initially set to seven minutes to midnight, where midnight represented the end of times by a nuclear war. The idea of the clock was to illustrate the threat to human existence on Earth with the development of the nuclear age and the Cold War (Rabinovitch 1947). Therefore, it makes sense that the furthest we have been from midnight (17 minutes) on the Doomsday Clock was 1991, when the United States and the Soviet Union signed the first Strategic Arms Reduction Treaty (START I) and the Soviet Union subsequently dissolved (The Clock Moves 1991). Nevertheless, the instability in the former Soviet republics, the Soviet scientist 'brain drain', and the continuous military spending on nuclear weapons development steadily decreased the minutes left before midnight (Moore 1996).

The scientists employed by the oil company Esso, now ExxonMobil, were working on one of the oil company's most ambitious projects: understanding how much CO_2 the oceans were sequestering. Between 1979 and the mid-1980s, several letters and memos shared between Esso executives, managers, and scientists contained predictions about climate change that would not become widely popular until the 2010s. An investigation from InsideClimate News in 2015 revealed that Exxon was aware of climate change as early as 1977 (Banerjee et al. 2015).

Studying the carbon cycle made it possible to determine how carbon moved between the air, the oceans, and the continents, so it could be predicted how much CO_2 was accumulating in the atmosphere. The accumulation of carbon in the atmosphere would reduce the polar caps, which reflect part of the sun's rays through the Albedo Effect. Shrinking the polar ice caps would accelerate global warming and the rise in sea level. This would cause the humid regions to become more humid and the dry regions to become desert. The drastic change in the ecosystems would produce changes in the agricultural conditions of the planet, epidemics, and an increase in pests, affecting the most vulnerable sectors of the planet's population. Human migrations would increase massively as living conditions would be severely affected. Based on computer models of the time, by 2010 there would be 400 parts per million (ppm) of CO_2 in the atmosphere, 280 ppm above the amount of CO_2 before the Industrial Revolution. By 2019, the level of 415 ppm would be reached, the highest concentration of atmospheric carbon in the last 400,000 years. The problem would only be contained if 80 per cent of the oil reserves were left underground. Were the trend in the accumulation of atmospheric CO_2 not to decrease substantially by 2030, the changes would be irreversible and catastrophic.

The above scenarios were all studied by Exxon scientists 40 years before climate change became popularised by the global media. Predictions back then have been confirmed almost verbatim in memos exchanged in Exxon offices. Then, in 1982, the Exxon leadership began to change its narrative and question the results of its researchers. They started a campaign to encourage scepticism about global warming and climate science. Exxon went from using its resources to be on the frontier of climate research to financing a denial campaign that would prevent the passage of laws to discourage the use of fossil fuels.

The end of the 20th century coincided with the articulation of palaeoclimatology and climate change science as coherent fields of study. The scientific consensus that climate change posed a threat to human existence started to coalesce during the 1970s after the environmental concerns in the 1960s grew into proper studies of environmental decay. In the 1980s, many scientists considered climate change a global risk. In 1988, the growing concerns translated into creating the Intergovernmental Panel on Climate Change (IPCC) inside the United Nations. The IPCC aimed to gather interdisciplinary teams to assess anthropogenic climate change's scientific basis and its impacts on political, economic, and social terms. In 1992, the accumulation of evidence worldwide established the United Nations Framework Convention on Climate Change (UNFCCC) in Rio de Janeiro. The IPCC's warnings and the urgent call for action resonated with the *Bulletin of Atomic Scientists*'s assessment of the dangers posed to civilisation by climate change, and included climate change as a threat alongside nuclear annihilation in 2007. As of 2007, the lack of political action on preventing climate change has been the principal catalyst in adding minutes to the Doomsday Clock, which was set to 100 seconds to midnight in 2020.

The Apocalyptic Global North

While anthropogenic climate change was considered a prediction in 1979, it became an observed fact in the early 21st century, and the consensus amongst climate researchers became established (Walther et al. 2005). However, outside of the field of climate research, several scientists have sought to add doubt about the certainty around the climate change science that their peers consider now a fact. For example, American biochemist Arthur Brouhard Robinson funded the Oregon Institute of Science and Medicine (OISM) in 1980 and has made part of its mission to study the impact of the environment on health and welfare and disaster preparedness. The OISM was behind the Oregon Petition of 1997 that urged the United States to reject the Kyoto Protocol on global warming, considering that reducing greenhouse gases could harm the environment and reduce the advancement of science. Furthermore, the OISM engaged in a strategy to dilute the pool of experts when it circulated a petition in 2008 addressed to people with scientific training to assess the climate change consensus. The petition read that there was "no convincing scientific evidence that human release of carbon dioxide, methane, or other greenhouse gases is causing [...] heating of the Earth's atmosphere and disruption of the Earth's climate".[2]

Moreover, it claimed that "there is substantial scientific evidence that increases in atmospheric carbon dioxide produce many beneficial effects upon the Earth's natural plant and animal environments." In this exercise, it gathered the signatures of 9,029 scientists with a PhD degree. However, it addressed the petition to anyone with scientific training to weaken the idea of the consensus that climate researchers had achieved. Arthur Robinson was elected senator of Oregon's District 2 in 2020 and ran a campaign based on the statement that climate change was a hoax.

Thus, this group of denialist scientists is not necessarily part of the scientific consensus, but its members stand amongst the actors who are preventing climate-change-related policies from being put forward. The scientists who have recognised that climate change poses a threat to human existence, either definite or in the form of sustained hardship, have urged that policies need to be deployed to redesign the current technology to avoid its most deleterious effects. Non-denialist scientists have described climate change in the same way theologians have described the Apocalypse. This discourse is always along the lines that climate change threatens our well-being and prosperity and that if we do not act soon to prevent further climate change, then humanity will suffer. The idea that climate change is not real has dwindled over the past decades, but the 'apocalyptic' nature of climate change is the element that has been put into doubt. Furthermore, other greenhouse gas sources such as volcanoes have been studied as possible culprits, suggesting that non-human activities may have been significant and sustained enough to generate a change. However, since the Industrial Revolution, rises in CO_2 have been coincident with massive increases in fossil-fuel burning.

On the other hand, the fixity of species and the fixity of continents were both scientific consensuses, and many scientists have exploited this fact to suggest that climate change may also be one such temporary consensus, and that it can be overturned as new empirical evidence is put forward. Politicians have known about anthropogenic climate change for as long as scientists have. Through the 1970s, scientific committees started to alert the US government of the threat posed by climate change. By 1980, scientists and politicians knew that greenhouse gas emissions could cause climate change and that the latter would be coupled with the generalised change in land use and deforestation. In 1988, atmospheric physicist James Hansen testified before Congress that the change that many scientists in the committees were speculating to start by the year 2000 was already detectable. Environmental concerns were a priority for many Republicans in the United States. George H.W. Bush called on world leaders to translate the written document of the UN Framework Convention on Climate Change (UNFCC) of 1992 into concrete action to protect the planet. Climate change was even part of George H.W. Bush's presidential campaign.

In late 1988, the Washington, DC-based George C. Marshall Institute, a political think-tank, became one of the most prominent voices challenging the scientific evidence on anthropogenic climate change. The George C. Marshall Institute has not only challenged the scientific evidence on global warming but has also proposed alternatives to the outcomes of what they frame as a hypothetical scenario. Interestingly, the think-tank's claims can be traced to three physicists, Robert Jastrow, head of Goddard Institute for Space Studies, solid-state physicist Frederick Seitz, who worked on the hydrogen bomb, and William Nierenberg, nuclear physicist and director of the Scripps Institution of Oceanography. The three physicists founded the George Marshall Institute in 1984 to counter the scientific boycott against the Strategic Defence Initiative (SDI or "Star Wars") encouraged by the Reagan administration.

The George C. Marshall Institute was the leading actor pushing environmental scepticism and climate change denial, and although part of it closed in 2015, it became part of the CO_2-Coalition. The George C. Marshall Institute amplified minor disagreements between scientists and emphasised oddities and exceptions to the general principles of the consensus, and then it presented them as cutting-edge science. For many, climate change is real, but it does not pose the threat that scientists claim it does, as humans will rapidly adapt to these changes. Another branch, environmental sceptics, questions the science around climate change and the evidence that supports its anthropogenic origin. Climate change denialists consider that the whole concept is a hoax and that the scientific consensus has a socialist agenda behind it.

The most common individualist speeches are around individual action, although some of the changes proposed necessarily require policy-making to become sustainable. Going car-free, for instance, requires better public transport infrastructure and cyclist-centred urban design, which individuals on their own cannot produce. For instance, most public transport in the United States is built around a suburb-to-downtown commute, making otherwise short suburb-to-suburb commutes impossible through public transport. Metro lines in the largest cities in the US converge in the centre, unlike urban transport designs in European and Latin American cities with several intersection points in a more grid-like pattern. For the individual action of going car-free in the United States, it is necessary to bring local and federal resources to improve the infrastructure. Electric kick scooters such as those made by Bird and Lime were implemented to allow last-mile transportation for commuters using public transport services.

In January 2022, the *Bulletin of the Atomic Scientists* updated their Doomsday Clock with the headline "At Doom's Doorstep: 100 Seconds to Midnight" (Yue 2022). The press release warns that the end of civilisation is nigh due to nuclear risks, the response (or lack thereof) to climate change, the emergence of new diseases, and the damage that disinformation is causing to society with regard to new disruptive technology and public health policies (Yue 2022).

The Peripheral Global South

The Global North has generated environmentalism linked to large cities, post-materialist values, and large non-governmental organisations such as Friends of the Earth and the World Wildlife Fund. On the other hand, the Global South's environmentalism is based on poor and rural communities and the local organisations that large companies often threaten. Thus, scientists may not want to stop exploiting resources but redistribute such resources' exploitation. Moreover, the scientific academies in countries of the Global South essentially replicate the discourses and practices in the Global North. Here, however, we will refer only to those approaches that are somewhat on the periphery inside these 'more northern' scientific academies.

The Mexican scientific institutions offer a wide range of approaches that could be classified as reformism, namely, the assumption that climate change is a threat to human existence that cannot be averted unless deep institutional reforms occur. In this reformist framework, the current systems are the ones that have triggered climate change, namely industrialisation, colonisation, and capitalism, and our efforts should be put into adapting to the consequences, making societies more resilient, and dismantling the systems that originated the problem.

The scientists in this group believe that the capitalist system has generated the conditions for climate change, and that the mitigation measures proposed by institutions, such as the United Nations, the World Bank, amongst many others, are focused on eradicating the environmental degradation caused by poverty while neglecting the effects caused by the consumerism and the world's large-scale industries. In this regard, reformists suggest that the causes of climate change are better understood as systemic, suggesting that colonialism, imperialism, agro-industrial modernisation, inadequate technology, and political and economic structures that favour unequal distribution of wealth are the culprits of climate change. Therefore, climate change can only be avoided if the whole system is dismantled. This reformist discourse is what denialists base their fearmongering on when they claim that climate change is an instrument being used to implement socialism. At any rate, reformism is the current framework within which many scientists in Mexico are responding to climate change (Toledo and Barrera-Bassols 2017).

It is important to emphasise that reformists do not posit that climate change is not a threat; instead, they argue that this threat is an unavoidable consequence of capitalism. This statement is at the core of the reformist thesis: the strategies used to mitigate climate change are irrelevant if they do not address the capitalism as their origin. Although the threat may be global, the rural and poor communities should not bear the responsibility to prevent an 'end' they did not cause. In this regard, the idea of sustainable development is a discourse that puts the responsibility on actors that have none, depriving these actors of the development that the Global North has achieved based on the exploitation of the Global South.

The national policies that aim to bring industrialisation to their regions are an element that speeds up climate change as steadily as greenhouse gas emissions. However, agro-ecologists argue that the crisis many scientists fear will be caused by climate change has already started due to modernising agricultural techniques, the free market, and the loss of genetic resources due to a reliance on few varieties of outcrops and the displacement of traditional farmers. Therefore, traditional agriculture is one of the main elements needed to avert the threat posed by climate change. This element of the agro-ecologists' position includes some of the principles that the IPCC recognised in 2014 as being needed to promote adaptation to climate change in Africa, one of which is the "integration of scientific, local, and indigenous knowledge in developing adaptation strategies".

The goal of adaptation, for instance, is that, through international strategic and coordinated efforts, it is possible to slow down the climate crisis so as to give time for human societies to adapt to the change. On the other hand, mitigation seeks to reduce the causes of climate change in an attempt to stop it altogether.

Climate change adaptation also considers the climate crisis unavoidable but emphasises the possibility of delaying it and building more resilient communities. Adaptation requires understanding local needs and preparing for the most detrimental effects that climate change will have on the communities. The central tenet in adaptation is that the technology needed to face the consequences of climate change exists.

In the Global South, climate change adaptation has become the most common response to the crisis as the effects of global warming have started to be felt in those regions and as these regions are generally the ones with a more uneven distribution of the adaptive capacity. World Bank (2010) estimated that adaptation to climate change would cost billions of dollars in annual expenditure, with most of the money invested in sub-Saharan Africa, in the decades to come. The adaptation framework results from the pessimism that comes from the frequent failure of international policies to stop or reduce climate change and a loss of trust in the institutions that put them into place. Furthermore, even if the policies seeking to stop greenhouse gas emissions were succeed, climate change has already been set into motion, and its consequences will be felt in the coming decades regardless. However, this discourse of reformism considers that it is possible to adapt to climate change with current technologies and that current economic and political systems do not need a profound ideological change. Nevertheless, as human and monetary resources are finite, priority should be put into projects that aim to adapt to climate change in the hope that this adaptation will have the concomitant effect of delaying or avoiding worst-case scenarios over projects that aim to stop greenhouse gas emissions. However, climate change adaptation requires a certain level of dependence on regional and local governance.

Nepal, one of the most vulnerable countries to climate change, has created the National Adaptation Programme of Action (NAPA) as part of its national response, identifying nine projects requiring urgent and immediate national adaptation priority. NAPA and some recent successes are considered bottom-up approaches to tackling climate change, with up to 80 per cent of the financial resources deployed on the ground.

Scientists all over Africa are also shifting their attention to developing community-based strategies to improve people's livelihoods in rural areas already affected by climate change. For example, in Ghana indigenous agricultural adaptations involve managing the soil and the crops, adjusting quantities of feed-to-feed livestock, storing enough feed during humid seasons to be given to animals during the lean season, and keeping local breeds that are already adapted to the climate of northern Ghana. This response to climate change opposes institutionalism, which is about using current institutions to implement mitigation measures.

Understanding the different discourses as expressions of apocalypticism can help us understand the different approaches to climate change taken in the Global South and Global North. The Apocalypse has already started for the Global South, or at the very least is the closest it has ever been, and this has led to people making more radical suggestions involving total reform. In the Global North, where the Apocalypse is still relatively faraway, scientists' responses are based more on survivalism, and emphasise the power of individuals and authorities. Nevertheless, as the effects of climate change start to be felt worldwide, the Global South and Global North may start to coincide in their responses to the reformist framework. The current trend to completely depart from a carbon-based economy and to emphasise the importance of institutional reforms, such as those promoted by Extinction Rebellion, shows that the Global North's understanding of apocalypticism is converging with that of the Global South, which is reformist. Furthermore, the more radical the apocalyptic scientists become, it is to be expected that the denialists will also radicalise, since the reformists' contend that the main culprit of our climate emergency is capitalism.

Conclusion

Neocatastrophism posits that catastrophes are an accumulation of incremental changes, which will be sustained and which will hinder the ability of the communities to adapt to these changes the more they are left unaddressed. The individual changes that the environmental activism of the late 20th century suggested as the solution to environmental decay, and later pandered to by the large oil companies during the 2000s and 2010s, have not deviated us from the path that the scientists from ExxonMobil laid out in the 1980s. Acknowledging the different ways in which communities understand and address the concept of an end of times is crucial to creating the bridges needed to reach international cooperation. The press release of the *Bulletin of the Atomic Scientists* that updated the Doomsday Clock to 100 seconds to midnight in 2022 highlights the lack of results delivered by the 2021 United Nations Climate Change Conference (COP26) (Yue 2022).[3] On November 13, 2021, Mohamed Adow, Director of Energy at the think-tank

2

THE SHAPES OF APOCALYPTIC TIME

Decolonising Eco-Eschatology

Jakub Kowalewski

In a recent article, Delf Rothe identifies three main strands of contemporary environmentalism: eco-catastrophism, which calls for emergency management on a global scale; eco-modernism, which imagines experiments aimed to mitigate climate change; and planetary realism, which emphasises resilience in the face of a disaster which is already taking place. Importantly for Rothe, these three types of ecological discourses are "deeply influenced by a linear temporality and a common orientation towards the threat of the end of time" derived from Christian eschatology (Rothe 2020: 145). The various genres of environmentalism share one presupposition – a belief in time as a single, unidirectional line tending towards an end, which mirrors the Christian view of history as "a flow or movement from a starting point (the creation) towards a final event in the divine plan (the eschaton)" (Rothe 2020: 156). This leads him to conclude that "the discourse on the Anthropocene is essentially eschatological" (Rothe 2020: 147).

However, the belief that historical time is a single line leading to an apocalyptic event generates two serious interrelated problems (P1 and P2) for any environmental eschatology, which, together, constitute what I call a 'decolonial critique' of eco-eschatology:

> P1: The linear view of time centred on a present climate crisis or a future ecological catastrophe "disregards that many people in the majority world have already lived through the ecological catastrophe brought about by European colonialism and its repercussions" (Rothe 2020: 146). For example, as Déborah Danowski and Eduardo Viveiros de Castro point out, "for the native people of the Americas, *the end of the world already happened* – five centuries ago" (Danowski and Viveiros de Castro 2017: 104). However, the exclusive focus on the environmental disaster as a future or present 'end' of linear history blinds us to 'ends of history' experienced by colonised communities in the past.

DOI: 10.4324/9781003189190-3

Power Shift Africa (2022) About us: Power Shift Africa. https://powershiftafrica.org/about-us/ [Accessed January 25, 2022].

Rabinovitch, E. (1947) If the UN Atomic Energy Commission fails. *Bulletin of the Atomic Scientists*, 7(3): 169–170. https://doi.org/10.1080/00963402.1947.11459076.

Raup, D.M., Sepkoski, J.J. (1982) Mass extinctions in the marine fossil record. *Science*, 215 (4539): 1501–1503.

Sagan, C. (1983) Nuclear war and climatic catastrophe: some policy implications. *Foreign Affairs*, 62 (2): 257–292.

Schuchert, C. (1932) Gondwana land bridges. *GSA Bulletin*, 43(4): 875–916. https://doi.org/10.1130/GSAB-43-875.

Smith, P., Beaumont, L., Bernacchi, C.J., Byrne, M., Cheung, W., Conant, R.T., Cotrufo, F., Feng, X., Janssens, I., Jones, H., Kirschbaum, M.U.F., Kobayashi, K., LaRoche, J., Luo, Y., McKechnie, A., Penuelas, J., Piao, S., Robinson, S., Sage, R.F., Sugget, D.J., Thackeray, S.J., Way, D., Long, S.P. (2021) Essential outcomes for COP26. *Global Change Biology*, 28(1): 1–3. https://doi.org/10.1111/gcb.15926.

Teller, E. (1947) How dangerous are atomic weapons? *Bulletin of the Atomic Scientists*, 3(2): 35–36. https://doi.org/10.1080/00963402.1947.11455839.

Toledo, V., Barrera-Bassols, N. (2017) Political agroecology in Mexico: a path toward sustainability. *Sustainability* 9(2): 268. https://doi.org/10.3390/su9020268.

Turco, R.P., Toon, O.B., Ackerman, T.P., Pollack, J.B., Sagan, C. (1983) Nuclear winter: global consequences of multiple nuclear explosions. *Science*, 222 (4630): 1283–1292.

Tydnall, J. (1861) On the absorption and radiation of heat by gases and vapours and on the physical connection of radiation, absorption, and conduction. *Philosophical Magazine*, 22(146): 169–194;273–285.

Walther, G.-R., Hughes, L., Vitousek, P., Stenseth, N.C. (2005) Consensus on climate change. *TRENDS in Ecology and Evolution*, 20(12): 648–649. https://doi.org/10.1016/J.TREE.2005.10.008.

Whiston, W. (1696). *A New Theory of the Earth, from its Original, to the Consummation of all Things, where the Creation of the World in Six Days, the Universal Deluge, and the General Conflagration, as laid down in the Holy Scriptures, are Shewn to be perfectly agreeable to Reason and Philosophy*. Benjamin Tooke.

World Bank (2010) Economics of adaptation to climate change: Synthesis report. https://documents1.worldbank.org/curated/en/646291468171244256/pdf/702670ESW0P10800EACC-SynthesisReport.pdf.

Yue, L. (2022) At doom's doorstep: it is 100 seconds to midnight. *Bulletin of the Atomic Scientists*, January 20. https://thebulletin.org/doomsday-clock/current-time/.

Arrhenius, S. (1896) On the influence of carbonic acid in the air upon the temperature of the ground. *The London, Edinburgh and Dublin Philosophical Magazine and Journal of Science*, 41: 237–276.

Baker, V.R. (1998) Catastrophism and uniformitarianism: logical roots and current relevance in geology. In: Blundell, D.J., Scott, A.C. (Eds.), *Lyell: The Past Is the Key to the Present*. Geological Society, 171–182.

Banerjee, N., Song, L., Hasemyer, D. (2015) Exxon's own research confirmed fossil fuels' role in global warming decades ago. *InsideClimate News*, September 16. https://inside climatenews.org/news/16092015/exxons-own-research-confirmed-fossil-fuels-role-in-global-warming/.

Barberousse, A., Samedi, S (2010) Species from Darwin onward. *Integrative Zoology*, 5(3): 187–197. https://doi.org/10.1111/j.1749-4877.2010.00204.x.

Benton, M.J. (1990) Scientific methodologies in collision: the history of the study of the extinction of the dinosaurs. *Evolutionary Biology*, 24: 371–400. https://cpb-eu-w2.wpmucdn. com/blogs.bristol.ac.uk/dist/5/537/files/2019/08/1990Collision.pdf.

Benton, M.J. (1995) Diversification and extinction in the history of life. *Science*, 268(5207): 52–58. https://www.jstor.org/stable/2886491.

Booker, M.K., Thomas, A.M. (2009) *The Science Fiction Handbook*. Wiley-Blackwell.

Callendar, G.S. (1938) The artificial production of carbon dioxide and its influence on temperature. *Quarterly Journal of the Meteorological Society*, 64 (275): 223–240. https://doi. org/10.1002/qj.49706427503.

Clouse, R.G. (1974) John Napier and apocalyptic thought. *The Sixteenth Century Journal*, 5 (1): 101–114. https://doi.org/10.2307/2539589.

Granada, M. (2011) After the nova of 1604: Roeslin and Kepler's discussion on the significance of the celestial novelties (1607–1613). *Journal of the History of Astronomy*, 42(3): 353–390. https://doi.org/10.1177/002182861104200305.

Haenssgen, M.J., Lechner, A.M., Rakotonarivo, S., Leepreecha, P., Sakboon, M., Chu. T.-W., Auclair, E., Vlaev, I. (2022) Implementation of the COP26 declaration to halt forest loss must safeguard and include Indigenous people. *Nature Ecology and Evolution*, 6: 235–236. https:// doi.org/10.1038/s41559-021-01650-6.

Hegel, G.W.F. (2014) *Hegel's Philosophy of Nature: Volume II*. Routledge.

Matthew, W.D. (1921) Fossil vertebrates and the Cretaceous-Tertiary problem. *American Journal of Science*, 5(2): 209–227. https://doi.org/10.2475/ajs.s5-2.10.209.

McKenna, M.C. (1972) Possible biological consequences of plate tectonics. *BioScience*, 22(9): 519–525. https://doi.org/10.2307/1296311.

McLean, D.J. (1978) A terminal Mesozoic 'greenhouse': lessons from the past. *Science*, 201 (4354): 401–406. https://doi.org/10.1126/science.201.4354.401.

Moore, M. (1996) On the scale. *The Bulletin of the Atomic Scientists*, 52(1): 2. https://doi.org/ 10.1080/00963402.1996.11456578.

Morrone, J.J., Crisci, J.V. (1995) Historical biogeography: introduction to methods. *Annual Review of Ecology and Systematics*, 26: 373–401. https://doi.org/10.1146/annurev.es.26. 110195.002105.

"A new era" (1991) [Editorial]. *The Bulletin of the Atomic Scientists*, 47(10): 3.

Oettinger, R. (2003) Thomas Murner, Michael Stifel, and songs as polemic in the early reformation, *Journal of Musicological Research*, 22 (1–2): 45–100. https://doi.org/10.1080/ 01411890305919.

Pearson, R.G., Santon, J.C., Shoemaker, K.T., Aiello-Lammens, M.E., Ersts, P J., Horning, N., Fordham, D.A., Raxworthy, C.J., Ryu, H.Y, McNees, J., Akçakaya, H.R. (2014) Life history and spatial traits predict extinction risk due to climate change. *Nature Climate Change*, 4(3): 217–221. https://doi.org/10.1038/nclimate2113.

Power Shift Africa, founded in 2018 and whose mission is "to drive public debate on climate and energy and amplify African climate voices" (Power Shift Africa 2022), tweeted that "For the first time we have a COP decision calling for efforts towards the phase-down of coal and fossil fuel subsidies", suggesting that during the previous 26 meetings the role of the fossil fuel industries was not mentioned. Adow's tweet summarised why many scientists have considered COP26 and its predecessors to be failed attempts at international cooperation (e.g. Arora and Mishra 2021; Haenssgen et al. 2022; Smith et al. 2021).

The discourse proposed by Global North institutions is in contraposition with many peripheral discourses in the Global South, where the climate crisis is part of reality and not a distant threat. As more societies in the Global North are exposed to this fact, we can expect that this apocalyptic discourse should drift away from capitalist realism and target the oil industry's lobbying. It is essential that scientists, science communicators, and journalists start to imagine the alternative to an end of times to prevent nihilist or denialist attitudes from being used as part of this lobbying. The current climate crisis needs to be understood as a sustained chain of changes that will generate suffering in societies around the world in a transgenerational manner. For many societies in the Global South, the Apocalypse has already started, and they are suffering the consequences of the sustained failure of world governments to rein in the oil industry. We need to address the climate crisis as a long-standing catastrophe that will wear down the resilience of human society rather than as a specific 'event' for which we can repent later on or at avoid. The potential suffering and worsening of living conditions of millions of people should not have to be taken seriously only after being looked at through the lens of the Apocalypse.

Notes

1 The collision of a meteorite to explain the extinction at the end of the Mesozoic was known as the "Alvarez Hypothesis". The English writer Arthur Clarke used the Alvarez Hypothesis as the inspiration for a piece in *Time* magazine (1992) that would then become *The Hammer of God* (1993), where an asteroid is set to collide with the Earth. In the acknowledgements, Clarke explains that the "Alvarez Hypothesis" was confirmed a few months after he had finished his book. Steven Spielberg bought the rights for *The Hammer of God* and helped to make it into the movie *Deep Impact* (released in May 1998). In the documentary *Tales from the Script* (2009), one of *Deep Impact*'s writers, Bruce Joel Rubin, revealed that one of the producers at Disney took notes on the movie and started the production of *Armageddon*, a film released a few months after *Deep Impact*. Armageddon is the name of the location where a battle will take place that will trigger the end of times in the New Testament (Revelation 19:19).
2 The *Oregon Petition* website is still online as the "Global Warming Petition Project" and can be reached at www.petitionproject.org.
3 It is known as COP26 because it was the 26th Conference of Parties to the UNFCCC.

References

Arora, N.K., Mishra, I. (2021) COP26: more challenges than achievements. *Environmental Sustainability*, 4: 585–588. https://doi.org/10.1007/s42398-021-00212-7.

P2: The single timeline, which is expressed, for instance, in a narrative about future human extinction and which is common to contemporary eco-apocalyptic discourses, gives the impression that "'we are all in this together'" (Rothe 2020: 146). This, in turn, depoliticises the environmental emergency by obscuring the historically and geographically specific effects of climate change. As Anupama Ranawana and James Trafford put it, the faux 'anti-political universalism' of a single apocalyptic narrative "actively conceals both how climate crises are temporally and spatially distributed, and how they are symptoms of ongoing imperialist practices" (Ranawana and Trafford 2019).

The aim of this chapter is to suggest that eco-eschatology – that is to say, environmentalism attentive to 'the end of the world' – does not *necessarily* reproduce the Western-centric standpoint which leads to P1 and P2. In fact, as Danowski and Viveiros de Castro (2017) make clear, and as I will further demonstrate in the last section of this chapter, the eschatological focus on the 'end of the world' can effectively lend itself to a decolonial project. Eco-eschatologies, therefore, don't have to be abandoned altogether, although their understanding of time must be rethought in light of the decolonial critique presented above. In order to do so, I offer a theoretical corrective to eco-eschatologies constructed on the basis of a singular and linear temporality by proposing an alternative epistemological model of eco-eschatological time capable of addressing both P1 and P2.

First, I will argue that the model of historical time found in apocalyptic literature is not a line but a *spiral* which combines linear and cyclical elements. Such an understanding of time would respond to P1 by recognising the connection between the past, present, and future apocalypses and the constitutive role of past 'ends of the world' for an eschatological history. Second, I will demonstrate that apocalyptic discourse presupposes multiple timelines, whose relationship can be understood as a dislocated and non-contemporaneous historical totality. I will sketch the latter with the help of the work of Louis Althusser (1969, 2015) in order to show how such a model of time can address P2. I will conclude this chapter by suggesting that a two-fold understanding of eschatological time – as a spiral and as a non-contemporaneous totality – can help us to, on the one hand, inform decolonial theory and, on the other, devise *tactics* and *strategies* for a decolonised environmental politics.

★★★

In her book on apocalypse in Japanese science fiction, Motoko Tanaka rightly observes that a lot of 'research on world eschatologies refers to the major distinction between linear historical apocalypses and cyclical traditional apocalypses'. In contrast to the popular characterisation of the Judeo-Christian apocalypse as a final event of a linear history,

cyclical apocalyptic narratives envision no absolute end; many of these stories and myths envision apocalyptic destruction at the end of each cycle, yet assume that the world will be restored in the next rotation. Cyclical eschatology has its end, but implicit in this ending are both restoration and rebirth.

(Tanaka 2014: 16)

The distinction between these two models of time might initially lead an eco-eschatologist to substitute the linear conception of history with a single end (rejected on the basis of the decolonial critique) with the cyclical image of historical time with multiple ends. However, I believe that cyclical time cannot offer a viable eco-eschatological alternative to linear time.

If we conceive of historical time as a cycle, we are led to two contradictory conclusions regarding the end of the world. First, there is no singular apocalypse which fulfils time; rather, history's end repeats itself at and as the end of each cycle, generating a *multiplicity* of 'world ends'. Second, and in contrast to our first claim, if history is a cycle, it must be *the same end* which is repeated each time a cycle reaches its conclusion. If this weren't the case, and *end X* and *end Y* were, in fact, qualitatively distinct, then we could present the relationship between these two ends as a linear process from *end X* to *end Y* – thus abandoning the commitment to a purely cyclical model of time. Consequently, the multiplicity of 'world ends' generated by strictly cyclical history is only *apparent* – in fact, it is the same, single end which has taken place in the past and will take place in the future. As such, cyclical eschatology repeats the belief in a single end characteristic of its linear counterpart.

Furthermore, when considered from an existential point of view, cyclical history can generate a need to aim for a qualitatively distinct future. As Mircea Eliade puts it in his study on the eternal return in religions,

> all these numberless aeons also have a soteriological function; simply contemplating the panorama of them terrifies man and forces him to realize that he must begin this same transitory existence and endure the same endless sufferings over again, millions upon millions of times; this results in intensifying his will to escape, that is, in impelling him to transcend his condition of 'living being' once and for all.
>
> *(Eliade 1959: 115–116)*

However, to truly escape cyclical history, and to transform our existential condition "once and for all," the cycle must be broken and time must become a line, since it is only a linear model of time which can make sense of an *irreversible* change between two distinct temporal points necessary for an existential 'escape'. Thus, even if history is cyclical, it makes a subjective demand for an end conceived on the basis of a linear temporality.

From an eco-eschatological perspective, therefore, cyclical time is not a feasible replacement for linear time. First, on the cyclical view of time, we can acknowledge past ends of the world only by equating them with present and future apocalypses – thus creating a universal narrative of an eternally returning extinction event. Such a single narrative, however, cannot capture the temporal and spatial *differentiation* of the climate crisis. In other words, although the cyclical model of history can account for past ends of the world (thus addressing P1), it can do so only by reducing the *specific differences* between the multiple ends, and thus running into P2, that is, the faux universalism of a single extinction story centred on *the same*, recurring apocalypse. Second, the existential demand to overcome cyclical time refocuses our attention on a qualitatively distinct end which has to be located in the future, which, in turn, risks reintroducing P1 with its blindness to past ends of the world.

★★★

Since both purely linear and purely cyclical models of eco-eschatological time are flawed, in what follows I will suggest a third shape of time: a *spiral* which combines linear and cyclical elements. Such a model of historical time can be reconstructed on the basis of medieval and contemporary apocalyptic literature. In my reconstruction, I will first draw on the works of 12th-century monk Joachim of Fiore, to then turn to the 20th-century philosopher Jacob Taubes.[1]

Joachim is perhaps best known for his division of history into three stages or *status*: the age of the Father, corresponding to the events of the Old Testament; the age of the Son, marked by the domination of the Catholic Church; and the age of the Holy Spirit, which Joachim envisaged as a time of monastic orders.[2] Joachim frames the events of the Apocalypse as a transition from the age of the Son to the age of the Holy Spirit. Although the progressive movement through the three ages is undoubtedly built on a linear model of time, in Joachim's conception of history the linearity of history is often articulated in terms of cycles.

Joachim, a prolific illustrator, famously pictures historical time as three intertwined circles. Bernard McGinn describes Joachim's drawing in the following way:

> Three interlocking rings demonstrate how the mystery of the Trinity relates to the course of history. The first circle in green belongs to the Father and forms the time of the Old Testament. The middle blue circle, that of the Son, interlocks with both extremities – the median that joins the extremes. The final flaming red circle of the Holy Spirit indicates the double procession of the Third Person by intersection with both the green and the blue circle [...] the noninterlocking area of this last circle does suggest a coming special era within history, not unrelated, but still superior to what has gone before.
>
> *(McGinn 1979b: 104)*

One of the possible reasons why Joachim thinks of the relationship between the three stages of history as interlocking circles is that he views history as structured around a repetition of meaning – or *concordance* – found in the parallels between people and events of the Old and the New Testaments.[3] For Joachim, history moves forward in a cyclical movement which repeats the significance of particular characters and events across the past and the present *status*. As Marjorie Reeves explains:

The medieval approach to history which sought in each episode an inner meaning which linked it by concord with events of another era is, of course, quite foreign to us. It is as if each happening had a vertical point of reference, a 'thread' in the hand of God who combined threads into patterns on the inner side of history, whereas we look only for the horizontal connections and the pattern of visible cause and effect spun along the time span [...] Thus the three chief patriarchs, Abraham, Isaac and Jacob, are in concord with Zacharias, John the Baptist and Christ, and, of course, typify the three Persons of the Trinity and, in consequence, the three *status* of history.

(Reeves 1999: 10–11)

Importantly, for Joachim, the concordance between characters and events of the present and past *status* offers a possibility of anticipating the shape of the future age. This suggests that the notion of concordance does not reduce the differences between given characters or events, merging them into one single entity. If it did so, history would be purely circular – since it would be constituted by the repetition of the same event or character – and consequently could not be moving towards the next, qualitatively distinct *status*. Concordance finds identities across history; however, what the parallels reveal are specific historical developments between repeated meanings, significant from the point of view of the era to come, because it is these developments (say, between Abraham, Isaac, and Jacob, and Zacharias, John the Baptist, and Christ) which allow us to anticipate the next *status*. In the words of McGinn:

These letter-to-letter comparison and parallels between the Testaments are not used merely to understand the past, but also, and far more daringly, to reveal the future [...] The Old Testament, the New Testament, and especially the Book of the Apocalypse, when illuminated by the typological understanding, can show the meaning of what is to come.

(McGinn 1979b: 102)

Thus, for Joachim, eschatological prediction presupposes a *spiral* movement of historical time, where historical meanings move *forward cyclically*.

Interestingly, a very similar spiral shape of historical time is present in Jacob Taubes's philosophical history of apocalypticism, which was published in 1947 as *Occidental Eschatology*. Taubes's text seems to be characterised by an apparent tension between an apocalyptic commitment to a linear view of time and the recognition of the cyclical return of the apocalyptic theme throughout history. On the one hand, Taubes – a self-proclaimed apocalyptic (Taubes 2004: 103) – states explicitly that for apocalypticism "history does not complete a circle" (Taubes 2009: 33) and that, instead, "In the once-was of creation history has its beginning, and in the one-day of redemption it comes to its end. The interim between creation and redemption is the pathway of history" (Taubes 2009: 13). However, this seemingly philosophical commitment to linear time is contrasted in the course of

Taubes's exposition by a repetition of apocalypticism as a historical phenomenon – as he makes clear, the expectation of the end of times is a motif which returns cyclically from antiquity to modernity. It is this precisely this repetition of apocalyptic theories and experiences which allows Taubes to draw parallels, for example, between ancient apostles and modern philosophers.[4]

The apparent tension in *Occidental Eschatology* between the linear and cyclical aspects of historical time can be easily resolved if we emphasise that Taubes is committed to thinking about both apocalypticism and history as *spirals*.[5] On the *spiral* model, the apocalyptic motif is both repeated and altered throughout the forward march of time. Furthermore – in a manner reminiscent of Joachim – Taubes uses the cyclical parallels between various figures in the history of apocalypticism, and the developments between them, to predict a new era in the history of eschatology (admittedly, in rather obscure terms).[6] Taubes, therefore, seems "in concord" with Joachim in at least three ways: (1) they are both apocalyptic thinkers interested in the history of the end of the world; (2) they both emphasise the repetition and development of specific meanings throughout this history, which demonstrates their commitment to the spiralic character of time;[7] and (3) for both Taubes and Joachim, the analysis of the spirality of historical time serves to anticipate the next *status*.[8]

Importantly for our purposes, Joachim and Taubes demonstrate that the apocalyptic tradition offers a model of historical time which can help us to address P1.

First, in contrast to a purely linear time with its exclusive interest in the present or future end, the spiralic shape of history enables us to account for past ends of the world in its concordance with ends to come. In other words, present and future climate disasters should be understood as related to, and oriented by, ecological catastrophes of the past. Second, and in contrast to a purely cyclical understanding of time, we are not condemned to continually repeat the same end. On the contrary, the parallels between the 'world ends' can help us to identify historical developments, and thus to illuminate the ways to "break the cycle" by anticipating a qualitatively distinct *status* demanded by our existential situation. For example, past extinction events can be seen as offering insights into the character and the conditions of possible future extinction events; but in so doing, they also point towards changes necessary for a habitable Earth for communities most affected by the destruction of their ecosystems. In short, the spiralic model of history allows us to address P1 because it views past apocalypses as necessary both for understanding present and future environmental catastrophes and for putting forward transformative political solutions.

★★★

As long as we conceive of history as single timeline – even if this timeline is a spiral – we face P2, that is, the problem of the universal extinction narrative, which depoliticises the environmental disaster by concealing the unequal distribution of the climate crisis. In this section, therefore, I will suggest that eco-eschatology can be articulated in terms of multiple temporalities capable of capturing the differentiated climate catastrophes, and that such a model of historical time can serve as a useful tool for radical politics.

In the previous section, my discussion of apocalypticism operated with a number of distinct temporalities. They can be categorised in terms of three 'levels', which themselves include a multiplicity of temporal elements.

1. *Existential time (level 1)*: this type of temporality can be understood as pertaining to one's lived experience. For example, as we have seen, the awareness of one's existential condition generated by cyclical time, produces a subjective demand for a new future – the grasp of time as a cycle of endless suffering may result in an experience of terror and a will to escape. Similarly, the experience of apocalyptic prophets produces in them a subjective certainty that they live in the end times (Taubes 2009: 32). Since this level is constituted by individuals' experiences, it is essentially multiple.

2. *Historiographical time (level 2)*: this type of temporality is structured by historical events or characters, in so far as they play a role in history. Both Joachim and Taubes employ this type of time, for example, when the former demonstrates the concord between the Old and the New Testaments, and when the latter discusses the relationship between ancient, medieval, and modern eschatologies. Here, apocalyptic predictions are not grounded in a subjective experience, but in the analysis of past and present historical events and characters, and it is the latter which ensure the multiplicity constitutive of this level.

3. *Cosmological time (level 3)*: this type of temporality poses the question of the end of the world from extra-human and often extra-terrestrial outlooks. Examples of cosmological temporality can be found in religious beliefs about the death and rebirth of the universe, geological discussions concerned with major extinction events which pre-date humanity (Brannen 2017), and in theoretical explorations of the consequences of the eventual solar catastrophe (Brassier 2007). Here, the multiplicity within the level is introduced by the particular character of the extra-human or extra-terrestrial timelines (religious, geological, astronomical, etc.).

How are we to conceive of the relationship between levels 1–3, such that we can successfully address P2?

On one reading – which I believe is erroneous – existential time would be situated within historiographical time, while both would be grounded in cosmological time. The difference between the levels would be constituted by the *scale* of each temporality – from subjective, through historiographical, to cosmological. On this reading, the three levels could be visualised as gradually expanding circles of a *continuous* and *contemporaneous* temporal space in a shape of a cone, with existential time on top, historiographical time in the middle, and cosmological time at the bottom. Thus, when Taubes writes that "History does not just happen objectively in macrocosm, it also comes about in man as microcosm. Subject and object receive their identity from history" (Taubes 2009: 13–14), this reading would lead us to situate the subjective microcosm *within* the objective macrocosm.

However, as I will show below, the above view is problematic for two reasons: first, the difference between the existential, historiographical, and cosmological levels is not simply one of scale, which means that the temporal space shared by these levels is not continuous. Second, the understanding of the different timelines as part of a contemporaneous space constituted by a shared present depoliticises theories of history committed to such a view.

After suggesting the relationship between microcosm and macrocosm, Taubes observes a possibility of dislocation which characterises subjective time in relation to worldly temporality and which can be found in memory. For Taubes, memory has a capacity to resist worldly time by its power to prevent events from disappearing:

> Memory uncouples an event from the stream of time. An event can be released from the time element in this way because it is set fast and does not disappear in the course of time [...] time is conquered by memory. Because memory stands *outside* time, it can be aware of its transitory nature.
>
> *(Taubes 2009: 14)*

However, if memory – that is, an experience proper to the subjective microcosm or existential time – can be (at least partly) exterior to both historiographical and cosmological temporalities, then the difference between the three levels cannot be one of scale within a continuous temporal space. Rather, the latter is constituted by an essential dislocation between the respective times of the microcosm and the macrocosm.

Furthermore, a similar non-identity can be found in the relationship between historiographical and cosmological time. Recently, Dipesh Chakrabarty has drawn a distinction between 'global' and 'planetary' temporalities: while the former refers to human history, the latter involves 'vast processes of unhuman dimensions':

> The global, as I have said, refers to matters that happen within human horizons of time, the multiple horizons of existential, intergenerational, and historical time, though the processes might involve planetary scales of space. Planetary processes, including the ones that humans have interfered with, operate on various time tables, some compatible with human times, others vastly larger than what is involved in human calculation. Thus air and surface water have 'short recycling times,' as do many metals, but soils and ground water take 'thousands of years' to replenish themselves [...] The two modes of thinking represent two different kinds of knowledge and, for humans, two different ways of comporting themselves to the world within which they find themselves. The global with humans at its center is ultimately all about forms and values [...] But the planetary as such [...] cannot be grasped by recourse to any ideal form. There is no ideal form for the earth as a planet or of its history or for the history of any other planet.
>
> *(Chakrabarty 2019: 24–25)*

As Chakrabarty suggests, despite the intersection between the global (or historiographical) and the planetary (or cosmological) timelines, there is a more fundamental difference between these respective temporalities, namely, the generation of two types of knowledges about and comportments in the world. Consequently, the temporal space between historiographical and cosmological temporality can be represented as dislocated according to the divergent epistemic and axiological effects produced by the respective macrocosmic timelines.[9]

One may argue that the dislocation of the three timelines makes them discontinuous, but this does not prevent them from being *contemporaneous*. The gradually expanding circles of a cone may not perfectly overlap, but they still express a common present, running vertically through all the horizontal levels. The existential timeline would then be located within *the same moment* as parts of historiographical and cosmological timelines, and it is this universal present which would account for the interconnectedness between the three temporalities recognised above by both Taubes and Chakrabarty.

Interestingly, it is Louis Althusser who offers reasons as to why temporal contemporaneity should be rejected. In *Reading Capital*, the French Marxist argues that philosophies in which "all the elements of the whole always coexist in one and the same time, one and the same present, and are therefore contemporaneous with one another in the present" (Althusser 2015: 241) make it impossible to gain any knowledge of the future; this, in turn, makes such philosophies politically redundant:

> the ontological category of the present prevents any anticipation of historical time, any conscious anticipation of the future development of the concept, any *knowledge* of the *future* [...] The fact that there is no knowing the future prevents there being any science of politics, any knowing that deals with the future effects of present phenomena.
>
> *(Althusser 2015: 242)*

The moment we situate all temporal levels within a contemporaneous temporal space constituted by a universal moment, we, in effect, assert that "*nothing can run ahead of its time*": if the three apocalyptic temporalities express a shared apocalyptic present, none of them can point *beyond* the current catastrophe. This, in turn, makes it impossible to devise any reliable plans for political action: to anticipate possible future developments and with them opportunities for change, at least some temporalities would have to reach *outside* of the all-encompassing present disaster, offering access to future alterations (or lack thereof), which, in turn, would illuminate the direction of other timelines – which is something the above-mentioned three levels, in as much as they are contemporaneous, cannot do. Thus, if the knowledge of the future is in principle unavailable, we can only "divine it as a presentiment" (Althusser 2015: 242), instead of attempting to prepare for and shape what is coming. The belief in a universal present, therefore, depoliticises eco-apocalypticism.

Now, Althusser offers an alternative model of historical time which, I believe, can be useful for conceptualising the relationship between the different levels of apocalyptic temporalities in a manner which enables us to address P2.

Althusser asserts the dislocation, and the concomitant relative autonomy, of multiple timelines.[10] However, he also recognises the interconnectedness of divergent temporalities:

> The fact that each of these times and each of these histories is *relatively auton-omous* does not make them so many domains which are *independent* of the whole: the specificity of each of these times and of each of these histories [...] is based on a certain type of *dependence* with respect to the whole.
>
> *(Althusser 2015: 247)*

For Althusser, therefore, the temporal levels create a particular historical *conjuncture* in which they co-exist in a relatively independent way and in which they are nonetheless determined by "the current mechanism of the whole" (Althusser 2015: 254). Thus, on the Althusserian model the relations of backwardness, forwardness, survival, or unevenness of different temporal levels are determined not by the base time or the universal present, but by their relatively autonomous positions in a given historical totality (Althusser 2015: 254).

Importantly, and in contrast to continuous and contemporaneous history, the Althusserian conjuncture is *dislocated* and *non-contemporaneous*. First, we should not expect to find the element constitutive of one level in other levels: "The present of one level is, so to speak, the absence of another, and this co-existence of a 'presence' and absences is simply the effect of the structure of the whole in its articulated decentricity" (Althusser 2015: 252). A similar point is illustrated by Siegfried Kracauer (1995) and Carlo Ginzburg (1993) in their respective discussions of the non-identity between macro- and micro-histories. Kracauer, for example, uses a cinematic analogy to demonstrate how the diverse effects of 'close-ups' and long shots in film are mirrored by historical 'close-ups', which are "apt to suggest possibilities and vistas not conveyed by the identical event in high magnitude history"[11] (Kracauer 1995: 126). Furthermore, it is possible that the different levels run ahead or behind each other. If we take the existential and the historiographical timelines as an example, we can observe that an individual can be "ahead of their time" or "behind the times," and that historical events can be missed or anticipated by individuals (think here of a partisan in a forest who continues to fight because the news of the peace treaty hasn't reached them yet; or of an army strategist capable of responding to the effects of a battle before it takes place). As Vittorio Morfino puts it, commenting on relevant passages in *Reading Capital*, time

> is the articulation of a plurality of durations and at the same time the guarantee of the impossibility of hypostatisation of one rhythm in relation to the others [...] It is, in a certain sense, the guarantee that time has no secret, that its fundamental structure is that of *non-contemporaneity*.
>
> *(Morfino 2014: 15)*

In an eschatological context, the Althusserian model of history would suggest that an analysis of ends of the world belonging to different levels would yield heterogeneous findings (analogous to the difference between macro- and micro-histories), and that the ends of the world would happen at different points in each timeline. However, the differentiated apocalypses would nonetheless belong together within an apocalyptic historical conjuncture, which would determine their positions in relation to one another.

The rather abstract characterisation of the temporal levels as constitutive of a dislocated and non-contemporaneous historical conjuncture has concrete consequences for politics – illustrated by Althusser with an example of the Russian Revolution. In 'Contradiction and Overdetermination', Althusser suggests that the October Revolution was possible because "Russia was, *simultaneously at least a century behind the imperialist world, and at the peak of its development*" (Althusser 1969: 97). Lenin's success, therefore, consisted in grasping and exploiting the relationships of temporal forwardness, backwardness, survival, and unevenness of multiple heterogeneous timelines articulated in a decentred historical totality,[12] as "the *objective conditions* of a Russian revolution," in order to "forge its *subjective conditions*, the means of a decisive assault on this weak link in the imperialist chain" (Althusser 1969: 98).[13] The political potential of the dislocated and non-contemporaneous model of time, therefore, consists in representing a complex historical totality – with its relationships of backwardness, forwardness, survival, and unevenness between particular timelines – in a way which can inform campaigns aiming at radical change.[14]

I believe that the account of historical time I sketched in this section is able to address P2, because (1) it replaces a single eschatological timeline with multiple apocalyptic temporalities articulated in terms of three levels: existential, historiographical, and cosmological; (2) it avoids, by resisting subordinating heterogeneous timelines to a universal present or a base time, constructing universal extinction narratives – which makes it well-equipped to capture the unequal differentiations of the climate crisis;[15] and (3) it can inform environmental politics by representing apocalyptic times and narratives as a historical totality in a manner conducive to radical political transformations.

★★★

I would like to conclude this chapter by considering three possible objections to my answers to the 'decolonial critique' of eco-eschatology.

First, the understanding of history as a spiral (which allowed us to address P1) and as a decentred totality (sketched in my response to P2) offer two seemingly incompatible 'shapes' of historical time, which present history as either a *diachronic process*, where the intersection between past, present, and future generates the cyclical yet forward movement, or as a *synchronic structure* (albeit capable of breaks and mutations) where all elements constitute a whole. If this is the case, not only is eco-eschatology unable to offer a unified model capable of addressing both P1 and P2, but its solutions to P1 and P2 reveal it to be internally contradictory – because they demonstrate that eco-eschatology is committed to two irreconcilable models of historical time.

Second, while it may be possible to reconcile the eschatological understanding of time with a decolonial project, it is not clear whether eschatology adds anything new to decolonial theory. Why should we painfully reinterpret largely compromised eschatological concepts, if we can construct new notions more immediately grounded in decoloniality? Similarly, why should we struggle to map old apocalyptic ideas onto the climate catastrophe, if modern decolonial philosophies equip us with a plethora of categories suited to address the contemporary environmental crisis by informing decolonial eco-politics?

Third, Althusserian philosophy can be classified as 'Marxist-Leninist', 'structuralist', 'Spinozist', etc., but it is not *ecological* or *eschatological* in any straightforward sense. Consequently, turning to Althusser to address a problem with eco-eschatology may seem like 'cheating'; more importantly, however, an Althusserian answer to a critique of eco-eschatology may suggest that the eschatological tradition itself is incapable of generating theoretical resources necessary to deal with P2. In other words, far from advancing eco-eschatology, a reliance on Althusser would disclose the essential limits of the eco-eschatological project.

Let me briefly address the third objection first.

A number of authors have recently explored the pertinence of Althusserian philosophy to environmentalism (e.g. Margulies 2018; Tsoneva 2016). I will, therefore, restrict myself to briefly suggesting the link between Althusser and the eschatological tradition. As Warren Montag has argued, Althusser's project is incompatible with traditional eschatologies committed to a final event of history (Montag 2016) – and with it to Marxism which secularises this belief. But, as I have tried to show in this chapter, we can rethink eschatology in terms of multiple ends of the world, which would bypass the Althusserian problem with traditional eschatologies. More significantly, I believe that Althusser's philosophy has theoretical affinities with the apocalyptic tradition, which concern a shared approach to history. As we have seen above, Althusser wants to found radical political change on the knowledge of the future, anticipated on the basis of a given historical conjuncture. Importantly, for him, the structure of the historical totality – and with it, the future – is not immediately legible, which necessitates a construction of a concept of history on the basis of the meaning hidden behind past and present historical facts (Althusser 2015: 248–251). As such, the Althusserian theory of history bears a striking resemblance to the theory of history operative in Joachim and Taubes, where the knowledge of the next era is grounded in the disclosure of the *hidden* concordance and the construction of parallels between the Old and the New Testaments or between different eschatologists. In addition, both Althusser and Taubes recognise the importance of dislocation and the relative autonomy of individual temporalities. On my reading, therefore, Althusser is 'in concord' with both Joachim and Taubes, since they share a motivation for, and approach to, historical analysis, which may explain the ease with which the Althusserian model of time could be adapted to this chapter's explicitly eco-eschatological problematic.

The first objection concerning the two seemingly incompatible models of time can be answered by pointing out (1) that the synchronic totality consists of diachronic processes and that (2) the concept of a conjuncture can be a tool for conceptualising the relationship between spiralic developments taking place in individual timelines constitutive of a historical totality. In other words, eco-eschatology has two *complementary* ways of thinking about historical time at its disposal: while the first one focuses on the spiralic histories found in existential, historiographical, and cosmological levels, the second one captures the complex co-existence of the spiralic timelines in a given historical whole.[16]

The second objection can be addressed by pointing out that the two-fold, eco-eschatological approach to history can be beneficial for both decolonial theory and decolonial politics.

First, some decolonial thinkers employ categories and theoretical models which possess an undoubtable eschatological inflection. In *The Politics of Decolonial Investigations*, Walter D. Mignolo argues – in a Joachimite manner – that "we, on the planet, are experiencing a change of epochs": a movement from the "second nomos" understood as "unipolarity in international relations and the hegemony/dominance of Western modernity" and the "third nomos" marked by "de-Westernization and decoloniality" (Mignolo 2021: 484–485). Here, for example, the decolonised eschatological model of time – with its focus on the ends and advents of timelines and eras – can provide the theoretical framework capable of identifying the synchronic and diachronic elements of what Mignolo calls "the change of epochs." Second, eschatological history which recognises the multiple ends of the world can offer a temporal supplement to decolonial philosophies which explore the existence *of* and *in* the 'pluriverse' (Escobar 2017; Ortega 2016) – for instance, by framing the dynamic relation between the different worlds in terms of ends and survivals of distinct timelines.

Furthermore, the two-fold approach to eschatological history maps onto the distinction between *tactics* and *strategy*, which makes it useful for decolonial eco-politics. The focus on the individual spiral histories enables us to come up with tactical political solutions shaping the development of a given timeline. By contrast, the meta-perspective representing the historical conjuncture allows us to devise an overall strategy capable of transforming a given totality. Furthermore, the synchronic grasp of diachronic processes can be helpful for evaluating particular tactical solutions, since it can question the latter's role in bringing about radical change. For example, as long as they are rethought as spiral (i.e. as long as they recognise the significance of past ends of the world for their respective present- and future-oriented politics), the three strands of ecological politics identified by Rothe – eco-catastrophism, eco-modernism, and planetary realism – can furnish an effective tactic for an individual timeline. However, the decision on the usefulness of their application (or on the usefulness of the resistance to their application) can only be made from a strategic point of view grounded in the synchronic meta-perspective. In addition, the concept of a conjuncture may reveal a need for new tactics, which would effectively respond to the political demands unseen from a diachronic position. Decolonised eco-eschatology, therefore, doesn't only recognise the "heterogenous temporality of the Anthropocene" (Rothe 2020: 163), but it can also generate tactics and strategy, which apocalyptically aim at a radical change in the historical whole and an emergence of a new *status*.

Acknowledgements

I would like to thank Darshan Cowles, Ben Randolph, and Simon Thornton for their help in the preparation of this chapter.

Notes

1 While the *content* of Joachim's and Taubes's respective eschatologies can be accused of Eurocentrism, as I show below they develop a *formal* account of history compatible with decoloniality.

2 See McGinn (1979a: 133–134). The division of history into three stages has been highly influential. As Jacob Taubes points out, "the model of antiquity – Middle Ages – modern age is nothing but a secular extension of Joachim's prophecy of the three ages of the Father, the Son, and the Holy Spirit" (Taubes 2009: 82). We can also map the Marxist periodisation of history into feudalism, capitalism, and communism, as well as Auguste Comte's theological, metaphysical, and positive stages of society onto the three *status* of Joachim's. For the discussion of the importance of Joachim for modern philosophy, see Lynch (2019: 38–48).

3 Alternatively, Marjorie Reeves suggests that, perhaps, it was the "annually returning order of the Church's liturgy" which inspired Joachim's usage of cyclical elements in his account of history (Reeves 1980: 270).

4 "The dialectic of Paul's history of salvation is both *quantitative in terms of world history* (Hegel) and *qualitative in existential terms* (Kierkegaard)" (Taubes 2009: 63). In another work, Taubes explains the parallels between the texts of St Paul and Walter Benjamin by suggesting a concordance of their experiences (Taubes 2004: 72–74).

5 "Apocalypticism and Gnosis inaugurate a new form of thinking [...] The logic of the dialectic, whether in apocalypticism and Gnosis or in the works of Hegel, is not circular but spiral. The 'bending backward' characteristic of the dialectic does not progress back to the thesis in a circular manner but broadens out into spiral toward the synthesis [...] Dialectic logic is a logic of history, giving rise to the eschatological interpretation of the world" (Taubes 2009: 35). For a more detailed discussion of the relationship between apocalypticism, Gnosis, Hegel and Taubes, see Bielik-Robson (2014), especially chapter 5.

6 "A new epoch is beginning, which introduces a new aeon that is post-Christian in a more profound sense than that of the calendar. This epoch, in which the threshold of Western history is crossed, regards itself primarily as the no-longer of the past and the not-yet of what is to come [...] For the coming age is not served by demonizing or giving new life to what-has-been, but by remaining steadfast in the no-longer and the no-yet, in the nothingness of the night, and thus remaining open to the first sings of the coming day" (Taubes 2009: 193).

7 One interesting question, which I don't have the space to explore here, is whether historical meanings – and with them, history as such – for Joachim and Taubes are ideal or material, or if they constitute a third, ideal-material category.

8 The concordance between Joachim and Taubes is far from accidental – *Occidental Eschatology* contains a relatively long section devoted to Joachim, which demonstrates Taubes's in-depth engagement with the thought of the medieval monk.

9 It would be interesting to read the notion of dislocation into Joachim's texts – such an interpretation would then split the history of the three *status* into heterogeneous histories

taking place on the levels of characters, events, and the Godhead. This, in turn, would complicate the concept of a *single status*.

10 "Each of these different 'levels' does not have the same type of historical existence. On the contrary, we have to assign to each level a *peculiar time*, relatively autonomous and hence relatively independent, even in its dependence, of the 'times' of the other levels" (Althusser 2015: 246–247).

11 In a similar fashion, Ginzburg notes: "A battle, strictly speaking, is invisible, as we have been reminded (and not only thanks to military censorship) by the images televised during the Gulf War. Only an abstract diagram or a visionary imagination [...] can convey a global image of it. It seems proper to extend this conclusion to any event and with greater reason to whatever historical process. A close-up look permits us to grasp what eludes a comprehensive viewing, and vice versa" (Ginzburg 1993: 26).

12 Among the examples constitutive of the historical conjuncture in 1918 listed by Althusser, we find the historical contradictions between feudalism and "large-scale capitalist and imperialist exploitation"; the industrial cities and "the medieval state of the country side"; the class struggle within the ruling classes and "the '*advanced*' character of the Russian revolutionary élite [...] *far ahead of any Western 'socialist' party in consciousness and organisation*" (Althusser 1969: 96).

13 Furthermore, the dislocated and non-contemporaneous model of historical totality can also help us make sense of the existence of pre-revolutionary elements after October – these surviving pre-revolutionary elements belong to timelines not yet transformed by the Revolution (Althusser 1969: 115–116).

14 For an example of political apocalypticism based on non-contemporaneous historical structures (with references to pop culture), see Williams (2011).

15 Although I cannot develop this claim here, I believe that it is possible to answer the second part of Ranawana and Trafford's (2019) critique, namely, that a single apocalyptic narrative conceals not only temporal, but also *spatial* differentiation by linking the multiple ends of the world found in individual timelines to particular locations in space.

16 Importantly, the fact that the individual timelines share a spiral shape does not undermine their relative autonomy, since they remain both formally and qualitatively distinct: the spiral relationship between past, present, and future has a different form and quality in the case of one's psychological life than in the case of a spiral recurrence of historical or cosmic events. For example, the repetition of a traumatic memory – a personal end of a world – is both formally and qualitatively distinct from a repetition of an extinction event.

References

Althusser, L. (1969) *For Marx*. Trans. B. Brewster. Harmondsworth, UK: Penguin Books.

Althusser, L. (2015) *Reading Capital: The Complete Edition*. Trans. B. Brewster and D. Fernbach. London: Verso.

Bielik-Robson, A. (2014) *Jewish Cryptotheologies of Late Modernity: Philosophical Marranos*. London: Routledge.

Brannen, P. (2017) *The Ends of the World: Volcanic Apocalypses, Lethal Oceans and Our Quest to Understand Earth's Past Mass Extinctions*. New York: Simon and Shuster.

Brassier, R. (2007) *Nihil Unbound: Enlightenment and Extinction*. London: Palgrave Macmillan.

Chakrabarty, D. (2019) "The Planet: An Emergent Humanist Category". *Critical Inquiry*, vol. 46. pp. 1–31. https://doi.org/10.1086/705298.

Danowski, D. and Viveiros de Castro, E. (2017) *The Ends of the World*. Trans. R. Nunes. Cambridge: Polity Press.

Eliade, M. (1959) *Cosmos and History: The Myth of the Eternal Return*. Trans. W.R. Trask. New York: Harper & Brothers.

Escobar, A. (2017) *Designs for the Pluriverse: Radical Interdependence, Autonomy, and the Making of Worlds*. London: Duke University Press.

Ginzburg, C. (1993) "Microhistory: Two or Three Things That I Know about It". Trans. J. Tedaschi and A.C. Tedaschi. *Critical Inquiry*, vol. 20. pp. 10–35.

Kracauer, S. (1995) *History: The Last Things before the Last*. Princeton, NJ: Markus Wiener Publishers.

Lynch, T. (2019) *Apocalyptic Political Theology: Hegel, Taubes and Malabou*. London: Bloomsbury.

Margulies, J. (2018) "The Conservation Ideological State Apparatus". *Conservation and Society*, vol. 16. pp. 181–192. https://doi.org/ 10.4103/cs.cs_16_154.

McGinn, B. (1979a) *Visions of the End: Apocalyptic Traditions in the Middle Ages*. New York: Columbia University Press.

McGinn, B. (1979b) *Apocalyptic Spirituality*. New York: Paulist Press.

Mignolo, W.D. (2021) *The Politics of Decolonial Investigations*. London: Duke University Press.

Montag, W. (2016) "Althusser and the Problem of Eschatology," in *Althusser and Theology: Religion, Politics, and Philosophy*, Agon Hamza (ed.). Leiden: Brill, pp. 31–46.

Morfino, V. (2014) *Plural Temporality: Transindividuality and the Aleatory Between Spinoza and Althusser*. Leiden: Brill.

Ortega, M. (2016) *In-Between: Latina Feminist Phenomenology, Multiplicity, and the Self*. Albany: SUNY Press.

Ranawana, A. and Trafford, T. (2019) 'Imperialist Environmentalism and Decolonial Struggle', *Discover Society*, August 7. https://archive.discoversociety.org/2019/08/07/imperialist-environmentalism-and-decolonial-struggle/.

Reeves, M. (1980) "The Originality and Influence of Joachim of Fiore". *Traditio*, vol. 36. pp. 269–316. https://doi.org/ 10.1017/S0362152900009260.

Reeves, M. (1999) *Joachim of Fiore and the Prophetic Future*. Guildford, UK: Sutton Publishing.

Rothe, D. (2020) "Governing the End Times? Planet Politics and the Secular Eschatology of the Anthropocene". *Millennium: Journal of International Studies*, vol. 48. pp. 143–164. https://doi.org/10.1177/0305829819889138.

Tanaka, M. (2014) *Apocalypse in Contemporary Japanese Science Fiction*. London: Palgrave Macmillan.

Taubes, J. (2004) *The Political Theology of Paul*. Trans. D. Hollander. Stanford, CA: Stanford University Press.

Taubes, J. (2009) *Occidental Eschatology*. Trans. D. Ratmoko. Stanford, CA: Stanford University Press.

Tsoneva, J. (2016) "From the 'International of Decent Feelings' to the International of Decent Actions: Althusser's Relevance for the Environmental Conjuncture of Late Capitalism," in *Althusser and Theology: Religion, Politics, and Philosophy*, Agon Hamza (ed.). Leiden: Brill, pp. 168–181.

Williams, E.C. (2011) *Combined and Uneven Apocalypse: Luciferian Marxism*. Winchester, UK: Zero Books.

3

QUEER ECOLOGIES AND APOCALYPTIC THINKING

Elizabeth Pyne

Apocalyptic Provocations

Sometimes it pays to state the obvious – so one parent's heart-shaped cardboard sign held aloft at a climate protest says: 'I care about the climate coz I care about my kids'. The sentiment is also regularly mobilized, in inverted form, to indict a failure of care and stewardship on the part of those responsible – so one young person's block-lettered accusation of the adults who are meant to be in charge reads: 'You say you love your children, but you are destroying their future'. The constellation of planetary conditions, the future, and the figure of the child is familiar, even reflexive, in the symbolic repertoire around discussions of climate change and sustainability. If it is now acknowledged that the environmental crisis is also a socio-political crisis and that assessments of the health of Earth's bio-physical systems inevitably have a temporal dimension (i.e., sustainability), then appeals animated by their relation to children are an apt illustration of this convergence. This juncture makes for a powerful platform to articulate the urgent threat posed by climate warming.

Greta Thunberg has gained notoriety among the ardent youth voices clamoring for climate action in the past several years. The Swedish activist came on the scene in 2018 with weekly protests to urge her government to adopt policies in line with the Paris Agreement. She is hardly alone in her fierce advocacy; equally outspoken peers organizing across the world have helped to rally a global movement. Using the hashtag #FridaysForTheFuture, Thunberg and scores of young people have engaged in school strikes, most notably in a coordinated effort across some hundred-odd countries on March 15, 2019, in which an estimated 1.4 million students took part. In its plea for the future – their future, her future – Thunberg's rhetoric remains striking for its uncompromising evaluation of a dire situation and a correlatively blunt challenge. Most quoted are the closing lines of her January 2019 address at the World Economic Forum in Davos:

DOI: 10.4324/9781003189190-4

Adults keep saying we owe it to the young people to give them hope. But I don't want your hope, I don't want you to be hopeful. I want you to panic, I want you to feel the fear I feel every day. And then I want you to act, I want you to act as if you would in a crisis. I want you to act as if the house was on fire, because it is.[1]

The ethical force Thunberg summons draws on the sense that things are coming to an end – indeed, screeching to a halt. The time for hope is over. The time for quietly adhering to pedagogical routine and appropriately circumscribed political participation is no more, since a linear progression into adulthood can no longer be assumed. Her and others' calls to recognize a social order interrupted by climate crisis and systems of life that require radical overhaul – lest a truly impossible future finally arrive – accentuate a fluidity between actual and potential children in their expressions of threat. 'What will you tell your children?', Thunberg asks, projecting politicians' present failure forward. Others judge the burden of care, cost, and risk to be grounds for forestalling any actualization of this imaginative exercise. In contexts where lifestyles are carbon-intensive, recent climate commentary includes think pieces and op-eds – often anguished, sometimes angry, occasionally smug – that adapt the sentiment with which I began: 'I'm choosing not to have kids because I care about the environment'. This genre extends in fascinating ways into the youth movement. In 2019, 18-year-old Canada-based activist Emma Lim pledged not to have kids until governments begin taking meaningful action on climate change; several thousand people signed on, using the hashtag #NoFuture-NoChildren. The negative framing of this campaign, as in Thunberg's rhetoric, may be seen as a radicalized continuation of climate rhetoric on behalf of children whose fate is a metonym for the future. Yet, upping the ante by way of negation, so as to more forcefully communicate severity and urgency, also seems to subvert an obvious proposition (protecting the future for children) with an unthinkable one: a future without children – maybe no future at all.

★★★

By some appraisals, the degree to which Thunberg's and Lim's advocacy is awash with an apocalyptic spirit indicates where the environmental movement has gone wrong – in emotional dramatics and catastrophizing narratives (Nordhaus and Shellenberger 2007). For the present volume, however, talk of apocalypse is appropriate to the scale and range of contemporary planetary shifts and to the profound nature of the questions thereby raised: grave, but uncertain too, neither giving license to inaction nor foretelling an all-consuming destruction (Skrimshire 2010). That the refusals Thunberg and Lim express – no hope, no future – are simultaneously opposed to *and* on a continuum with articulations of what it means to show regard *for* the future locates us in the peculiar quandaries around ends (i.e., cessations) and ends (i.e., aims or objectives) that mark this moment of eco-social exigency. As several other chapters in this collection explore, responses to climate change channel the elusive dynamic of redemptive transformation in ways that touch on the religious roots of the concept of apocalypse, one which says: 'in what is coming everything will change and yet that future will not be wholly discontinuous with the past, thus something of the present must persist'. Might one read these protests around children and the question of a sustainable future as indicative of this paradoxical quality of apocalyptic unveiling?

Perhaps, and I am intrigued by the possibility that their doing so puts a queer twist on the sanitized poster-child politics that infuses environmental causes no less than other social causes. What are we to make, in other words, of the surge of negativity wherein the future is seen only, if at all, through what is refused?[2] For there is no doubt that climate change discourse regularly posits an equation of affective investment, the wellbeing of children, and ecological futurity (Ensor 2012; Seymour 2013; Sheldon 2016). More often than not, its hortatory force relies on what queer theorist Lee Edelman (2004) has termed 'reproductive futurism', a dynamic in which politics — that is, universal conceptions of value and social life — are organized around the figure of 'the Child'. While queer families with children exist and queer people may well desire better futures, Edelman's argument hinges on a claim about the structuring position of the 'queer' as a threat to the intelligibility of the social order, the repudiation of which is necessary to its ongoing reproduction. Therefore, a built-in hostility to queer sexuality is not unique to reactionary politics but applies as well to liberalism's ameliorative figurations of a 'more perfect social order' (Edelman 2004: 4). Queers cannot be made to 'fit' and should not, Edelman maintains, try to square the circle of misrecognition; instead, he encourages acceding to the mantle of unintelligibility, disorder, and negativity foisted on queer sexuality and abandoning the assimilation that politics demands. He thus advances a paradoxical ethics of defiance: 'no future'.

In an era of climate anxiety, the question of the future and, with it, the putatively self-evident good of bio-social reproduction is newly invigorated with a planetary charge. Some have argued that Thunberg occupies a position of enunciation squarely in the midst of this rhetorical framing (Kverndokk 2020), which would suggest that the way she and Lim warn of the negation of the future by climatic threat only reinforces the value of reproducing the (heteronormative) same. Other observers, however, find that Thunberg confounds the linear temporality of reproductive futurism in a manner that is 'rather queer' (Baraitser 2020: 505). My interest is not to settle a correct interpretation, but instead to highlight how situating her intransigent rage and that of the youth-led #NoFuture-NoChildren alongside Edelman's polemical critique reveals the proximity of two registers of apocalyptic thinking. Insofar as the figurations of hope, destruction, and futurity that permeate discussions of environmental apocalypse are inevitably entangled with issues of nature and social reproduction, they also display the relevance of queer theory and specifically, I argue, of the antifuturist/antisocial strand sometimes regarded as queer theory's apocalyptic turn (Giffney 2016; Woltersdorff 2012).

Too-frequent reductions of 'apocalypse' — to stand in for a blockbuster conflagration or, latterly, vaguely-but-fatalistically looming ecological cataclysm — have motivated efforts to reclaim biblical and historical accuracy by a strict definition of terms and restricted theological usage. However, something would be lost in trying to excise the ambiguity that unfolds between and at the edges of queer theory and the 'apocalyptic'. Queer theory's searing debates about 'sex and negativity' distill key questions from across humanities scholarship concerning the status of social critique and its relationships to issues of agency, identity, ethics, and politics, and the possibility of changing death-dealing systems for the better (Wiegman 2017). As recent theological treatments of queer temporalities demonstrate, these are questions which at the same time double back on, cross, and reframe Western modernity's supposedly settled boundary between secular and religious thought (Brintnall et al. 2018). Reverberations of apocalypse feature prominently in explorations of these intersecting queer times and their disorientations relative to 'seemingly commonsensical categories such as past, present, and future', which – no less than sexuality, gender, and race/ethnicity – are 'culturally constructed and [...] intimately bound up with the (il)logics of desire and power' (Brintnall et al. 2018: 3). As we have seen, prominent examples of recent climate activism both hew to and challenge such commonsensical renderings of temporality in what they propose for the sake of the children and of the planet. Prompted by these multiple and ambiguous configurations of care, reproduction, and the future, I contend that efforts to interpret the meaning of environmental apocalypse have an integral partner in the apocalyptic thinking found in varieties of queer critique.

This first section thus situates the relevance of queer theory with respect to this book's overarching concern with climatic and ecosystemic disaster. However, my identification of apocalyptic negativity as a stimulating point of overlap between queer and ecological studies cannot be taken as read. As I describe in the next section, leading voices in the field known as queer ecology have tended either to disregard or distance themselves from the association of queerness with (apocalyptic) negativity. While there is much to commend in the pathways these authors propose, there are also serious drawbacks. The third section outlines a critical rendering of queer ecology and its posture towards the future in conversation with a selection of recent works that have engaged queer negativity in an ecological register. Then, following from my initial discussion here of queer theory's theological imbrications, I return in the final section to the question of what is at stake in insisting on this queer eco-negativity's specifically apocalyptic character when it comes to approaching the ambiguities embodied in the child-future-nature triad.

Queer Ecologies

For several decades now, scholars from a number of disciplines have explored how and with what consequences understandings of queerness might be brought together with a range of ecological categories. Nature and the natural are foremost among them. The development of the self-described field of queer ecology reflects two main impulses at this intersection. On the one hand, it sets out to challenge the heteronormativity of scholarly trends, from evolutionary thought to ecocriticism, which frequently naturalized conventional heterosexual patterns of human desire and sexuality by reference to their nonhuman animal counterparts and wider biological environments (Mortimer-Sandilands and Erickson 2010). But while some ecologically inclined thinkers make productive use of the denaturalizing bent of certain queer strategies, many have also been keen to show that an environmental awareness must contest the ascendance of deconstructive and psychoanalytic approaches in queer theory and the humanities more broadly. Such thinkers resist the foregrounding of not only denaturalization but, with it, negativity, antisociality, and antirelationality in contemporary queer theory – themes fueled largely by the work of Edelman and Leo Bersani – as impediments to understanding sex and gender in the context of ecological belonging and planetary pressures. Thus, they argue, the merger of 'queer' with 'ecology' needs to transform not only certain assumptions about the natural world but also the sense that negative projects exhaust what it means to think queerly about nature.[3]

The divergence of queer ecology from antisocial queer theory is particularly evident when it comes to the status of the future. Appraisals of Edelman's *No Future* by Nicole Seymour (2013) and Diane Chisholm (2010) are indicative. Seymour, unlike Chisholm, grants that Edelman's piercing analysis of 'reproductive futurism' is relevant to diagnosing problematic tendencies within environmental advocacy – namely, an often heteronormative vision of the future that environmentalism seeks to protect. Nevertheless, both judge the negativity animating his verdict of 'no future' to be lacking ecologically speaking and effectively bankrupt in an era of climate change. Their related critiques include three main charges. First, and most directly, they maintain that Edelman's work veers into nihilism and indifference when what we desperately need right now are articulations of hope, visions of possibility, and nudges towards responsibility for the future. Second, this rendering of queerness is anthropocentric, failing sufficiently to connect human affairs with the nonhuman world. Third, its attack on a pro-life regime dictated by reactionary social forces neglects life's creative and multiple ways of becoming; so, for instance, in the observation that evolution includes nonreproductive sex, Chisholm finds an alternative pathway of queer desire that Edelman disregards. These criticisms suggest that Edelman remains locked in an unnecessarily oppositional framing that dispenses too neatly with the more-than-political possibilities circulating in nature, or something like it.[4] Because he only ever locates queer difference *against* – against nature; against what is; against what we know and that for which we can hope – he expresses hostility towards the future that these authors find unwarranted and irresponsible.

Conversely, the queer futures sketched by authors such as Seymour and Chisholm refer to and in various ways get their grounding in an ontology. Chisholm's interest in the queer swerve of evolutionary processes is one example of a broad interest in materialist and scientific approaches to queer theory, a significant portion of which identify the environmental crisis as a motivating factor for getting beyond a too-narrow focus on language, society, and the human. Timothy Morton, for one, writes of how 'queerness [...] is installed in biological substance as such' in terms of the strange intimacies and blurry ontologies of living matter (quoted in Seymour 2020: 109). In a field-defining article, he concludes: 'Ecology and queer theory are intimate. It's not that ecological thinking would benefit from an injection of queer theory from the outside. It's that, fully and properly, ecology is queer theory and queer theory is ecology: queer ecology' (Morton 2010). Karen Barad (2012) has labored to show how 'nature's queer performativity', as she puts it, inheres at the atomic level. Her posthumanist forays into new materialism via quantum physics have been hugely influential for queer thinking about the way nonnormative social relations have a place – even perhaps a privileged place – in a universe where queerness is constitutive, abundant, and constantly surprising. Such work, and much more besides, poses a stark rejoinder to any perspectives, whether scientific, literary, or theological, that conscript nature or the natural into 'straight' paradigms.[5] It not only depathologizes queer identity but claims that those heteronormative efforts fight a losing battle against the very fiber of the vibrantly and queerly material world.

These queer ecological insights are understood as a potent antidote to the forces that wreak environmental havoc and climate disaster – that is to say, they are key to making possible a planetary future. One such connection winds its way through the celebration of nonnormative forms of affiliation and care. For Morton and Seymour with him, queerness as a practice of 'loving the strange stranger' has salutary ecological implications insofar as it analogously describes nonanthropocentric ethical orientations: love, care, and responsibility for those beyond our own species (Seymour 2013). *Contra* Edelman and the antisocial camp, the move here is to unsettle, expand, and in *that* sense queer the domain of sociality. The former's opposition of queer ethics to the future is taken as the unfortunate consequence of his totalizing negativity around the figure of 'the Child' as object of political fantasy. Queer ecology of the variety discussed thus far instead seeks to dislodge that symbolic center by considering the array of human beings' surprising (and often repressed) relations, from the animal to the atom, the biotic to the quantum. In this move to dethrone the normative and exceptional human by way of its pluriform sexualities and more-than-human connections, work in queer ecology enhances environmentalists' rebuttal of Western modernity's nature–culture dualism, a separation widely indicted as the main driver of ecological degradation and climate change (Luciano and Chen 2015).

Many of those who present queerness as a figure traversing a world that has erroneously been divided into the 'social' and the 'ecological' strive not to settle for a romantic optimism. As much as they laud the wonderfully weird possibility of queer becoming, some queer ecologists mourn what has been lost and try to face head-on landscapes of destruction and extinction (O'Dell-Chaib 2017). Indeed, for a thinker like Donna Haraway, the apocalyptic tenor of so many calls for ecological responsibility is problematic precisely because it evokes binary certainty: annihilation or salvation. In *Staying With the Trouble*, she asks: 'How can we think in times of urgencies without the self-indulgent and self-fulfilling myths of apocalypse?' (Haraway 2016: 35). Haraway is not alone in the way her allergy to apocalypse aligns with a parallel criticism of the negativity that pervades antisocial queer theory – a point to which I will return. A revealing illustration is found in her queer-ecological rebuke of the heteronormative imperative to reproduce, which is captured in the controversial slogan 'make kin not babies' (Haraway 2016: 137). Despite an apparent overlap, it differs markedly from Edelman's critical posture towards the future because of the presence in Haraway's work of a version of queer ontology I have charted in this section. For Haraway, the negativity associated with queer critique of the oppressive status quo is muted, redirected, and ultimately revealed to be enfolded in and buoyed by this other sense of queer: multispecies entanglement, unruly becoming, dogged possibility.

This positive ontological leaning is the basis for many queer ecologists' divergence from the antisocial thesis regarding the meaning of queerness as it concerns the future. In their efforts to harness the salutary ecological implications of non-normative sexualities, such thinkers are not only directly suspicious of the negativity animating 'no future' but, more fundamentally, seek to render such a positioning superfluous. In other words, if their work finds that there can be a future for which we hope, then this is because the future *will* be queer; it *must* be, in some sense, because life and matter themselves are already superlatively queer.

While a more thorough review of this creative work would turn up fascinating insights, it would be unlikely to make more than passing contact with the apocalyptic provocations surfaced in the first section's opening images of climate protest and the refusal of the reproductive same. For as we have seen, queer ecologies find an abundant alternative to the apparent double bind in which the future threatened by anthropocentric ecocide can only be imagined in heteronormative terms: futurity tied to planetary, biological, and molecular queerness. These queer ecological futures thus issue a reproach precisely where I have proposed a striking consonance regarding the discourse of ends, of refusal, among young climate activists and within queer theory. In so many words, the suggestion is that this youthful negativity, like that of queer theory writ large, might, in time, come around.

'Everything Queer, Nothing Radical?'[6]

But in the proliferation of speech about a queerness that so vastly exceeds any oppositional queer politics, others ask whether the critical edge of queer theorizing dissipates. Sophie Lewis, for instance, elaborates her disappointment at seeing Haraway's long negotiation of the tense interstices among feminism, socialism, and materialism resolve, unhappily, into 'an apolitical notion of transspecies Gemeinschaft' (Lewis 2017: 7). Of particular concern to Lewis is Haraway's leveraging of a posthuman twist on queer nonreproductivity – that is, her injunction to make queer kin through cross-species relationships, rather than make more human babies – which Lewis views as a 'decisive turn towards a primitivism-tinged, misanthropic populationism' (Lewis 2017: 5). Despite Haraway's protestations to the contrary, Lewis alleges that Haraway does not do enough to distance her position from the 'neoimperialism, neoliberalism, misogyny, and racism' that pervade population reduction discourses (Lewis 2017: 11). While Haraway continues to gesture towards the need for direct resistance to systems of exclusion and oppression, those aspects of her text are unsubstantiated, Lewis contends, because the insurgent character of 'queer' kinship is assumed rather than argued. In this way, and crucially for my earlier reading of queer ecologies, Lewis zeroes in on a predicament faced by those who seek to root the political thrust of queer theory in a broader queer universe:

> the queer biological efflorescence [Haraway] valorizes is already here − she says − everywhere. We have to actively generalize it anyway. This feels like a flat ontology; less a matter of '[staying with the] trouble' (or struggle) than infinite regression [...] Why 'kinnovate' if traditional families are already queer? How does kinnovating break down the structural apparatuses of slow violence?
>
> *(Lewis 2017: 9−10)*

Lewis ultimately sees an unacceptable tradeoff: whatever Haraway's work contributes by imagining the future outside of a heteronormative reproduction of family and kinship, it surrenders something decisive with respect to a robust vision of eco-social justice for dispossessed human beings.

Jordy Rosenberg shares the conviction that however necessary the ecological rejoinder to anthropocentric theorizing, including that relating to sexuality, the principal paradigms of queer ecology are insufficiently descriptive of and antagonistic to the specific features of social reproduction that cause suffering and immiseration (Rosenberg 2014; see also Watson 2017). Like Lewis, Rosenberg calls into question the futurist orientation that is said to be enabled through recognition of planetary, cosmic, and molecular queerness. Situating work by thinkers such as Morton and Claire Colebrook as representative of a wider theoretical trend, Rosenberg writes:

> The ontological turn [...] reshapes an old paradigm, a primitivist fantasy that hinges on the violent erasure of the social: the conjuring of a realm − an 'ancestral realm' − that exists in the present, but in parallax to historical time. A *terra nullius* of the theoretical landscape.
>
> *(Rosenberg 2014)*

As with Lewis in regard to Haraway, Rosenberg is unpersuaded by the figuring of queerness as inherently productive (i.e., of possibility, of the future, of life), and suggests that the social contradictions to which queer studies traditionally orients itself are sidestepped rather than overcome; thus:

> We can see that the ontological turn is looking for a line of flight − for a way out of capitalist logics and repetitions. We might wish to take this flight without asking questions, but can we? No-one wants to be the messenger − to be the one to say that queerness itself, in itself, isn't generative, doesn't create things, cannot break us out of this present and into a different future.
>
> *(Rosenberg 2014)*

But for Rosenberg, it is essential to bear this message if queer thinking is to interrupt overlapping global systems of violence. Taking simultaneous aim at the categories of ontology and temporality in Morton's writing, Rosenberg maintains that in the move to 'grasp biology as a kind of *sheer* queerness [...] the molecular becomes the vehicle for the cleaving of ontology from politics – investing it with a dual temporalization that is simultaneously a dehistoricization' (Rosenberg 2014). This is not a neutral mistake but one which is found, in Rosenberg's extended and complicated argument, to occlude the historical operations of 'settler [colonial]-forms of dispossession under neoliberal regimes of finance capital' (Rosenberg 2014). From this vantage, Morton's doubly transitive equation between 'queer theory' and 'ecology' (mentioned above) is an illusory suturing of the social and the ecological that leaves the resulting 'queer ecology' unable to speak meaningfully about how to achieve the fragile and hard-fought material conditions of a world transformed. For Rosenberg, who puts forward a fundamentally Marxist reading, a future that merits the designation 'queer' must be consciously wrested from its opponents and collectively built.

In a recent review of the field, Seymour acknowledges this depoliticizing tendency in queer ecology, even as she reminds readers of significant exceptions to it. Still, her welcome push for critical specificity in response to the problem traced in this section – she asks: 'if we can "queer" everything, why call anything queer?' (Seymour 2020: 109) – only glances at the thoroughgoing negativity of the antisocial thesis, one which proposes the incommensurability of queerness with future-oriented political projects. But another vein of critical engagement with queer ecology grapples with this distinct meaning of queer negativity and the challenges that flow from it. The theorists I turn to now describe a more basic confrontation with the ambivalence of the world, of human subjects, and of all our transformative labors.

Elizabeth Wilson and Steven Swarbrick both dissect the queer tonalities of Barad's work, which was identified above as an influential reference point in new materialism. These thinkers aver that the radical claim in antisocial queer theory consists in its defense of the 'foundational destructiveness in sexuality' (Wilson 2018: 22). Though the association of this spiraling death drive (in antisocial queer theory's Lacanian vocabulary) exclusively with 'deviant' sexual practices is in error, queer people may for that reason be better able to embrace the irreconcilable status of all sexuality as a site of nonmeaning lodged in the subject, constantly undoing the subject's search for meaning, identity, belonging, and continuity across time (Edelman 2004; Swarbrick 2019). If queer theory is to be employed across social and ecological contexts, then it is this negativity, Wilson maintains, that 'needs to be clearly formulated in any queer account of nature', yet it is precisely that which recedes in Barad's account of a fluid and aleatory universe; thus Wilson 'want[s] to find a way to articulate more fully what Barad gestures towards but seems unable to entirely countenance: that negativity, never under our control, has a permanent place in the spacetimemattering of the world' (Wilson 2018: 22, 24).

For similar reasons, Swarbrick and Wilson are suspicious of the 'queer ethics of repair' that emerges in Barad's posthuman forays. In short, their worry is that the reparative impulse to redeem the human from its destructive flaws is essentially transposed onto the intra-active web of 'life' or 'nature' or 'matter' which, it is supposed, overcomes by overflowing: in life's queer vibrations, one finds no warrant for violations against marginalized human beings or for unsustainable human activities on planet Earth, but instead an unraveling of both. Yet while Wilson and Swarbrick each acknowledge, in ways we can now recognize, that this depathologizing recuperation of queerness has political and environmental teeth, in their view it requires an unacceptable sacrifice of queer theory's most radical insights: that the subject's coherence and wholeness is indebted to a repudiation of its lack; that nature is only ever with *and* against human efforts, never a stable foundation for ethics or politics; that the stability of the 'life' or future we seek to protect is a selective fantasy. These elements indicate a 'reproductive' process within queer ecologies that subtly excises negativity – a maneuver to which queer theory, as these thinkers understand it, must carefully attend. Thus, both Wilson and Swarbrick draw on Edelman's antifuturist and antisocial reading of queerness to challenge Barad's and others' liberative gestures (even as they continue to criticize the obviously oppressive dynamics those thinkers strive to upend).[7] Swarbrick contests the elision of negativity or antagonism as such with capitalism's regime of enforced destruction; 'Nature's queer negativity', he writes, 'does not simply befall Mother Earth, though it does take on historically specific and uncanny forms – "fossil capitalism" being the most pressing of those forms.' Lest this be read as equivocating around the violence of the present, he clarifies that nature's negativity 'fits neither the capitalist's image of life as endless accumulation nor the posthumanist's image of life as endlessly adaptive network of living beings' (Swarbrick 2019). Wilson alleges that even work which, in dialogue with Edelman, explores the nonhuman in terms of the 'inhuman' queer 'tends to under-read the negativity and confusion that queer entails, and so it renders nature, and the politics we might extract from it, more palatable than perhaps they should be' (Wilson 2018: 21; see Luciano and Chen 2015). 'Life without the optimism of repair' verges on the unbearable, but it is not, for Swarbrick, bereft of a 'paradoxical agency' that would allow the ecological subject to, in Haraway's words, stay with the trouble (see Wilson 2018: 27–28).

★★★

The array of responses to queer ecologies in this section indicate, I would argue, that queer negativity is not simply a nonstarter for understanding the intersection between social reproduction and environmental futures. However, I must underscore the ambiguity that emerges with respect to the character of this negativity, its operations, and its objects of critique. The first pair of assessments (Lewis and Rosenberg) and the second (Wilson and Swarbrick) sit uneasily alongside each other. While both find fault with queer ecologists' unsatisfactory circumvention of certain crucial antagonisms or contradictions, it is not immediately clear how a materialism focused on mobilizing minoritarian subjects to construct an alternative future can be reconciled with a concentration on nature's abiding negativity. The latter presents the queer practice of saying 'no future', and meaning it, as a fearful prospect, to be sure, yet also as an enduring and a most urgent one. This is a significant departure from the tendency of other queer ecologists to view the theoretical fixation on antifuturity as a distraction and a liability. It is also, to some extent, a challenge to the humanist-utopian bent of Lewis's and Rosenberg's critiques of those projects.

These thinkers all grapple with queer negativity as a critical discourse of ending and unmaking and, at the same time, with the manner in which a critique of the existing world can open onto possibilities for altering or perhaps transforming its patterns of relation. Divergent appraisals of the function of 'apocalypse' in connection with these efforts demonstrate that this category serves as a crucial index for ways of perceiving the dynamic between negativity and possibility. Neither Lewis nor Rosenberg regard talk of apocalypse to be justified or useful for queer politics, whereas the apocalyptic negativity of Edelman's work becomes the medium through which Wilson and Swarbrick undertake their demanding revision of queer ecology. Their differences on this score highlight decisive questions about where and how queer thought should exert pressure on the environmental equation of child–future–nature.

Queer Negativity and Environmental Apocalypse: A Revealing Nexus

The claim that apocalypticism manifests a negativity which shuts down possibility of one kind or another is a recurrent one. What is perhaps surprising is the degree to which similar worries appear even among interlocutors working constructively with Edelman's futural negativity. As ecologically inflected interest in antisocial queer theory has recently gained some traction, theorists from several disciplines have found a valuable resource in his critique of reproductive futurism and heard anew the stinging force of 'no future'. While some adopt and even amplify the apocalyptic association of his charge (Giffney 2016), a significant faction distances itself from it. Although I am arguing for the importance of the apocalyptic as a point of articulation between queer theory and ecological thought, it is instructive to know what is at stake for those seeking to disentangle their accounts of queer negativity from any apocalyptic knots. In contrapuntal fashion, the following three examples help me clarify the explanatory capacity of queer futural negativity in the context of environmental apocalypse.

Heather Davis sees a literalizing of Edelman's nonreproductive queer theory as she tracks the role of plastic – identified as a nonhuman 'toxic progeny' – in pollution, extinction, and biological mutability. She does not endorse the petrochemical regime fueling these changes, but nevertheless observes that its results instantiate a 'queering' of life on Earth. Davis contends that 'through the saturation of the world with the advents of modern chemistry, in the multiple forms of endocrine disruption, Edelman's queer future is no longer a particular political position, but rather increasingly bleeds into biological reality' (Davis 2015: 240–241). She tries to embrace the ambivalence of 'toxicity', taking it as both a regrettable contingency and a condition that signals a new form of life. It is in this vein that she also separates her deployment of negativity from Edelman's. Even as she worries about naïve celebrations of queer futurity in the face of irreparable loss, she indicts the 'apocalyptic nihilism that subtends Edelman's argumentation' (Davis 2015: 242). That version of 'no future' is, she holds, ultimately unable to account for the asymmetrical loss of a future among human beings and other species. The specifically *apocalyptic* tenor of futural negativity therefore comes to shoulder the burden of a totalizing nihilism that is guilty of a privileged white masculinism in its neglect of multiply differentiated subjects.[8] Davis is right that queer theory in an ecological register ought to meaningfully address intersections among race, gender, class, and geopolitical status. Whether queer negativity as Edelman formulates it can account for the differential production and deprivation of, and divestment from, the future is a wider debate (Bliss 2015; Caserio et al. 2006), but I resist the claim that an apocalyptic mode of thinking necessarily attenuates its ability to do so.

A second example evinces another disavowal of apocalypse within an argument that affirms the ongoing diagnostic value of antisocial queer theory in ecological terms. In *The Child to Come: Life after the Human Catastrophe*, Rebekah Sheldon argues for a planetary extension of Edelman's mapping of biopower: 'Locked in tropic correspondence, the planet inflects and deepens the child's association with nature, the child lends its humanity to the planet, and the vulnerable innocence historically associated with the child enshrouds Earth' (Sheldon 2016: 26–27). 'No future' detects and defies the prevaricating emptiness of the planetary-political system that reproduces itself, since, '[i]n the name of the future, we must be protected from the future' (Sheldon 2016: 36). Sheldon's problem with apocalyptic discourse emerges in the context of this system. She echoes Swarbrick's and Wilson's worry about pacifying nature's queer negativity in order to uphold the stasis of 'life' and the 'value of civilization' and likewise identifies a persistent redemptive impulse in environmentalism – the pastoral fantasy. Yet for her, an apocalyptic imagination partakes of the cover-up rather than challenging it. She writes:

> [T]hinking the catastrophic requires the apprehension that all systems are unstable and groundless, without necessity and with no truth other than their own capacity to continue operating. By contrast, apocalypse, which also labors in the temporal register, designates that which has always already been awaiting our discovery, now at the end of the quest literally unveiled. So, apocalypse requires a self-similarity beyond duration, lurking within all the ephemera of the passing hours.
>
> *(Sheldon 2016: 46–47)*

Effective ecological engagement with queer negativity is, for Sheldon, hindered by the false sense of epistemic certainty lent by the dipolar harmony-apocalypse structure. It is that opposition, rather than ecological negativity as such, which forecloses the possibility of something genuinely different. In this way, she aims to clarify the meaning of queer (non)futurity for environmentalism: 'The issue is not that there is no future but rather that there is no sure way of orienting toward that future, either to save it or to survive it' (Sheldon 2016: 179–180). That ecological perspectives must venture a clear-eyed recognition of that which is without whither or why is imperative. I have suggested that apocalyptic queer negativity pursues this very task, what Sheldon calls 'thinking the catastrophic', with particular vigor. More than a semantic adjustment is required, however, to address the fundamental problem she identifies.

The assertion that apocalyptically tinged critique settles into a rigid oppositional structure and thereby does a disservice to the future's uncertainty is both familiar and persistent. This structure becomes the focal point in Sarah Ensor's attempt to mediate the tension between 'ecofuturists' and antisocial queer theorists. She alleges that the 'two camps' commit a mirror-image error with the result that

> [t]he antisocial theorist's rejection of futurity is […] ultimately no more radical and no less normative than is the steadfast promotion of child-rearing – in large part because it continues to concretize and externalize the future, to treat it as the grammatical object of our transitive acts.
>
> *(Ensor 2012: 412)*

Edelman's polemic is her primary illustration of a transitive negation of the future whose intensity cannot finally belie its investment in futurity. Yet his initial figuration of queerness in the language of intransitivity inspires Ensor's alternative vantage on a queer environmental future. She cites his association of this intransitivity with the death drive, which 'can only insist, and every end toward which we mistakenly interpret its insistence to pertain is a sort of grammatical placeholder, one that tempts us to read as transitive a pulsion that attains through insistence alone the satisfaction no end ever holds' (Ensor 2012: 412, citing Edelman 2014: 22). Though she claims that his 'no' to the future submits to this temptation, she also sees what might have been in his titular formulation:

> 'not a directed act but rather a gesture of evasion – one that would allow "the future" to exist while positioning the queer subject in a slanted relation to it'; this posture characterizes a 'queer relationship to futurity [which] is intransitive not because of how it refuses but rather because of how it facilitates a notion of the future (and of futurity) outside the realm of objects, outside the push and pull of acceptance or refusal, both outside and beyond our capacity to control'.
>
> *(Ensor 2012: 413–414)*

This is the move that, for Ensor, redraws the relation between queer negativity and possibility towards an epistemically chastened futurity that is neither positive nor negative, finally. She also ventures that this intransitive negativity would better accommodate ecological subjects' personally and structurally differential relation to the future, noting: 'Perhaps the question is not the future, yes or no, but the future, which and whose, where and when and how' (Ensor 2012: 414).

Ensor's argument regarding Edelman's failure to hold the line relies on an analysis of his syntax and form that I do not dispute, but I want to stay with what she identifies as his aspiration. While my interest is not exegetical, Edelman has been a central figure in this chapter, and so it is worth noting that others interpret his queer negativity as more consistent in regard to an insistently intransitive approach (Brintnall 2018). Moreover, some seize on precisely this quality as the apocalyptic dimension of his work (Lynch 2019; Tonstad 2016b). They resource his queer ethics in dialogue with the language of apocalypse as a tool for rendering judgment on the world as it is and cultivating a disposition that learns not to settle for reformism's half measures while dealing thoughtfully with the risk of a quietist pessimism that perpetuates privilege (Brintnall et al. 2018; Tonstad 2018). Such work can include countering Edelman on behalf on an aspiration in queer negativity that is rightly called 'apocalyptic'.

Those who make the language of apocalypse the fault line for separating varieties of negativity identify significant pressure points, but to permanently consign an apocalyptic imaginary to its flattening and dichotomizing uses and abuses has no more warrant than efforts to retrieve its revelatory potential and self-critical character. And this is not only because the negativity within apocalyptic discourse can be evaluated as a performative enactment of possibility (Woltersdorff 2012) in ways that are functionally equivalent to the negativity-futurity dynamic said to be foreclosed by apocalypticism. More pointedly, I contend that using the label 'apocalyptic' to discriminate 'bad' from 'good' forms of negativity or critique (thus reproducing a structural antagonism) is itself an operation that may be exposed by the revelatory distillation engendered in *apokalypsis* (Tonstad 2018). Theologians practiced in the custodianship of terrorizing concepts know well the friction entailed in projects of retrieval (Keller 2005). Turning the apocalyptic reflex inward indicates that what is unveiled is equally a judgment on the world and on the one who judges, leading to an affirmation of the unfinished labor of revealing. When interpreted in the intransitive sense just adduced, '[a]n apocalyptic mindfulness' understands that "the removal of the veil" [...] means not closure but *dis-closure*' (Keller 2021: 3). It is in this sense akin to a spiritual practice, 'a guide and discipline for living in the end' (Lambelet 2021: 483). Within the orbit of queer theory, to speak of persistence raises the question of desire and where desire crosses into the 'optimism of repair'. Does my intransitive reading of apocalyptic queer negativity trade an annihilative 'ending' for a redemptive 'end'? The rich exchange between queer theory and Christian theology may reach an impasse at this point (Tonstad 2019). What I have tried to draw out by examining environmentalism's child–future–nature equation is the extent to which the dilemmas posed by the notion of environmental apocalypse are not merely analogous to but interwoven with that exchange at nearly every turn.

Taken together, the collection of scholarly voices surveyed returns us to our opening scenes: What species of negativity animates moments of concomitant rejection of future and children in a climate movement largely intent on protecting the future for children? What do these activists' renderings of environmental catastrophe have to do with queer theory's critique of reproductive futurism? In each case, what exactly is refused and what, if anything, can be said about whatever would take its place? I have argued that, in negotiating the challenges of a historicized ecological materialism, tensions concerning the place of queer futural negativity coalesce in a particularly intense way around the category of apocalypse. A queer apocalyptic mindfulness exposes sentimentalizing interpretations of those cardboard protest signs, yet looks for an opening where the youth movement wields platitudes of future-oriented care at the vanishing point of sincerity and irony, while also being conscious of their uneven exclusions (Stall-Paquet 2020). As an intransitive practice, it understands that every effort to carve a pure space for intersectional and intergenerational solidarity will fail; by the same token, it disallows a pat negation of these efforts because the negativity that drives the world's reproduction is present in the Earth-bound subject as well. At issue, then, is a negative mode that would resist oppressive systems and renounce pretensions ever to purge the negativity that animates resistance; seek the end of the world that reproduces violence and have the wisdom not to confuse the end of the world with the end of the Earth; and make visible how the possibility of possibility verges on collapse both in grasping the future and in rejecting it. Apocalyptic thinking helps queer ecologies tread closer to that razor's edge.

Notes

1 https://www.weforum.org/agenda/2019/01/our-house-is-on-fire-16-year-old-greta-thunberg-speaks-truth-to-power/.
2 Thunberg asks: 'And why should I be studying for a future that soon will be no more, when no one is doing anything whatsoever to save that future?' (Baraitser 2020: 500).
3 The tension I describe tracks with a broad-brush narrative regarding a material or onto-logical 'turn' in humanities scholarship, which follows and to various degrees revises its earlier linguistic turn.
4 Among the queer ecologists mentioned, there is disagreement about whether the material world can be interpreted using the fraught category of 'nature' or whether it is better approached without it, as Timothy Morton (2009) argues in *Ecology without Nature: Rethinking Environmental Aesthetics*.
5 For Whitney Bauman, 'Religion is much queerer than we had ever imagined. Nature is as well' (Bauman 2018: 11).
6 I borrow this phrasing from the title of Linn Marie Tonstad's (2016a) perceptive article on strategies of queering in Christian theology.
7 Wilson uses Edelman's notion of 'homographesis', while Swarbrick deploys his figuration of the *sinthome*.
8 Briohny Walker (2019) and Seymour (2013) reach conclusions similar to Davis's on slightly different grounds. Walker initially counterposes her Edelmanian rejection of the future from theorists who, '[w]hile making valuable use of Edelman's work [...], resist his searing nega-tivity and refusal of the future itself' (Walker 2019:145). However, she too backs away from this negativity, alleging a pitfall in his work to which Davis and Seymour also allude:

    To suggest, as Edelman seems to, that all futurity is capitalist, Western, and hetero-reproductive, is to silence the resistance, perseverance and futurity of other ways of being. In addition to dismissing alternative queer futures and the textures of anti-capi-talist resistance projects, this oversimplification of futurity has colonial connotations. (Walker 2019: 148)

References

Barad, K. (2012) 'Nature's queer performativity (the authorized version)', *Kvinder, Køn og forskning /Women, Gender and Research*, 1–2, 25–53. https://9pdf.org/document/y9635lxw-nature-s-queer-performativity.html.

Baraitser, L. (2020) 'The maternal death drive: Greta Thunberg and the question of the future', *Psychoanalysis, Culture & Society*, 25, 499–517. https://doi.org/10.1057/s41282-020-00197-y.

Bauman, W. (2018) *Meaningful flesh: reflections on religion and nature for a queer planet*. Goleta: Punctum.

Bliss, J. (2015) 'Hope against hope: queer negativity, Black feminist theorizing, and reproduction without futurity', *Mosaic*, 48(1), 84–98. https://doi.org/10.1353/mos.2015.0007.

Brintnall, K. (2018) 'The politics of revelation: unveiling negativity in the work of Lee Edelman and Georges Bataille', *CrossCurrents*, 68(2), 309–322. https://doi.org/10.1111/cros.12313.

Brintnall, K., Marchal, J., and Moore, S. (eds.) (2018) *Sexual disorientations: queer temporalities, affects, theologies*. New York: Fordham University Press.

Caserio, R., Edelman, L., Halberstam, J., Muñoz, J. and Dean, T. (2006) 'The antisocial thesis in queer heory', *PMLA*, 121(3), 819–828. http://www.jstor.org/stable/25486357.

Chisholm, D. (2010) 'Biophilia, creative involution, and the ecological future of queer desire', in Mortimer-Sandilands, C. and Erickson, B. (eds.) *Queer ecologies: sex, nature, politics, desire*. Bloomington: Indiana University Press, 359–381.

Davis, H. (2015) 'Toxic progeny: the plastisophere and other queer futures', *philoSOPHIA*, 5(2), 231–250. https://www.kabk.nl/en/lectorates/design/toxic-progeny-the-plastisphere-and-other-queer-futures-by-heather-davis.

Edelman, L. (2004) *No future: queer theory and the death drive*. Durham, NC: Duke University Press.

Ensor, S. (2012) 'Spinster ecology: Rachel Carson, Sarah Orne Jewett, and nonreproductive futurity', *American Literature*, 84(2), 409–435. https://doi.org10.1215/00029831-1587395.

Giffney, N. (2016) 'Queer apocal/o/ptic/ism: the death drive and the human', in Giffney, N. and Hird, M.J. (eds.) *Queering the non/human*. London: Routledge, 55–78.

Haraway, D. (2016) *Staying with the trouble: making kin in the Chthulucene*. Durham, NC: Duke University Press.

Keller, C. (2005). *God and power: counter-apocalyptic journeys*. Minneapolis: Fortress.

Keller, C. (2021) *Facing apocalypse: climate, democracy, and other last chances*. Maryknoll, NY: Orbis.

Kverndokk, K. (2020) 'Talking about your generation: "our children" as a trope in climate change discourse', *Ethnologia Europaea*, 50(1), 145–158. https://doi.org/ 10.16995/ee.974.

Lambelet, K. (2021) 'The lure of apocalypse: ecology, ethics, and the end of the world', *Studies in Christian Ethics*, 34(4), 482–497. https://doi.org/10.1177/09539468211031352.

Lewis, S.A. (2017) 'Chthulu plays no role for me', *Viewpoint Magazine*, May 8. https://viewpointmag.com/2017/05/08/cthulhu-plays-no-role-for-me/.

Luciano, D. and Chen, M. (2015) 'Has the queer ever been human?', *GLQ*, 21(2–3), 183–207. https://doi.org/10.1215/10642684-2843215.

Lynch, T. (2019) *Apocalyptic political theology: Hegel, Taubes, and Malabou*. London: Bloomsbury.

Mortimer-Sandilands, C. and Erickson, B. (2010) 'Introduction: a genealogy of queer ecologies', in Mortimer-Sandilands, C. and Erickson, B. (eds.) *Queer ecologies: sex, nature, politics, desire*. Bloomington: Indiana University Press, 1–48.

Morton, T. (2009) *Ecology without nature: rethinking environmental aesthetics*. Cambridge, MA: Harvard University Press.

Morton, T. (2010) 'Guest column: queer ecology', *PMLA*, 125(2), 273–282. https://doi.org/10.1632/pmla.2010.125.2.273.

Nordhaus, T. and Shellenberger, M. (2007) *Break through: from the death of environmentalism to the politics of possibility*. New York: Houghton Mifflin.

O'Dell-Chaib, C. (2017) 'Biophilia's queer remnants', *Bulletin for the Study of Religion*, 46 (3–4), 18–23. https://doi.org/10.1558/bsor.33167.

Rosenberg, J. (2014) 'The molecularization of sexuality: on some primitivisms of the present', *Theory & Event*, 17(2). https://www.muse.jhu.edu/article/546470.

Seymour, N. (2013) *Strange natures: futurity, empathy, and the queer ecological imagination*. Urbana: University of Illinois Press.

Seymour, N. (2020) 'Queer ecologies and queer environmentalisms', in Somerville, S. (ed.) *The Cambridge companion to queer studies*. Cambridge: Cambridge University Press, 108–121.

Sheldon, R. (2016) *The child to come: life after the human catastrophe*. Minneapolis: University of Minnesota Press.

Skrimshire, S. (ed.) (2010) *Future ethics: climate change and apocalyptic imagination*. London: Continuum.

Stall-Paquet, C. (2020) 'The women pledging not to have kids until meaningful action on climate change is made', *Elle*, 9 March. https://www.ellecanada.com/culture/society/the-women-pledging-not-to-have-kids-until-meaningful-action-on-climate-change-is-made.

Swarbrick, S. (2019) 'Nature's queer negativity: between Barad and Deleuze', *Postmodern Culture*, 29(2). https://doi.org/10.1353/pmc.2019.0003.

Tonstad, L. (2016a) 'Everything queer, nothing radical?', *Svensk teologisk kvartalskrift*, 92(3–4), 118–129.

Tonstad, L. (2016b) *God and difference: the trinity, sexuality, and the transformation of finitude*. London: Routledge.

Tonstad, L. (2018) *Queer theology: beyond apologetics*. Eugene, OR: Cascade.

Tonstad, L. (2019) 'Response to Brintnall', *Syndicate*, October 7. https://syndicate.network/symposia/theology/queer-theology/.

Walker, B. (2019) 'Precarious time: queer Anthropocene futures', *Parrhesia*, 30, 137–155. http://www.parrhesiajournal.org/parrhesia30/parrhesia30_walker.pdf.

Watson, M. (2017) 'Raging hallelujah', *Science as Culture*, 26(2), 271–275. https://doi.org/10.1080/09505431.2017.1292232.

Wiegman, R. (2017) 'Sex and negativity; or, what queer theory has for you', *Cultural Critique*, 95, 219–243. https://doi.org/10.5749/CULTURALCRITIQUE.95.2017.0219.

Wilson, E.A. (2018) 'Acts against nature', *Angelaki*, 23(1), 19–31. https://doi.org/10.1080/0969725X.2018.1435368.

Woltersdorff, V. (aka Logorrhöe, L.) (2012) 'Apocalypse NOW! Radical negativity and the performativity of ending in queer theory', *InterAlia*, 7. https://interalia.queerstudies.pl/issues/7_2012//05_apocalypse_now.htm.

4

SLOW CATASTROPHE

A Concept for the Anthropocene

Jonathon Catlin

1 Introduction: Climate Change and the Unthinkable

In his 2016 book, *The Great Derangement: Climate Change and the Unthinkable*, Amitav Ghosh asks how global warming, despite the palpably accelerating frequency, scale, and destructiveness of natural disasters around the world, has for so long remained an 'unthinkable' subject for leading artists, intellectuals, and politicians. Why is it, he asked, that as soon as a novel or film ventures to depict climate catastrophe or ecological collapse, it is dismissed as science fiction, even while such 'fantasies' increasingly become reality as weather-related disasters cause unprecedented devastation around the world? Ghosh concluded that admitting the reality of global warming would be so threatening to modern capitalist society's prevailing enlightenment ethos emphasizing progress, free markets, moral individualism, and the advantage of capital over the rights of human beings that the necessary collective responses to these disasters are not yet even thinkable, much less actionable.

This chapter argues that making climate catastrophe *thinkable* requires that its apocalyptic dimensions first have to be made *imaginable* through both narrative and conceptual labours. Drawing upon contemporary climate fiction and critical theory, I argue that global warming must be grasped as a social totality through the constellation of rhetorical registers and conceptual frames I term *slow catastrophe*. In between progressive *crisis optimism*, which frames the climate emergency as an opportunity for social transformation, and *apocalypse pessimism* verging on climate fatalism, which frames it as unavoidable or as an apolitical cycle of violence, this chapter argues for conceiving global warming as a single, ongoing *slow catastrophe* rooted in the extractivist and productivist social order of late capitalist society. Apocalyptic narratives and images can support such a conception so long as they politicize climate catastrophe and render it contingent and potentially avoidable as

DOI: 10.4324/9781003189190-5

the end of *this* social world but also as the beginning of another decarbonized and more just one.

2 Toward 'A New Structure of Feeling': Making Climate Apocalypse Imaginable

The American science-fiction writer Kim Stanley Robinson (2020) has stepped up to fill the imaginative void Ghosh identified following the wager of writers such as Ursula K. Le Guin and Donna Haraway that science fiction can make for uniquely insightful political theory (Burns 2008). Robinson's 2020 novel *The Ministry for the Future* opens with an apocalyptic scene set in the near future: a heat wave sweeps over India, causing deadly wet-bulb temperatures. The nation's power grid fails under strain, wiping out air conditioning, and no cool places are left. Over twenty million people die, and this triggers a revolution.

In India, the governing far-right BJP is blamed for the catastrophe and thrown out of power, ushering in a new, progressive pro-climate government that transitions India to clean energy and large-scale geo-engineering. The United Nations launches the Ministry for the Future to coordinate global climate action and represent the interests of future generations. A traceable cryptocurrency called 'carbon coin' pays those who hold oil reserves not to burn them; it helps eliminate tax havens, spur investment in green technologies, and redistribute global wealth. At the same time, survivors of the heat wave turn to eco-terrorism in a 'War for the Earth' (Robinson 2020: 229), assassinating carbon profiteers and using micro-drones to shoot down passenger flights on 'Crash Day', sink diesel cargo ships, and blow up dirty power plants until the pressure on these industries forces them to turn to renewables (which are soon cheaper anyways). Under the 'Half Earth' initiative, populations are relocated and land is ceded back to nature. Rising ocean levels from polar ice cap melt is slowed by pumping water back on top of glaciers to re-freeze. 'Project Slowdown' halts the capitalist imperative for endless growth. Ultimately, a combination of globally coordinated political will and technological breakthroughs leads to a drop in the real amount of greenhouse gasses in the atmosphere. However, vast damage has been done – 'Irreversible and unfixable catastrophe' for extinct species and millions of human lives lost (Robinson 2020: 293). The planet is not triumphally saved, but total extinction is averted. Characters ultimately recognize that 'there is no other home for us than here [...] That people can take fate in their hands' (Robinson 2020: 563). In a moment of collective danger, a revolutionary vanguard realizes that '*we* are the driving force in history' and resolves to 'put the crisis to use' (Robinson 2020: 458, 459). Indians in particular 'remain horrified by the memory of the heat wave, galvanized, and if not united then nearly so, in a broad coalition determined to re-examine everything, to change whatever needs changing' (Robinson 2020:125).

Robinson's narrative is apocalyptic: it draws the reader in and shocks her to attention by vividly depicting a landscape piled with heat-exhausted corpses as far as the eye can see. Yet this apocalypse is not set in a different world or at the end of time. It could happen almost anywhere, and it could happen tomorrow. As the science-fiction writer William Gibson has quipped: 'The future is already here – it's just not very evenly distributed' (Kennedy 2012). Ghosh puts this claim a bit more pointedly: 'The Anthropocene has reversed the temporal order of modernity: those at the margins' – especially in the Global South – 'are now the first to experience the future that awaits all of us' (Ghosh 2016: 63). It is not an accident that Robinson sets this apocalypse in India, a site of rapid capitalist growth and carbonization.

Ministry's contemporaneity exemplifies Samuel Delany's claim that 'science fiction is not about the future; it uses the future as a narrative convention to present significant distortions of the present' (Delany 2009: 26). As Shelly Streeby has shown, speculative fiction should not be dismissed as a utopian escape into the future; rather, it should be seen as a genre especially important to Indigenous writers and writers of color who for decades have been at the forefront of imaginative work on climate justice (Streeby 2018: 33). For Afrofuturist Octavia Butler, science fiction is not about the future as such, but about preparing us now to 'think about it, foresee it' (quoted in Streeby 2018: 25). Donna Haraway similarly argues that the imaginary worlds of fiction can help us learn to face ongoing catastrophes: 'staying with the trouble requires learning to be truly present, not as a vanishing pivot between awful or edenic pasts and apocalyptic or salvific futures, but as mortal critters entwined in myriad unfinished configurations of places, times, matters, meanings' (Haraway 2016: 2). And even when science fiction does utilize utopian elements, this is not necessarily escapist. As Alfred Schmidt has observed: '[Walter] Benjamin was right when he wrote that in view of the crimes men daily commit against themselves and against external nature [...] there is good sense even in the most eccentric fantasies and extravagant utopias' (Schmidt 2014: 163).

The decisive driver of the great transformation that averts extinction in *Ministry* is 'a new structure of feeling, underlying politics as such' (Robinson 2020: 358). After witnessing the Indian heat wave, diverse constituencies affected by global warming are radicalized and called to action. And indeed, studies have shown that people impacted by extreme weather are more likely to believe in and be concerned about global warming (Bergquist et al. 2019). Short of actual experience, however, climate fiction and reporting can make this abstract process tangible and conceivable. Through this *politicized apocalypse*, Robinson offers us a psychological portrait of how humanity might avert extinction: it depends upon overcoming our 'universal cognitive disability, in that people had a very hard time imagining that catastrophe could happen to them, until it did' – the problem whereby 'until the climate was actually killing them, people had a tendency to deny it could happen' (Robinson 2020: 349). In particular, in the wake of the heave wave, much of the Global North initially believes that 'It was mostly poor people, in particular poor people of color. It couldn't happen in the North. It couldn't happen to prosperous white people' (Robinson 2020: 349). But, of course, it eventually does. Los Angeles, for example, is flooded and washed away into the sea. Robinson wagers that such apocalyptic narratives have a role to play in bringing the Earth back from the brink. However, as Zygmunt Bauman (2014) has asked: 'Does one need catastrophe to happen in order to admit its coming? A chilling thought, indeed'.

The American writer Jonathan Safran Foer puts the problem in slightly different terms: global warming is a 'conceptual event' and thus 'a crisis that we know about but don't believe in' enough to spur us to take action as others have done before in the face of historic threats such as the Second World War (Foer 2019: 46, 53). Similarly, the head of the Ministry for the Future is accused by a survivor of the heat wave of 'pretend[ing] not to know what the future is bringing down on us' and doing too little; abstract scientific knowledge fails to motivate urgent action (Robinson 2020: 95). Another character worries that 'we can't imagine the suffering of the people of the future, so nothing much gets done on their behalf' (Robinson 2020: 172). Foer and Robinson implicitly share the view of Eva Horn's *The Future as Catastrophe* (2018) that, in such cases of 'catastrophe without event' (Horn 2018: 8), impending catastrophes become imaginable and actionable only when they are given concrete instantiation through relatable stories (Catlin 2021). Ghosh has similarly suggested that 'to think about the Anthropocene will be to think in images' (Gosh 2016: 83). Having constructed such images and rich narratives, Robinson concludes: 'Arranging this situation' – averting extinction – 'is left as an exercise for the reader' (Robinson 2020: 58).

In *Ministry*, the horrific images and 'the sheer numbers' from the Indian apocalypse 'recalled the Holocaust, which left a huge hole in civilization's sense of itself' – 'And yet still they burned carbon. […] Everyone alive knew that not enough was being done, and everyone kept doing too little' (Robinson 2020: 227–228). While this near *universal* moral shock is insufficient on its own, it is the cognitive precondition for the necessary global actions to be taken. A woman from Zimbabwe says to an emergency global climate meeting: 'We are all in a single global village now. We share the same air and water, and so this disaster has happened to all of us' (Robinson 2020: 24). But progress is not linear. Building on the Paris Agreement, the 'Trembling Twenties' give way to the 'zombie years' of the 1930s before the final 'Great Turn' (Robinson 2020: 123, 227). After this catastrophe, however, climate denialism simply loses plausibility, and it becomes impossible not to recognize that such 'natural catastrophes' are really 'anthropogenic catastrophes' (Robinson 2020 :250). Global catastrophic experience makes it impossible to ignore the premise with which we began: addressing climate emergency requires renouncing human *omnipotence* over nature while also creating new forms of *agency* on both species-wide and planetary scales.

3 The Permanence and Banality of Climate Catastrophe

While the popular recognition of carbon-fuelled human history as catastrophic ultimately achieved in *Ministry* would be novel, the idea of history itself as a 'permanent catastrophe' – more of a *Denkbild*, or thought-image, than a theory – is not new; it was developed by the Frankfurt School critical theorists Walter Benjamin and Theodor Adorno starting in the 1920s. As Benjamin wrote in his *Arcades*

Project: 'The concept of progress must be grounded in the idea of catastrophe. That things are "status quo" *is* the catastrophe. It is not an ever-present possibility but what in each case is given' (Benjamin 1999: 473). This idea culminated in Benjamin's famous last testament, 'On the Concept of History' (1940), in which he wrote: 'Where a chain of events appears before *us*', the angel 'sees one single catastrophe which keeps piling wreckage upon wreckage and hurls it in front of his feet. [...] What we call progress is *this* storm' (Benjamin 2006: 392). Following Benjamin's death, and after the atrocities of the Second World War, Adorno elaborated this idea further: 'After the catastrophes that have happened, and in view of the catastrophes to come. [...] The world spirit [...] would have to be defined as permanent catastrophe' (Adorno 1973: 320). This radical view of the 'ongoingness' of catastrophe presents ecological devastation as an inherent feature of capitalist modernity, not a bug.

At the same time, the early work of Benjamin (2019) and Adorno (1977) argued that no worldly process can be truly 'permanent' amidst the dialectical interplay of nature and history. The notion of 'permanent catastrophe' thus bears a deliberately provocative and polemical thrust; as Adorno wrote, sometimes the truth lies in such exaggerations (Adorno 2005: 29). On a rhetorical level, polemics can, of course, be politically mobilizing. At the same time, however, this concept can also be redeemed for rational social criticism. The catastrophe at play here is not human life as such. As Adorno wrote, to make 'common cause with the catastrophe' is not genuine criticism but reactionary Spenglerian pessimism (Adorno 1998: 99). Though climate catastrophe may *appear* 'permanent' in its ideological forms, it is not inevitable, but a historical product of a contingent form of life: *extractivist and productivist capitalist modernity*.

Adorno and Max Horkheimer argued that the triumph of instrumental rationality in 'enlightened' modernity led to the domination of nature and human beings rather than advancing genuine freedom (Adorno and Horkheimer 2002), and they also saw this principle at work in prevailing currents of Marxism: 'In Marx the principle of the *domination of nature* is actually accepted quite naïvely [...] so that we might say that the image of a classless society in Marx has something of the quality of a gigantic joint-stock company for the exploitation of nature' (Adorno 2008: 58). However, Horkheimer and Adorno also saw that alongside these tendencies developed possibilities for overcoming catastrophe. Adorno later reflected that after Auschwitz and amidst the Cold War spectre of nuclear annihilation he could only conceive of 'progress' in its most minimal form as 'the prevention and avoidance of total catastrophe' (Adorno 2004: 143). Here 'permanent catastrophe' is revised in a pragmatic direction and intended to be overcome through a heightened consciousness of catastrophic threats. Bruno Latour has more recently called for the *end of all goal-oriented or teleological philosophies of history* and a shift to the minimal goal of maintaining terrestrial 'habitability': 'A *good enough planet*, not the Promised Land. That's what everyone on this planet suddenly dreams of' (Latour and Chakrabarty 2020: 435). Dipesh Chakrabarty (2021: 202) calls this the shift from *Homo sapiens* to *Homo prudens* – a transformation that will, however, require a great deal of courage.

My aim in reviving this catastrophic lens is to argue for moving away from what the disaster historian Scott Knowles (2018) has called 'event-thinking', or conceiving catastrophes as isolated accidents rather than as ongoing *processes* that play out along the lines of, and often exacerbate, existing inequalities and injustices. J.R. McNeill and Peter Engelke similarly describe global warming as a 'great acceleration' of existing trends rather than as a novelty or departure (McNeill and Engelke 2016). Inspired by Robinson's depiction of the emergence of universal catastrophic consciousness in the face of extinction, I argue for viewing serialized climate-related disasters as a *single collective catastrophe* – but one that is not *permanent*, and hence ahistorical and inevitable, but rather *slow* and still potentially preventable.

John Palatella (2021) has criticized the profusion of apocalyptic rhetoric surrounding the Covid-19 pandemic, asking: 'If there have been so many apocalypses' throughout human history, 'how can there have been any at all?' Robinson (2021) answers that apocalypse must be brought – to use Latour's (2018) expression – 'down to earth'. Apocalypse is not something to be *deferred* until the end of times but rather *recognized* as taking place all around us all the time in forms both extreme and banal. *Apocalypse* – as *revelation* – does not have to entail the end of the world as such. As the Indian apocalypse illustrates, the end of one way of life can enable the birth of another from its ruins. As Priya Satia (2022) has written: 'human civilization as we know it *should* expire. History has been a nightmare from which we all must awake'. In *Learning to Die in the Anthropocene*, Roy Scranton (2015) similarly suggested that the death of consumerist, capitalist civilization does not have to entail the death of civilization as such. Instead, as Geoff Mann and Joel Wainwright have argued: 'Rather than accept that "civilization" is dead, we need to struggle to create one that is truly civilized' (Mann and Wainwright 2018: 12).

Robinson was trained in revolutionary thinking about the end of capitalism. He completed a PhD on the science fiction of Philip K. Dick under the critic Fredric Jameson, who was himself schooled in the catastrophic consciousness of Western Marxist figures like Benjamin and Adorno. Robinson updates Jameson's most famous idea for the Anthropocene: 'Easier to imagine the end of the world than the end of capitalism: the old saying had grown teeth and was taking on a literal, vicious accuracy' (Robinson 2021: 25). He wagers that the collective, everyday experience of catastrophe might create cracks in the status quo of 'neoliberal' ideology, 'the structure of feeling in our time; we can't think in anything but economic terms' (Robinson 2021: 75). In contrast to Margaret Thatcher's notorious quip that 'there is no such thing as society', Robinson writes: 'when the taps run dry, society becomes very real' (Robinson 2021: 169).

Yet catastrophe is not only to be found in times of extremity. The subject of the American writer Andrew Durbin's 2017 novel *MacArthur Park* is 'the banality of catastrophe' (Rabinbach 1997: 12). Inspired by Durbin's experience of living in a post-Katrina New Orleans and interest in Rob Nixon's work on the challenges of representing 'slow' environmental violence (Siemsen 2017), the novel follows a protagonist from the sinking island of Miami Beach to New York City during Hurricane Sandy, in which much of the city was flooded and lost power (except for the glowing Goldman Sachs headquarters), watching the sublime destruction outside from the safety of a luxury apartment he is house-sitting, 'just at the edge of disaster' (Durbin 2017: 65). Climate catastrophe is simply something Durbin's protagonist learns to live with: he travels, does an artist retreat, goes to queer nightclubs. But everywhere he is followed by 'the weather' – 'weather as politics, weather as history' (Durbin 2017: 84). Reflecting on the way Hollywood disaster films often redeem mass death through individual heroism, he remarks that, in the West today, 'a catastrophe is an opportunity' (Durbin 2017: 21). At the same time, Durbin's novel is overshadowed by a melancholic 'notion of the ongoing catastrophe of the world' akin to the way 'a forest teems with insects, each gnawing through their environment, destroying it to make it new. [...] it's happening now. It being disaster' (Durbin 2017: 138–139). Catastrophe becomes mundane. We follow the life of Durbin's protagonist slip into what Andreas Malm (2020a) has called our new condition of 'chronic emergency', with endemic ecological and epidemiological threats persisting latently and erupting unexpectedly. Yet whereas Malm (2020a, 2020b, 2021), galvanized by the mass climate demonstrations of 2019, calls for a stance of 'war communism' and climate terrorism and a break with carbon capitalism, Durbin's protagonist, finds himself, like much of his generation, stuck circling around in catastrophic sameness.

4 From Redemptive Apocalypse to the End of the World

While Durbin's *MacArthur Park* lacks the revolutionary vision of Robinson's *Ministry*, taken together these works render important aspects of global warming tangible, both as banal catastrophe *and* as shocking apocalypse. However, just as the notion of *permanent* catastrophe should be revised to *slow* catastrophe, the traditional understanding of apocalypse must also be transformed. The etymology of *apocalypse*, from the biblical Greek, refers to a *revelation* or *uncovering*. However, Anson Rabinbach (1997, 2008) has argued that, after the twentieth-century limit catastrophes of Auschwitz and Hiroshima, the 'redemptive apocalypse' – the uncritical, messianic, or aestheticizing embrace of revolutionary violence – of interwar thinkers like Benjamin came to seem like a dangerous abuse of what Emmanuel Levinas (1988) called the 'useless suffering' of atrocities that were also '*conceptually* devastating' (Neiman 2002: 254). The historical ruptures of 1945 were also conceptual ruptures. After Auschwitz, the notion of *apocalypse* must resist the redemptive tendencies of religious eschatologies and their logic of turning useless suffering into poisoned salvation.

Apocalypse must also be de-theologized in another respect: while *apocalypse* is often taken to mean the end of *the* world, it can be salvaged for secular social criticism if it is qualified to mean the end of *a* world, namely the world of fossil capitalism. This *historical* rather than *theological* end of one order can also entail the beginning of another, as we saw in *Ministry*, and this reminds us that *catastrophe* was once a synonym for *revolution*. This is the premise of Timothy Morton's *Hyperobjects*: 'The end of the world' or 'the catastrophe', Morton claims, 'has already occurred' (Morton 2013: 7; Morton 2015): 'The very feeling of wondering whether the catastrophe will begin soon is a symptom of its already having begun' (Morton 2013: 177). But this does not entail 'an apocalypse' in the sense of 'a predictable conclusion' (Morton 2013: 108) or that it's 'too late' to save the planet (Morton 2020). While Morton's aim is to criticize ecological damage, they embrace the 'epistemological' catastrophe (overturning) or apocalypse (revelation) it has caused: 'Three cheers for the so-called end of the world, then, since this moment is the beginning of history, the end of the human dream that reality is significant for them alone' (Morton 2013: 108). If climate *apocalypse* can be freed from the sense of a *final* end, it might serve as the opening toward what William Connolly (2017) and Chakrabarty (2021) have called 'the planetary' – a conception of history and politics that moves beyond progress-oriented philosophies of history and the inadequate, anthropocentric framework of 'sustainability'.

5 The Trouble with Crisis Optimism

Revolutionary transition to a new social order has often been theorized with the concept of *crisis*, which denotes a period of uncertainty requiring decision, first in medical or juridical contexts, then when the legitimacy of political authority is contested (Koselleck 2006). In 1973, Jürgen Habermas defined 'legitimation crisis', in the narrow, functionalist view of systems theory, as junctures when a social order is not adaptive enough to solve a problem internal to it (Habermas 1988: 2). This conception has been developed, but also challenged, by two of Habermas' successors, the critical theorists Rahel Jaeggi and Nancy Fraser. Against purely functional accounts, Jaeggi argues that 'crises are both given and made': crisis also has a *subjective* dimension, and both are needed for an objective *problem* to escalate into a *crisis* (Jaeggi 2018, 2019). Building upon Habermas, Jaeggi argues that crises are moments when 'learning processes' fail and societies prove unable to adapt to and overcome problems through social transformation. This line of thinking suggests that if a social order can manage or persist through a crisis then it is not actually a crisis. But this reveals the limitations of the crisis framework: we must object: *endurable for whom, and for how long?* Countless social movements have turned *non-crisis* (unrecognized suffering) into *crisis* (recognized suffering), in the process exposing differentials of 'crisis-worthiness' along the lines of race, gender, class, etc.; Dara Strolovitch (2022) thus provocatively theorizes crisis as 'when bad things happen to privileged people'. The designation *crisis* has often meant crisis for those in power, and it thus abandons Benjamin's attention to the suffering of the socially marginal.

Collaborating with Jaeggi, Fraser presents late capitalism as a total 'institutionalized social order' wracked by intersecting crises of economy, ecology, social reproduction, and politics (Fraser and Jaeggi 2018, 2021). This view de-exceptionalizes discrete disasters of economics, climate, etc., and recasts them as mere 'flashpoints' of the same deeper, structural crisis. Fraser writes:

Present-day eco-politics unfolds within, and is marked by, an epochal crisis. A crisis of ecology, to be sure, but also one of economy, society, politics and public health – that is, a *general crisis* whose effects metastasize everywhere, shaking confidence in established worldviews and ruling elites. The result is a crisis of hegemony – and a 'wilding' of public space. No longer tamed by a ruling common sense that forecloses out-of-the-box options, the political sphere is now the site of a frantic search not just for better policies, but for new political projects and ways of living.

(Fraser 2021: 95–96)

For Fraser, consciousness of this 'general crisis' creates a Gramscian interregnum of authority, meaning that there is no going back to the carbonized neoliberal *status quo ante*: the old is dying, even if the new cannot yet be born. Fraser joins the forces of Karl Marx with Karl Polanyi, quipping that 'two Karls are better than one' (Fraser 2013a). On top of Marx's central *economic* insight concerning the exploitation of labor, Polanyi identified several additional *extra-economic* spheres in which capitalism does not pay the costs of its own reproduction, creating several additional crisis fronts: ecology (resource exploitation and toxic dumping), social reproduction (care work, education, health), and politics (inequality destabilizing democracy). Hence, 'Like a tiger that bites its own tail, neoliberalism threatens now, just as its predecessor did then, to erode the very supports on which capitalism depends' (Fraser 2013b).

Fraser emerges as a committed crisis theorist. In a 2019 seminar at the New School's Institute for Critical Social Inquiry, she proudly exclaimed: 'Look, *I'm a crisis girl!*' – a slogan her students emblazoned on a t-shirt for her. For Fraser, the strategic upshot of marrying the 'two Karls' is that one is left with *more* 'flashpoints of trouble and nodes of crisis' well beyond the traditionally male-dominated factory or firm. Theorizing capitalism not as a narrow economic system but as a more expansive 'institutionalized social order' radically expands possible political fronts for organizing and mobilizing new counter-hegemonic and intersectional coalitions that must engage diverse social movements from across the Global South (Fraser 2018).

To combat the 'hopelessness' of climate fatalism, the Canadian activist and author Naomi Klein proposes an 'inverted shock doctrine' in which emergencies become opportunities for action and social justice (Klein 2014: 406; Klein 2019a). Graham Jones has similarly advocated a 'shock doctrine of the left' that can prepare progressive organizations to 'enact rapid, irreversible change' in moments of crisis (Jones 2018: 7). For Klein, the question is not only what powers of resistance disasters sap, but also which others they activate: 'Climate change acts as an accelerant to many of our social ills – inequality, wars, racism – but it can also be an accelerant for the opposite, for the forces working for economic and social justice and against militarism', thereby serving as 'the catalyst we need to knit together a great many powerful movements' (Klein 2016). The political anthropology in Klein's (2019b) writings is close to that of Robinson's revolutionary view: 'Many more people are beginning to grasp that the fight is not for some abstraction called "the Earth". We are fighting for our lives'.

Leftist thinkers like Fraser and Klein who frame global warming as such an acute *crisis* present it as a rare *opportunity* to destabilize the status quo and transform the global economy in more ecological and egalitarian directions. However, as Christophe Bonneuil and Jean-Baptiste Fressoz have observed of such rhetoric, does not the word 'crisis' 'maintain a deceptive optimism' by suggesting 'a perilous turning-point of modernity, a brief trial with an immanent outcome, or even an opportunity'? In fact, they argue, the Anthropocene epoch we long ago entered into 'is a point of no return' (Bonneuil and Fressoz 2017: 21), not a period of indecision but the irreversible crossing of an 'ecological *threshold*' (Horn and Bergthaller 2020: 1–2).

To be sure, *crisis optimism* is to some extent performative and strategic rather than purely descriptive. In an influential 2009 article, Chakrabarty memorably presented climate emergency as a Benjaminian 'moment of danger' ripe with potential to generate for the first time truly *species-wide* solidarity and a new 'universal that arises from a shared sense of a catastrophe' (Chakrabarty 2009: 222). As Morton suggests, however, such messianic, Marxian views of climate catastrophe as revolutionary openings *to come* are temporally flawed if the catastrophe has already happened, as well as poisoned by the immense suffering climate catastrophe is already causing. As Mike Davis (2022) has also remarked: 'the "Anthropocene" with its hint of the promethean, seems especially ill-fitted to the reality of apocalyptic capitalism'.

Crisis optimism presents new openings for progressive political transformation and decarbonization. However, global warming has thus far defied this framing: The growing subjective sense of climate 'crisis' popularized by movements like Fridays for Future has thus far led neither to systemic failure nor to significant material change. Western societies have been able to go on essentially ignoring global warming by failing to rapidly decarbonize, and this has hardly challenged their legitimacy, stability, or habitability. Lacking the sense of historical totality we get from the notion of *slow catastrophe*, the crisis framework lacks a temporal horizon connecting past extraction, pollution, and violence, the present space of experience and action, and the future horizon of apocalyptic extinction.

6 From Permanent to Slow Catastrophe

Catastrophe comes from the Greek for 'overturning' or 'downturn' and long referred to the denouement in a drama. *Catastrophe* has never shed this creative valence: catastrophes are not first-order events but second-order *judgments* and *narratives* about material disasters. In the early modern period, *catastrophe* was a synonym for *revolution*, and the concept is still invested with the silver lining of potential transformation through 'a subversion of the order or system of things'.[1] This redemptive view of catastrophe is captured in Benjamin's analogy between anti-capitalist revolutionaries and passengers on the train of progress pulling 'the emergency brake', thereby bringing about the 'real state of emergency' he saw as the goal of revolution (Löwy 2005). After the anti-redemptive caesuras of Auschwitz and Hiroshima, thinkers from Adorno to Zygmunt Bauman (1989) brought this messianic view back down to Earth by theorizing the banality or permanent possibility of catastrophe. More recently, scholars have theorized long-term social harms as 'slow violence' (Rob Nixon) and a 'crisis ordinariness' (Berlant 2011, 2018). These refer to what Slavoj Žižek calls 'objective' or 'systemic' violence associated with social structures, as opposed to 'subjective' or 'symbolic' violence associated with discrete events or accidents (Žižek 2009: 1–2). The trouble is that media and hegemonic accounts of catastrophes often focus on subjective and symbolic events, which are presented as exceptions that take place against the background of a 'normal' state of things – what Klein has called 'random, free-floating bad events, drifting in the political ether, to be condemned by all people of conscience but impossible to understand' (Klein 2007: 120). The moralizing of individual actors and events reinscribes conventional notions of agency, but in the process it invisibilises structural, abstract, or more distant forms of ongoing violence.

Already in the late 1980s, the Afrofuturist writer Octavia Butler spoke of the 'slow disaster' of global warming not as a sudden 'incident' but rather as 'an ongoing trend – boring, lasting, deadly' (Streeby 2018: 32). Scott Knowles' (2020) more recent elaboration of this notion calls for a *Gestalt* shift in how we think about disasters. Instead of blaming 'natural' events like weather disasters, we should identify what makes them so devastating in the first place: the 'unnatural' social forces of poverty and long-term failures of infrastructure, prevention, and response. This framework accounts for how seemingly sudden disasters often emerge from less visible, but no less real, *slow* disasters; it also expands the scope of whose suffering is considered disastrous, unacceptable, and worthy of relief. Investing the concept of disaster with these dimensions of social visibility, political power, and historical perspective is the goal of the nascent field of 'critical disaster studies' (Remes and Horowitz 2021).

Slow catastrophe moves even further away from 'event thinking' by drawing upon the notion of 'permanent catastrophe' developed by Benjamin and Adorno, yet it differs in two important respects: it resists both the ahistorical emphasis on permanence evident in certain formulations of catastrophe in Adorno (the performative apocalypse pessimist) as well as Benjamin's 'undialectical' (as Adorno alleged) embrace of redemptive ruptures of potentially revolutionary violence (the revolutionary crisis optimist). Against these sometimes one-sided conceptions, a truly *critical* conception of catastrophe would have to remain *dialectical*, neither positing the tragic inevitability of catastrophe nor making common cause with it.

7 Against Making Common Cause with Catastrophe

The dialectical intervention of *slow catastrophe* is also warranted in light of an exchange between Fraser (the crisis optimist) and Étienne Balibar (the apocalypse pessimist) that took place at the New School's 2019 Institute for Critical Social Inquiry. We have already raised problems with Fraser's optimistic view. On the other side, in a lecture entitled 'Socialism before the Catastrophe', Balibar (2019) began from the thesis that 'the *catastrophe* has already happened', a pessimistic view he attributed to Adorno and Heidegger. This view, he readily acknowledged, risks succumbing to climate fatalism, presenting 'dialectics as permanent tragedy'. Only a 'collective invention of politics', he said, could respond to the reactionary and xenophobic responses to mass migration and genocide that are already resulting from climate-related displacement. He elaborates this new form of politics:

> Socialism was intrinsically linked to ideologies of progress and economic policies of growth, which led to ecological catastrophe. Now that the catastrophe is no longer a possibility, but an irreversible reality threatening us with more frequent crises and wilder forms of extreme violence, we need a different concept of time and political action, whether in the modality of regulation, insurrection, or civility.
>
> *(Balibar 2019)*

The politics of fear of which Balibar warns has also been called the 'armed lifeboat' or 'eco-ethnonationalism' model: developed nations engage in 'a war for the earth' (Kelly 2019: 119) and strengthen their 'fortresses' to protect their limited resources from increasing numbers of border-crossing climate refugees (Latour 2018: 18). Ultra-rich elites, meanwhile, retreat into 'bolt holes' safeguarded from catastrophe (Lawrence and Laybourn-Langton 2021: 58). While for crisis optimists global warming opens up genuine *political* opportunities, for apocalypse pessimists it fuels an *apolitical* Hobbesian cycle of unending violence that can only be broken by being brought into the sphere of genuine political contestation.

Two of the most popular apocalyptic depictions of global warming in English, by writer Jonathan Franzen (2019) and journalist David Wallace-Wells (2019), have corresponded with their relatively resigned political orientations. Lacking the revolutionary political imagination required to avert extinction results in climate fatalism that inaccurately represents global warming as an unalterable force out of human control. Franzen met strong criticism for his 2019 essay 'What If We Stopped Pretending?', which argued for facing up to the fact that global warming is not on track to be stopped and that therefore apocalyptic extinction looms ahead. He wrote that 'the constraints of human nature' prevent us from acting to avert climate catastrophe until it is too late – a notion we also saw raised but ultimately overcome through the 'concrete utopias' (Bloch 1995: 17) Robinson creates in *The Ministry for the Future*. Franzen's essay ends by encouraging readers to seek hope and comfort in initiatives like local community gardens, which surely have higher chances of success, but at the cost of abandoning the necessary structural transformation of the carbon economy and therefore accepting global catastrophe as inevitable. Franzen recognizes the potential of apocalyptic narratives like Robinson's, but he fails to jump back from the brink.

Wallace-Wells' non-fiction work *The Uninhabitable Earth* used gripping apocalyptic predictions to draw popular attention to a number of overlooked climate threats, from mass extinctions and dwindling water reserves to resource wars and genocides already producing millions of climate refugees. 'The path we are on as a planet should terrify anyone living on it', he wrote (Wallace-Wells 2019: 226). Yet such scare tactics also raise a question: at what point does apocalyptic imagery paralyze and demobilize readers rather than rally them to action? Wallace-Wells worries that we are 'collectively walking down a path of suicide', a *choice*, he claims, we might still avert (Wallace-Wells 2019: 220). Worse than the 'apocalyptic' tone of much of the literature he cites, he claims, is the 'climate apathy' and 'acclimatization' where 'the absolute moral unacceptability of the conditions of the world we are passing through' become normalized (Wallace-Wells 2019: 215–216). Yet his shocking book, an avalanche of alarming statistics and apocalyptic scenarios, falls victim to all the problems that Morton (2015) has identified with the numbing effects of 'information dumping'. Further, Wallace-Wells lacks the class-based framework for grasping what Latour (2018) has called the 'New Climatic Regime', in which global elites have for decades funded climate denial and opposed necessary societal transformation. Malm and Alf Hornborg thus reject the undifferentiated term 'Anthropocene' for its political vagueness (Malm and Hornborg 2014); global warming is not the result of democratic *anthropogenic* actions by all of humanity but the *sociogenic* forces of (mostly Western) capitalist elites controlling commodity production – a view that echoes calls to term our current epoch the 'Capitalocene' (Moore 2016) or 'Oliganthropocene' (Swyngedouw cited in Bonneuil and Fressoz 2017) in order to attribute responsibility where it is specifically due.

Lacking such a political frame, Wallace-Wells' apocalypticism remains uncritical: it plays upon fears rather than taking aim at power. While not as fatalistic as Franzen, Wallace-Wells 'makes common cause with the catastrophe' by depicting apocalypse without mobilizing us toward averting it through what Günther Anders, a leading interpreter of the existential threat of nuclear annihilation during the Cold War, called '*prophylactic* apocalypticism', the idea that 'we are apocalypticians only in order to be *wrong*' (Anders 1972: 179; see also Heidel 2018: 34; and Latour 2017: 217). Nathan Robinson has rightly concluded that climate catastrophe 'is actually not a fixable problem if you have Jonathan Franzen and David Wallace-Wells' politics', because their liberalism doesn't offer 'a theory of how political change actually happens' (Robinson 2019).

While Wallace-Wells can at first be seen to share the alarmism of Greta Thunberg's slogan 'We cannot solve an emergency without treating it like an emergency', his book, lacking a political framework for historical responsibility for global warming, ultimately risks climate fatalism and the neoliberal individualization of blame. Žižek (2019) has rightly warned that such diffuse 'ecological grief' (Craps 2020) can serve as a religious ideology, instilling a sense of universal guilt while also paralyzing action: 'We should avoid at all costs the trap of an "ecology of fear", a hasty, morbid fascination with looming catastrophe'.

8 Conclusion: Politicizing Slow Catastrophe

The competing framings of global warming we have considered are not sterile, analytic categories; they draw upon powerful rhetoric, metaphors, and images and tap into deep-seated anxieties, struggles, and hopes – what Robinson (2020: 358) called the 'structure of feeling, underlying politics as such'. The *crisis optimism* of

Fraser, Klein, and Chakrabarty importantly highlights the contingency of climate catastrophe and holds out hope that it might yet be averted through radical social, economic, and political transformation. At the same time, this sometimes romantic and one-sided view can blind us to problems raised by the apocalypse-pessimist view theorized by Balibar and capitulated to by Franzen and Wallace-Wells. Faced with these false options, Malm has called for an 'ecological Leninism':

> The problem with social democracy is that it has no concept of catastrophe – rather, it is premised on the opposite, namely the notion that we have time at our disposal and history on our side, meaning that we can move by incremental steps toward a socialist society. [...] It is therefore necessary to look to part of the socialist legacy that has an idea of catastrophe.
>
> *(Malm 2020a)*

With the threat of irreversible ecological damage fast approaching, we cannot afford the 'socialism before the catastrophe' Malm and Balibar fear any more than the quietism of liberal environmentalism, much less the ravages of eco-fascism. At this critical juncture, a dialectical concept of *slow catastrophe* theorizes the real danger of extinction while also maintaining a critical horizon for the possibility of radical social transformation to avert it.

I conclude by returning to the question this volume poses: how might *apocalyptic* framings of global warming serve as productive openings to new social orders not based on capitalist, carbon-based extractivism and productivism? I have argued that *apocalyptic* framings risk climate fatalism unless they are politicized and mobilizing in the manner achieved by Robinson's fiction. At the same time, *crisis-optimist* framings tend to underestimate the danger of the problem, misleadingly presenting catastrophe as a future revolutionary opening rather than a *fait accompli*, and to overstate the emancipatory opportunities it presents. In between the two lies the critical theorization of global warming as part of a larger *slow catastrophe* of capitalist modernity, a view which provides a dialectical middle ground that can guide both critique and activism. *Slow catastrophe* holds both the danger and opportunity of climate emergency in view, conceptualizing it as both sudden *and* slow, continuous *and* discontinuous, and structural *and* eventual, and thus envisions more affected stakeholders, more rallying points for action, and more potential ground for progressive social transformation.

Note

1 According to the *Oxford English Dictionary*.

References

Adorno, T.W. (1973) *Negative Dialectics*. Trans. E.B. Ashton. New York: Continuum.
Adorno, T.W. (1977) The Actuality of Philosophy. *Telos*, 1977(31): 120–133. https://doi.org/10.3817/0377031120.

Adorno, T.W. (1998) The Meaning of Working through the Past (1959) and Education after Auschwitz (1966). In *Critical Models: Interventions and Catchwords*, trans. H.W. Pickford. New York: Columbia University Press, 89–104 and 191–204.

Adorno, T.W. (2004) *History and Freedom: Lectures 1964–1965*. Trans. R. Livingstone. Cambridge: Polity.

Adorno, T.W. (2005) *Minima Moralia: Reflections from Damaged Life*. Trans. E. Jephcott. New York: Verso.

Adorno, T.W. (2008) *Lectures on Negative Dialectics, 1965–1966*. Trans. R. Livingstone. Cambridge: Polity.

Adorno, T.W. and M. Horkheimer (2002) *Dialectic of Enlightenment: Philosophical Fragments*. Ed. G.S. Noerr. Trans. E. Jephcott. Stanford, CA: Stanford University Press.

Anders, G. (1972) *Endzeit und Zeitenende: Gedanken über die atomare Situation*. Munich: C.H. Beck.

Balibar, E. (2019) Socialism before the Catastrophe. Lecture delivered at the New School's Institute for Critical Social Inquiry on June 10. https://www.facebook.com/events/467947987081200/.

Bauman, Z. (1989) *Modernity and the Holocaust*. Ithaca, NY: Cornell University Press.

Bauman, Z. (2014) Disconnecting Acts, Part II: Interview with Efrain Kristal and Arne De Boever. *Los Angeles Review of Books*, November 12. https://lareviewofbooks.org/article/disconnecting-acts-interview-zygmunt-bauman-part-ii/.

Benjamin, W. (1999). *The Arcades Project*. Trans. H. Eiland and K. McLaughlin. Cambridge, MA: Harvard University Press.

Benjamin, W. (2006). On the Concept of History. In *Selected Writings, Vol. 4, 1938–1940*, eds. H. Eiland and M. Jennings, trans. H. Eiland and E. Jephcott. Cambridge, MA: Harvard University Press, 389–401.

Benjamin, W. (2019) *Origin of the German Trauerspiel*. Trans. H. Eiland. Cambridge, MA: Harvard University Press.

Bergquist, M., A. Nilsson, and P.W. Schultz (2019) Experiencing a Severe Weather Event Increases Concern about Climate Change. *Frontiers in Psychology* 10: 220https://doi.org/10.3389/fpsyg.2019.00220.

Berlant, L. (2011) *Cruel Optimism*. Durham, NC: Duke University Press.

Berlant, L. (2018) Without Exception: On the Ordinariness of Violence: Brad Evans Interviews Lauren Berlant. *Los Angeles Review of Books*, July 30. https://lareviewofbooks.org/article/without-exception-on-the-ordinariness-of-violence/.

Bloch, E. (1995) *The Principle of Hope, Volume One*. Trans. N. Plaice, S. Plaice, and P. Knight. Cambridge, MA: MIT Press.

Bonneuil, C. and J.-B. Fressoz (2017) *The Shock of the Anthropocene*. New York: Verso.

Burns, T. (2008) *Political Theory, Science Fiction, and Utopian Literature: Ursula K. Le Guin and The Dispossessed*. Lanham, MD: Lexington Books.

Catlin, J. (2021) Catastrophe Now. *History & Theory*, 60(3): 573–584. https://doi.org/10.1111/hith.12231.

Chakrabarty, D. (2009) The Climate of History: Four Theses. *Critical Inquiry*, 35(2): 197–222. https://doi.org/10.1086/596640.

Chakrabarty, D. (2021) *The Climate of History in a Planetary Age*. Chicago: University of Chicago Press.

Connolly, W.E. (2017) *Facing the Planetary: Entangled Humanism and the Politics of Swarming*. Durham, NC: Duke University Press.

Craps, S. (2020) Introduction: Ecological Grief. *American Imago*, 77(1): 1–7. https://doi.org/10.1353/aim.2020.0000.

Davis, M. (2022) Thanatos Triumphant. *New Left Review Sidecar*, March 7. https://newleftreview.org/sidecar/posts/thanatos-triumphant.

Delany, S. (2009) Some Presumptuous Approaches to Science Fiction. In *Starboard Wine: More Notes on the Language of Science Fiction*. Middletown, CT: Wesleyan University Press, 25–34.

Durbin, A. (2017) *MacArthur Park*. New York: Nightboat Books.

Foer, J.S. (2019) *We Are the Weather: Saving the Planet Begins at Breakfast*. New York: Macmillan.

Franzen, J. (2019) What If We Stopped Pretending? *New Yorker*, September 8. https://www.newyorker.com/culture/cultural-comment/what-if-we-stopped-pretending.

Fraser, N. (2013a) Why Two Karls Are Better than One: Integrating Polanyi and Marx in a Critical Theory of the Current Crisis. Working Paper 01/2013. Jena: DFG-Kolleg-Forscher/Innengruppe Postwachstumsgesellschaft. https://www.econbiz.de/Record/why-two-karls-are-better-than-one-integrating-polanyi-and-marx-in-a-critical-theory-of-the-current-crisis-fraser-nancy/10011619079.

Fraser, N. (2013b) A Triple Movement? Parsing the Politics of Crisis after Polanyi. *New Left Review*, 81. https://newleftreview.org/issues/ii81/articles/nancy-fraser-a-triple-movement.

Fraser, N. (2018) Podcast: Beyond Economism. *The Dig*, September 12. https://www.thedigradio.com/transcripts/transcript-beyond-economism-with-nancy-fraser/.

Fraser, N. (2019) *The Old Is Dying and the New Cannot Be Born: From Progressive Neoliberalism to Trump and Beyond*. New York: Verso.

Fraser, N. (2021) Climates of Capital: For a Trans-Environmental Eco-Socialism. *New Left Review*, 127. https://newleftreview.org/issues/ii127/articles/nancy-fraser-climates-of-capital.pdf.

Fraser, N. and R. Jaeggi (2018) *Capitalism: A Conversation in Critical Theory*. Cambridge: Polity.

Fraser, N. and R. Jaeggi (2021) Climates of Capital. *Podcast: Critical Theory in Context*, July 3. https://ctic.podigee.io/1-climates-of-capital.

Ghosh, A. (2016) *The Great Derangement: Climate Change and the Unthinkable*. Chicago: University of Chicago Press.

Habermas, J. (1988) *Legitimation Crisis*. Trans. T. McCarthy. Cambridge: Polity.

Haraway, D.J. (2016) *Staying with the Trouble: Making Kin in the Chthulucene*. Durham, NC: Duke University Press.

Heidel, K. (2018). Leben im Anthropozän: Anmerkungen zur Wirklichkeit im 21. Jahrhundert. In *Leben im Anthropozän: Christliche Perspektiven für eine Kultur der Nachhaltigkeit*, eds. B. Bertelmann and K. Heidel. Munich: Oekom, 17–38.

Horn, E. (2018) *The Future as Catastrophe: Imagining Disaster in the Modern Age*. Trans. V. Pakis. New York: Columbia University Press.

Horn, E. and H. Bergthaller (2020) *The Anthropocene: Key Issues for the Humanities*. London: Routledge.

Jaeggi, R. (2018) *Critique of Forms of Life*. Cambridge, MA: Harvard University Press.

Jaeggi, R. (2019) Lecture at Cornell University's School of Criticism and Theory: Crisis, Progress and Critique, June 25. https://cdnapisec.kaltura.com/index.php/extwidget/preview/partner_id/520801/uiconf_id/31230141/entry_id/1_xxhs8jbn/embed/dynamic.

Jones, G. (2018) *The Shock Doctrine of the Left*. Cambridge: Polity.

Kelly, D. (2019) *Politics and the Anthropocene*. Cambridge: Polity.

Kennedy, P. (2012) William Gibson's Future Is Now. *New York Times*, January 13. https://www.nytimes.com/2012/01/15/books/review/distrust-that-particular-flavor-by-william-gibson-book-review.html.

Klein, N. (2007) *The Shock Doctrine*. New York: Metropolitan.

Klein, N. (2014) *This Changes Everything: Capitalism vs. The Climate*. New York: Simon and Schuster.

Klein, N. (2016) Let Them Drown: The Violence of Othering in a Warming World. *London Review of Books*, 38(11). https://www.lrb.co.uk/the-paper/v38/n11/naomi-klein/let-them-drown.

Klein, N. (2019a) *On Fire: The (Burning) Case for a Green New Deal*. New York: Simon and Schuster.

Klein, N. (2019b) The Green New Deal: A Fight for Our Lives. *New York Review of Books*, September 17. https://www.nybooks.com/daily/2019/09/17/the-green-new-deal-a-fight-for-our-lives/.

Knowles, S. (2018) What Trump Doesn't Get About Disasters. *New York Times*, September 13. https://www.nytimes.com/2018/09/13/opinion/trump-hurricane-maria-puerto-rico-disaster.html.

Knowles, S. (2020) Slow Disaster in the Anthropocene: A Historian Witnesses Climate Change on the Korean Peninsula. *Daedalus*, 149(4): 192–206. https://doi.org/10.1162/daed_a_01827.

Koselleck, R. (2006) Crisis. Trans. M.W. Richter. *Journal of the History of Ideas*, 67(2): 357–400. https://www.jstor.org/stable/2707264.

Latour, B. (2017) *Facing Gaia: Eight Lectures on the New Climatic Regime*. Trans. C. Porter. Cambridge: Polity.

Latour, B. (2018) *Down to Earth: Politics in the New Climatic Regime*. Trans. C. Porter. Cambridge: Polity.

Latour, B. and Chakrabarty, D. (2020) Conflicts of Planetary Proportion: A Conversation. *Journal of the Philosophy of History*, 14(3): 419–454. https://doi.org/ 10.1163/18722636-12341450.

Lawrence, M. and L. Laybourn-Langton (2021) *Planet on Fire: A Manifesto for the Age of Environmental Breakdown*. New York: Verso.

Levinas, E. (1988) Useless Suffering. In *The Provocation of Levinas: Rethinking the Other*, ed. R. Bernasconi and D. Wood, trans. R. Cohen. London: Routledge, 156–167.

Löwy, M. (2005) *Fire Alarm: Reading Walter Benjamin's 'On the Concept of History'*. Trans. C. Turner. New York: Verso.

Malm, A. (2020a) Interview: To Halt Climate Change, We Need an Ecological Leninism. *Jacobin*, June 15. https://jacobin.com/2020/06/andreas-malm-coronavirus-covid-climate-change?mc_cid=9aab712a86&mc_eid=581c6ff5ff.

Malm, A. (2020b) *Corona, Climate, Chronic Emergency: War Communism in the Twenty-First Century*. New York: Verso.

Malm, A. (2021) *How to Blow Up a Pipeline*. New York: Verso.

Malm, A. and A. Hornborg (2014) The Geology of Mankind? A Critique of the Anthropocene Narrative. *Anthropocene Review*, 1(1): 62–69. https://doi.org/10.1177/2053019613516291.

Mann, G. and J. Wainwright (2018) *Climate Leviathan: A Political Theory of Our Planetary Future*. New York: Verso.

McNeill, J.R. and P. Engelke (2016) *The Great Acceleration: An Environmental History of the Anthropocene since 1945*. Cambridge, MA: Harvard University Press.

Moore, J., ed. (2016) *Anthropocene or Capitalocene? Nature, History, and the Crisis of Capitalism*. Oakland: PM Press.

Morton, T. (2013) *Hyperobjects: Philosophy and Ecology after the End of the World*. Minneapolis: University of Minnesota Press.

Morton, T. (2015) IPCC: The Intraplanetary League of Concerned Critters. *Ecology without Nature*, November 8. http://ecologywithoutnature.blogspot.com/2015/11/my-editorial-in-todays-danish-politiken.html.

Morton, T. (2020) The End of the World Has Already Happened. *BBC Radio 4*, January 2. https://www.bbc.co.uk/sounds/series/m000cl67.

Neiman, S. (2002) *Evil in Modern Thought: An Alternative History of Philosophy*. Princeton, NJ: Princeton University Press.

Palatella, J. (2020) Earthly Anecdotes. *The Point*, April 12. https://thepointmag.com/exam ined-life/earthly-anecdotes/.

Rabinbach, A. (1997) *In the Shadow of Catastrophe: German Intellectuals between Apocalypse and Enlightenment*. Berkeley: University of California Press.

Rabinbach, A. (2008) From the Redemptive to the Non Redemptive Apocalypse in 20th Century German Thought. *Rivista di Filosofia*, 4: 199–205. https://www.academia.edu/ 75550211/From_the_Redemptive_to_the_Non_Redemptive_Apocalypse_in_20th_Century_ German_Thought1/From_the_Redemptive_to_the_Non_Redemptive_Apocalypse_in_20th_ Century_German_Thought.

Remes, J.A.C. and A. Horowitz (2021) *Critical Disaster Studies*. Philadelphia: University of Pennsylvania Press.

Robinson, K.S. (2020) *The Ministry for the Future*. New York: Orbit.

Robinson, N. (2019) In Conversation: Naomi Klein on Climate Crisis, Hope, and Jonathan Franzen. *Current Affairs Podcast*, September 12.https://open.spotify.com/episode/ 08221dNcpD5Hksl8vR4Fsx.

Satia, P. (2022) The Way We Talk about Climate Change Is Wrong. *Foreign Policy*, March 11. https://foreignpolicy.com/2022/03/11/climate-change-sacrifice-colonial-language-history-economics/.

Schmidt, A. (2014) *The Concept of Nature in Marx*. New York: Verso.

Scranton, R. (2015) *Learning to Die in the Anthropocene: Reflections on the End of a Civilization*. San Francisco: City Light Books.

Siemsen, T. (2017) Talking Disco and the Disaster Imaginary with Andrew Durbin. *LitHub*, September 21. https://lithub.com/talking-disco-and-the-disaster-imaginary-with-andrew-durbin/.

Streeby, S. (2018) *Imagining the Future of Climate Change: World-Making through Science Fiction and Activism*. Berkeley: University of California Press.

Strolovitch, D. (2022) *When Bad Things Happen to Privileged People: Race, Gender, and What Makes a Crisis in America*. Chicago: University of Chicago Press.

Wallace-Wells, D. (2019) *The Uninhabitable Earth: Life after Warming*. New York: Tim Duggan Books.

Žižek, S. (2009) *Violence: Six Sideways Reflections*. London: Profile Books.

Žižek, S. (2019) Yes, It Is a Climate Crisis: And Your Tiny Human Efforts Have Never Seemed so Meagre. *The Independent*, September 20. https://www.independent.co.uk/voices/amazon-fires-rainforest-capitalism-bolsonaro-climate-crisis-zizek-a9091966.html.

5

APOCALYPTICISM IN ISLAMIC ENVIRONMENTAL THOUGHT

The Anthropocene as a Theological Concept

Marita Furehaug

Introduction

Today, many scholars have noted the presence of apocalyptic scenarios in contemporary representations of environmental futures, whether portrayed in film and popular culture or in a variety of discourses on climate change (Breton and Hammond 2016; Crawford 2009; Fagan 2017). This framing in many instances is intended to galvanize action; however, some studies focused on the behavioral and psychological impacts of inducing fear as a means of motivation argue that apocalyptic scenarios tend to produce apathy, disempowerment, and skepticism (Fagan 2017: 229). While this apocalyptic framing of climate change shares commonalities with the religious concept of Doomsday, the environmentalist version involves no moment of transcendence or redemption. Further, it expresses a scientifically supported assessment of the environmental risks involved in maintaining the contemporary way of living. The proposed geological age dubbed the 'Anthropocene' incites alarming realities of humanity's planet-shaping abilities predicting a catastrophic environmental future by the hand of humankind, not God.

The Doomsday narrative is central in the Islamic faith system, and the Qur'an clearly warns humanity of an impending apocalyptic future, a recurring theme inciting images of cataclysmic and inconceivable events that will turn the natural order upside down (Lawson 2008). Images of natural disasters and catastrophic events take on a crucial role in the overall moral framework of Qur'anic discourse. Despite this central feature, apocalypticism and eschatology are not a major theme in most treatments of Islam and the environment (Gade 2019). In this chapter, Islamic eschatology is conceptualized, and central apocalyptic motifs identified. Further, I highlight the theological considerations the Anthropocene in its essence poses pertaining to God's omnipotence and the impact of human agency, along with the potential challenges of invoking eschatology through an environmental

DOI: 10.4324/9781003189190-6

perspective. Despite these issues, I wish to demonstrate that actors within the eco-Islamic field have formulated responses to the ecological crisis that illustrate that it is possible to align the Islamic worldview with the notion of the Anthropocene, and that eschatology and apocalypticism can in fact translate into environmental action. In this regard, studies aimed at exploring Muslim environmentalisms through fieldwork and interviews are highlighted to illustrate how apocalypticism informs Muslims engaging in environmental action.

Conceptualizing Islamic Eschatology

Todd Lawson argues that the Qur'an in its entirety can be viewed as belonging to the apocalyptic genre[1] and that apocalypticism in this sense forms the infrastructure of the holy text (Lawson 2008). Viewing the Qur'an as an apocalyptic text does not rule out its comprising other genres or literary styles[2]; however, it emphasizes the centrality of apocalyptic components and stress the eschatological worldview in the Qur'an and consequently the Islamic tradition. While eschatological content varies tremendously, the belief that in the fullness of time there will occur a fundamental transformation in the world, often described as the dawning of a new era, in which conditions are radically different than those in the present era, is usually shared (Donner 2017: 758). Thus, the purpose of eschatological schemes is to situate the subject (individual, community, or the cosmos) in the context of the new era. Fred Donner presents eschatological schemes in conceptual categories, and classifies Islamic eschatology as *linear, religious, other-worldly, moral, catastrophic,* and *future-oriented*. In addition, eschatologies are divided into *individual, communal,* and *cosmic*, whereby the Islamic one contains elements of all three. In a *linear* eschatology, every individual has one birth and proceeds through life to a single death, while the world as we know it proceeds from a single point in creation to a single definite end time, a 'last day'. The final fate after death depends on the actions during that lifetime (Donner 2017: 759). This entails a kind of urgency, as the individual's life involves a definite and inescapable 'deadline', namely, the ever-approaching death.

In the Qur'an, 'the last judgement' (*yawm al-din*), or 'the day of resurrection' (*yawm al-qiyama*) is God's final assessment of all creatures, one of the most central and recurring themes in text (Hasson 2005). Belief in this day, with the concomitant belief in paradise (*al-janna*) and hell (*jahannam*), and the 'weighing' of good and bad deeds are some of the most central elements in the Islamic faith. In the Qur'an, *yawm al-qiyama* is mentioned 70 times and *yawm al-din* 13 times; however, many terms or locutions have been identified by exegetes as synonyms. As an example, the medieval scholar Al-Ghazali (d. 505/1111) gives more than 100 names or epithets designating this day. Some of the synonyms listed give associations of the imminent horror to come, such as 'dreadful day' (*yawm 'azim*, Q 6:15; 10:15); 'the day of anguish' (*yawm al-hasra*, Q 19:39); 'the terror' (*al-waki'a*, Q 56:1; 69:15), and 'the great catastrophe' (*al-tamma al-kubra*, Q 79:34). Further, a number of synonyms also indicate the judging and the individual's part in their own destination (i.e. paradise or hell), such as 'the day of decision' (*yawm al-fasl*, Q 37:21; 44:40; 77:13; 14:38; 78:17); 'the day of reckoning' (*yawm al-hisāb*, Q 38:16; 26, 53; 40:27), 'the day of coming forth' (*yawm al-khurūj*, Q 50:52); 'the day when some faces are whitened, and some faces are blackened' (*yawma tabyaddu wujūhu wa-taswaddu wujūhu*, Q 3:106); and 'a day when no soul shall avail another' (*yawman lā tajzi nafsun 'an nafsin shay'an*, Q 2:123). This list is far from exhaustive; however, it indicates the frequent occurrence and centrality of Judgement Day. Additionally, numerous signs of the coming of this day occur as natural catastrophes as illustrated in Q 81:1–14, where the sun will be darkened, the stars will be thrown down, the mountains will be set moving, the seas will be set boiling (or will overflow), heaven will be stripped away, hell will be blazing, and paradise will be brought near (Hasson 2005). The apocalyptic sensorium of Judgement Day is interlinked with the other points in Donner's conceptualization.

While Islamic eschatology obviously is *religious*, as God in the Qur'an has decreed the end of time, the classification of *other-worldly* refers to scenarios played out in heavenly or other-worldly realms, such as the garden/paradise (*al-janna*) or hell (*jahannam*). These suprarational locations represent the destination of individuals and are not portrayed as being part of the current world; rather, they are depicted as perfect utopias or dystopias, and are representations of the afterlife. Linguistically, the Qur'an refers to two worlds: *al-ūla* (the first) and *al-ākhira* (the last) (Abdel Haleem 2017). Further, the Qur'an warns believers about being too attached to this world, and refers to the present world as a 'transitory world' (Zakzouk 2017: 39; Q 75:20). The Qur'an thus warns the reader from loving the first world (*al-ula*) too much and thereby disregarding the hereafter. However, the two worlds must be seen as interlinked and interconnected, since the actions in the first world determine the habitat of the afterlife. As Lawson points out, paradise connects other apocalyptic themes directly such as the triumph of good over evil, the final judgement, and the binary opposition to hell (Lawson 2017: 121). The descriptions and synonyms of paradise indicate a peaceful, beautiful, and blessed abode, revealing formulaic expressions of a lush landscape of ecological abundance (Gade 2019: 112) where inhabitants are portrayed as being in a state of immense psychological wellbeing with spiritual contentment (Abdel Haleem 2017: 61–64; Q 10:62; 15:47; 66:8; 54:54–55). The emphasis on this 'piece of mind' in a thriving natural environment can also be contrasted with the uncertainty, fear, and anxiety commonly felt by human beings in this (the first) world, and thus evoke images of an everlasting utopia. In cultivating what Lawson calls 'the apocalyptic sensorium', hell occurs in symmetrical antithesis to the descriptions of paradise, and emerges as a horrifying dystopia (Lawson 2017: 122).

Hell, too, has many synonyms and references describing it as an abyss with a barren and terrifying landscape with blazing and consuming fire, such as *al-nār* ('the fire'), *al-sa'ir* ('the blaze'), *al-jahim* ('the hot place'), *al-hutama* ('the consuming fire'), and *al-hawiya* ('the abyss'), to name a few (Thomassen 2009: 403, 407). The psychological state in hell, similarly, is highlighted, as inhabitants experience terror and humiliation (Thomassen 2009: 403–404; Q 3:178; 11:106; 35:36). These contrasting images and descriptions of the environmental and psychological states of hell and paradise form an apocalyptic or revelatory imagination that leads beyond the earthly world to a new (but not completely different) realm. Due to the element of a final (moral) judgement, Islamic eschatology is thus conceptualized as a *moral eschatology*. Muhammad Abdel Haleem (2017) and Mahmoud Zakzouk (2017) outline a theoretical framework for how to get to paradise from a Qur'anic point of view. By drawing extensively on quotes from the holy text, they are able to show the importance of good deeds, faith, and contemplation. Human accountability and spiritual connectivity thus emerge as central and all-encompassing notions. In addition, instructions throughout the Qur'an ask humans to reflect on and contemplate life, its origin, and its destiny, which are often invoked through the notion of signs (*ayat*). *Ayat* refer to both verses in the Qur'an and the numerous signs found within creation that point to God's glory and omnipotence.

Further, Donner differentiates between *individual, communal,* and *cosmic eschatologies*. Islamic eschatology contains elements of all three. It focuses on the fate of individuals, and emphasizes that your actions alone will be judged. However, there are Qur'anic instances where all the dead are described as being raised on Judgement Day as communities (Donner 2017: 767). Further, there is an emphasis on the interrelation between the *individual* and the *collective* also in Islamic law, especially when dealing with social responsibilities.[3] One intriguing observation regarding hell in this sense is that the Qur'an depicts it as a dynamic space of talking, whereby much of the dialogue is intergenerational, indicating the responsibility of the preceding generation for the subsequent one, which thereby reaps the consequences of the former's moral tendencies[4] (Gade 2019: 112). Lastly, Islamic eschatology is *cosmic*, as it presages the final fate of the entire world including images of environmental disasters and transformations (Donner 2017: 768).

Finally, Islamic eschatology can be described as *catastrophic* and *future-oriented*, as it anticipates a situation in which *all* souls are consigned at the same time for their final judgement at the end of time, which is sometime in the future. Because this single last judgement masks the end of the normal continuum of time, it is also described as taking place immediately following a cosmic cataclysm or a series of catastrophes that end the world as we know it and inaugurate the last judgement. However, the Qur'an contains references to the fate of past communities who have already suffered severe consequences, as in the stories of Sodom and Gomorrah and the Prophet Lot, the Prophet Nuh (Noah) and the flood, and the community of Thamud, to name a few. Donner argues that these stories can be considered eschatological as they are framed as lessons meant to inform those who are still living, that is, to warn them to avoid the errors of these communities already 'judged'. These are described as *realized eschatologies* because the fate of the community in question is already realized. *Partly realized eschatologies*, however, are of the utmost interest from an eco-apocalyptic perspective, as they refer to schemes in which the events of the present are portrayed as the beginnings of the end times or cataclysmic events associated with the last judgement. Such eschatologies combine an other-worldly orientation with the notion that the other-worldly fate of the subject group is not only imminent but already heralded by this-worldly events taking place in the present time. In the following section, I will highlight the parallel between Islamic eschatology and the Anthropocene as a secular eschatology and reflect upon potential questions this concept might raise for eco-theology.

Invoking the Anthropocene: Challenges and Implications for Theology

The Anthropocene refers to a new geological epoch which includes the scale and impact of human activity on Earth. The Anthropocene assumes that the supposed 'blind forces of nature' are brought about by human agency. Because the Anthropocene is closely linked to discourse on how the planetary condition will pose challenges to life on Earth, it is in effect a kind of secular eschatology that brings with it questions of finitude, irreversibility, and an absolute temporal ending (Rothe 2020: 147). Further, it seemingly challenges theological notions of an all-powerful omnipotent God, and instead ignites questions regarding the role, influence, and effect of human agency.

One relevant discussion that scientists are grappling with is establishing the *beginning* of the Anthropocene, thus, the border that demarcates the Holocene from the Anthropocene. The 'golden spike', in this context, refers to a marker in the fossil record indicating this boundary. While there is a consensus that carbon dioxide emissions, global warming, ocean acidification, habitat destruction, extinction, and widescale resource extraction are all indications that we have significantly modified our planet, the presence of these changes needs to be documented in the geological strata and fossil record before we officially declare the existence of the 'human epoch' (Pavid 2020). The marker would have to be significant enough to be detectable in rock layers thousands and even millions of years into the future (Pavid 2020). Some scientists have suggested that a possible boundary could be defined by the presence of radioactive particles in the soil around the world, a ubiquitous phenomenon observed after major events involving nuclear activity such as the Second World War and the Chernobyl nuclear disaster (O'Hare 2019). Others contend that the marker should be linked to the start of the Industrial Revolution in the 1800s, when human activity started to have a great impact on carbon and methane levels in the Earth's atmosphere (National Geographic 2019). Plastic is also a suggested marker, as Earth's is awash with plastic and millions of tons of it are still produced every day. Because plastic doesn't biodegrade, some evidence suggests that plastic is being deposited into the fossil record (Pavid 2020).

While consensus regarding the 'golden spike' has not yet been achieved, these examples of suggested markers attest to grave evidence that human activities in a multitude of ways will be detectable for a long time to come. Invoking the Anthropocene thus sends a powerful message that humanity is now a geological force of nature, to the extent that the planet is being altered not only because of natural causes but also because of human activity. This raises theological concerns, particularly pertaining to God's omnipotence and the role of human agency in narratives regarding the end times. In what ways is the Anthropocene challenging theology, and how is this expressed in Islamic eco-theological discourse? Can the Anthropocene be compatible with Islamic eschatology? These questions indicate the challenges the Anthropocene poses pertaining to philosophical, theological, and ethical aspects that will demand more elaborate answers in future eco-Islamic theology.

Further, the invocation of eschatological teachings in relation to the contemporary environmental situation is not without problems. Eschatology has been identified in some faith traditions as a potential reason for some of their adherents to intentionally (or unintentionally) disengage from environmental issues. In *Religion and Ecology*, John Grim and Mary Evelyn Tucker state that religions with a transcendent 'reality' can be preoccupied by an 'other-worldly' goal such as salvation in heaven or getting to paradise as opposed to hell in the afterlife, causing some actors to argue that the destructive treatment of this world is insignificant (Grim and Tucker 2014: 15). Within the logic of these worldviews, you may find religious practitioners that even suggest that degrading the environment hastens the end of the Earth and the return of a transcendent paradise. Examples of religious groups that actively deny the critical nature of environmental problems or reject the science of climate change exist; furthermore, as Laurel Kearns (2017) argues, for some Christians, and to a lesser extent for some Jews and some Muslims, climate change poses a challenge to their understanding of God. Accepting climate change, for these believers, requires doubt in God's omnipotence, and interferes with the idea of an all-loving, all-powerful God who is caring and who interferes in the lives of His adherents. For others, however, especially conservative Christians, Kearns continues, climate change is seen as part of the end times or a certain apocalyptic scenario, and they thus welcome it as a positive sign or feel that it doesn't matter because it is part of God's plan, one which includes the end of the world (Kearns 2017: 146). This degree of religious fatalism could certainly be applied to some Muslim circles, though I have not obtained any research linking Islamic eschatology to an exploitative treatment of the environment. One study by James Guth and colleagues (1995; see also Kearns 2017: 146) concluded that conservative eschatology (i.e., a belief in the literal reading of the Bible and that the Bible predicts that the world may end soon) was the strongest religious predictor of environmental apathy. The fact that the Qur'an is considered God's word and does emphasize the impending end of the world indicates the possibility of environmental apathy among some Muslim groups as well. Further, the Qur'an does warn the believer not to get too attached to this world. However, several authors have emphasized that 'this world' is the key to understanding the text's eschatological concerns.

These examples point to the potential (and actual) conflict between the notion of God's omnipotence, along with the promise of an 'other-worldly' reality, and the notion of the Anthropocene in contemporary discourse on climate change, which indicates that an alignment between the Anthropocene as a secular eschatology and religious eschatologies may prove challenging. However, the examples also showcase the negative consequences of disengagement that this theological strife can cause in the face of engaging with substantial and extensive environmental disasters, something that eco-theology at its core seeks to address. How does the Islamic eco-theological discourse deal with these issues? The following sections will first examine how authors from the eco-Islamic field draw on the Islamic tradition to emphasize human responsibility and destructive capabilities in an eschatological framework. Then, they will examine how two selected authors respond to the above-mentioned issues and bring together the Anthropocene and Islamic environmental thought.

Islamic Eschatology in the Eco-Theological Discourse

Several authors from the eco-theological discourse stress the ambiguity of the human character in Qur'anic discourse and humanity's potential for destruction (Abdul-Matin 2010; Gade 2019; Khalid 2019). Repeatedly, the Qur'an evidences the nature of people being forgetful, wasteful, complacent, ungrateful, hypocritical, and corrupt, even in the face of the presentation of the consequences of these dispositions (Gade 2019: 88). In an often-cited verse, the angels actually ask why God has created a being that will cause so much corruption on Earth. God does not answer them, but simply states: "I know what you know not" (Gade 2019: 87; Q 2:30). This ambivalent portrayal frames the human as a 'tested' being and further connects eschatological notions of the final accounting. In this regard, the commonly invoked notion of signs (*ayat*) is interesting. *Ayat* in the Qur'an encompass multifaceted notions of God's omnipotence and glory, as the term refers to verses in the Qur'an as signs for believers to reflect upon, on the one hand, and as signs of God's glory and the magnitude of his creation on all levels, on the other. This notion serves as a powerful theme in most eco-Islamic discourses and connects all of creation to the wonderous might, power, and omnipotence of God for those who reflect on and contemplate it.[5] However, some eco-theologies have also stressed that pollution, plastic waste impacting marine life, deforestation, and mass extinction are also examples of 'signs for those who reflect'; however, they are signs of humanity's destructive patterns and negative human behaviors (Abdul-Matin 2010: 12; Khalid 2019). In effect, these 'signs' become a way of confirming the destructive role of human agency in the phenomenal world as a negative parallel to the glorious signs of God's creation.

The eschatological dimension of the Qur'anic system of signs indicates another dimension that is significant when considering the contemporary environmental discourse. In this context, specific signs are prognostications of the end of the world, which are emphasized rhetorically as a predictive reminder to those in the present about inevitable consequences to be faced in the future (Gade 2019: 110). Many of the last 36 surahs of the Qur'an (Q 78–114) contain heavily apocalyptic material, and this section is commonly read, memorized, and recited in obligatory daily prayers. These surahs contain rich, detailed, and unique imagery of the natural world under radical transformation often in the form of natural disasters and phenomena such as earthquakes, eruptions, floods, and lunar/solar eclipses. For example, there are the images evoked in surah At-Takwir (Q 81:1–3) referring to the chaos of the 'last day' with descriptions of the sun darkening, stars falling, and mountains unmoored. Further, descriptions of frightening, incremental changes preceding it mark the start of the chain of inevitable events, such as water in the sea catching fire and the sky being stripped away (Gade 2019: 111; Q 81:5–11). Similarly, in surah Al-Infitar verses 1–3 we see that 'the sky breaks apart', the 'stars fall and scatter', and the 'seas are erupting' (Q 82:1–3). Immediately following are references to the day of recompense or the final judgement. According to Anna Gade, the apocalyptic framework of signs, changing signs in the phenomenal world, certain, and unknowable, points to ethical and environmental relationships and practices in the present moment that extend across space, time, and species (Gade 2019: 114). Further, both communal and individual responsibility for the state of the world are clear in the portrayal of the Qur'an's destructive signs, which God effects (Gade 2019: 111). However, Gade argues that it is precisely this connection of future worlds to come and the present moment that makes eschatology a powerful theme that mobilizes environmental thought and activism (Gade 2019: 111). In this sense, the eschatological dimension of 'signs' ties together the Islamic world view and its moral framework through the lens of environmental transformations. However, when considering the negative 'signs' of destructive human agency in light of the eschatological dimension of signs in the Qur'an, which God effects, the lines are not clear-cut. While there are eco-theologies that stress the ambiguous role of humans in the Qur'an and emphasize their destructive capabilities, is it possible to frame this in terms of the Anthropocene? Can Islamic eco-theology leave room for human agency destroying planetary conditions to this extent, or does the Anthropocene in its very essence challenge God's all-mighty omnipotence?

Framings of the Anthropocene in Islamic Eco-Theological Thought

As noted above, the Anthropocene construes a threat that is not only irreversible but also spatially and temporally absolute. While the Anthropocene is often mentioned in eco-Islamic discourse, few authors have dealt with the concept of the Anthropocene. The important exceptions are the writings of eco-theologian Fazlun Khalid and Islamic scholar Anna Gade. In a chapter entitled 'Surviving the Anthropocene', Khalid presents the idea that humanity is now itself a force of nature, a concept that doesn't seem to raise any theological questions for him. The Anthropocene is presented with the scientific discussions surrounding the 'golden spike', or marker of the Anthropocene. However, Khalid goes further and proposes a new marker – namely, the date on which usury/interest was legitimized by Henry VIII in 1545. This was an event that eventually led to the creation of the Bank of England in 1694, "enshrining the magical fractional reserve banking system which conjures money out of thin air and has kept the entire planetary population in its thrall ever since" (Khalid 2019: 181). This event, Khalid claims, profoundly changed the nature of the human relationship to the Earth and led to an acceleration of human history. This event, however, is an economic and political event in history, and cannot be detected in geological material. Further, this event supports Khalid's economic and political critique, highlighting that the fundamental problem is that our world system is based on the premise of perpetual growth and that to advert collapse there needs to be a radical restructuring of the economic system (Khalid 2019: 183). This means that while the introduction to the Anthropocene and the 'golden spike' is presented along the lines of scientific discussions, Khalid alters the premise of the discussion by introducing a historical moment that is linked to a highly socio-religious concept in Islam, namely interest/usury, which is forbidden and considered a great sin. By proposing a new marker, though this must be read as a symbolic marker framed religiously and ethically, Khalid implies that the Anthropocene as the 'human epoch' could in fact be playing out. Khalid refers to the Earth system collapse, and how humans have altered the Earth system qualitatively in ways that call into question our very survival over the coming few centuries (Khalid 2019: 182). By drawing on the authority of the 20,000 scientists who signed the 2017 document entitled "World Scientists Warning to Humanity: A Second Notice", Khalid presents the gloomy conditions of the planet while emphasizing the urgency of responding to these challenges. Khalid employs the Anthropocene to emphasize urgency and crisis, and the importance of responding accordingly. Khalid's political project of planetary response is framed economically and spiritually, calling for planetary emergency management through new global forms of 'Earth system stewardship' (*khalifah*), as urgent transglobal action is required to mitigate a dangerous shift in the Earth system. Even though Khalid's writings related to the Anthropocene reveal a discourse of humanity being on the tipping point of catastrophe, with cooperation and social change we can still avoid reaching the point of no return.

Gade argues that there exists an 'impending dooms-day atmosphere' in many secular writings, something which is further legitimized and backed up by an overwhelming majority of scientists drawing attention to the human impact on the Earth resulting in the Anthropocene (Gade 2019: 2). In this sense, she argues that the fundamental teachings on eschatology seem as ethically relevant today as in the seventh century. Further, Gade states that Muslim environmentalisms address head-on the notion of responsible and *human* existence as that of being among other creatures in the face of the imminent potential for the Earth's destruction (Gade 2019: 3). The notion of the Anthropocene is connected to Islamic eschatology in a way that does not challenge God's omnipotence or ultimate will, but rather assumes the potential of human destructive abilities, accountability, and the significance of human agency. In this sense, environmental discourses seen through the lens of Islamic eschatology completely recast the Anthropocene, as exemplified in the following comment made by Islamic scholar Marion Katz: "In a very different sense than we would understand today, traditional interpretations [...] suggest that, for Muslim thinkers, it was always the Anthropocene – environmental disaster was pervasively assumed to be the result of human wrongdoing" (Gade 2019: 116).

This seems to suggest that there is no contradiction pertaining to God's all-mighty power and the Anthropocene, where human activities determine planetary conditions; rather, the Anthropocene is part of the divine scheme. It should be noted that this way of reframing the Anthropocene makes it a holistic concept encompassing religious and ethical considerations beyond the scientific discussion regarding geological material. It suggests that through the Qur'anic discourse, in light of contemporary environmental awareness, Muslims can now only begin to realize what kind of destructive powers humans actually hold and what God in the Qur'an refers to. Thus, the concept is used to shed new light on Islamic eschatological teachings pertaining to the environment and human agency. The teachings from Islamic eschatology can also bring considerations regarding the notion of crisis, as Gade argues that the notion of crisis has always been embedded within the Muslim worldview as an eschatological reality. In other words, the image of crisis has always formed part of an ongoing ethical scenario, in which human impact forms an integral part of this moral system.

This entails that, in addition to images of crisis, Islamic eschatology can connect vivid and detailed depictions of the future transformation of the environment into a broader ethical framework in the moral present. Thereby, Muslim religious systems confront the hard facts of the unseen and unpredictable futures that are the result of human action in eschatological scenarios (Gade 2019: 116).

The general message in this framing of the Anthropocene is to embrace the realities of the uncertain and unpredictable future. Examples of Muslim environmentalisms, as presented through Gade's ethnographic fieldwork in Indonesia, illustrate an emphasis on mercy for creation, something that is only affirmed by engaging with nature in local settings. This is something that can also be extracted from Inga Härmälä's (2014) informants that engage in sustainable agriculture, or permaculture, that spending time in nature, paying attention to the 'signs' in nature, and drawing experience from these observations and reflections are essential. This indicates that local conditions should inform and guide environmental action as human beings need to adapt and change in the face of environmental transformations. Though only two authors are addressed here, these examples indicate that Islamic eschatology and eco-theology can display different framings of the Anthropocene in addition to encompassing the theological dimensions and concerns that may seem contradictory or challenging in the fields of religion and ecology.

Expressions of Apocalypticism in Muslim Environmentalisms

This final section explores how apocalyptic thinking is expressed in Muslim environmental activism through Islamic theoretical constructs and through a perceived shared experience of the eco-apocalypse. These examples illustrate that apocalypticism, for these communities, is a powerful theme that informs Muslims engaging with environmental concerns, and show that apocalyptic eco-theology can be successfully applied to Muslim environmentalism.

In Härmälä's master's thesis on transformative Islamic ecology, the topic of the end times and the connection between eschatology and ecological collapse emerged among some informants. According to them, Islam holds the answers to the contemporary ecological crisis, in which the human role in the environmental degradation is emphasized through interpretations of the Qur'an. Repeatedly cited was verse 30:40 stating that:

> Corruption has appeared throughout the land and sea by (reason of) what the hands of people have earned so He may let them taste part of (the consequence of) what they have done that perhaps they will return (to righteousness).
>
> *(Härmälä 2014: 43)*

In their interpretation, environmental degradation is equated with corruption and is a direct result of human agency. Further, emphasis was placed on the ways the Earth provides clear signs of this corruption through, for example, soil erosion, the destruction of food systems, lack of animal welfare, and climate change. These destructive signs are a means for reflection to understand the righteous way to engage with people and objects in the world. In this way, ecological degradation is invoked as a sign of the need to return to God. Here, the idea of signs could be argued to take the form of *realized eschatologies* as they are framed as lessons meant to inform those still living. This points to a shared terminology between discourses on the ecological crisis and Islamic eschatology, as Härmälä argues that "the eco-logical, economic, political, and climate crisis are interpreted by many Muslims as a sign that we have entered or are entering the end of times" (Härmälä 2014: 44). One informant was reluctant to use the word 'crisis', and instead called it a 'col-lapse' "because this system in which we are currently living in is designed to crash [...] It has been designed to collapse" (Härmälä 2014: 44). In fact, this idea encompasses the notion that even though human agency brings the destruction of the Earth, it is somehow part of a divine scheme. Following these statements and drawing on Donner's conceptualization, this idea is an example of a *partly realized eschatology*, where events or signs of environmental degradation in the present are portrayed as the beginnings of the end times or as being associated with the last judgement. The idea that the world is *designed* to collapse would essentially mean that you cannot change the outcome. But even though the informants stress that "everything is in the hands of Allah", that does not entail being passive, "because at the same time we need to deal with our own day of judgement. That way we must always show that we have acted based on hope and not based on despair" (Härmälä 2014: 48). What the informants refer to as the *signs* of the end times is interpreted as a call for environmental action along the lines of religious duty. This approach entails an acceptance of a creation (i.e., natural world) in transformation where humanity is not necessarily able to change the outcome; however, this view should not be translated into apathy or anxiety. Humans are still meant to 'listen' to the natural world, examine the signs, and cooperate with the soil, as Härmälä's informants emphasize.

While presenting examples from her ethnographic fieldwork in Indonesia, Gade illustrates that eschatological awareness and the consequences of humans as 'tested' creatures are a motivating rationale for environmentalism among pious Indonesians (Gade 2019: 113). Her claim is that when environmentalism is integral to religious piety it draws on tradition differently than when religion is enrolled in service of environmental projects. Rather than drawing on a selection of what have come to be known as 'environmental verses' in the Qur'an or key environmental conceptions, Gade stresses that when Muslim commitments to environmentalism are primarily for the sake of religious goals, apocalypticism along with material on the Prophet Muhammad's conduct and sayings emerge as central aspects (Gade 2019: 199). Prophetic teachings that accentuate having mercy (*rahma*) towards creation are especially underscored. One narrative widely circulated in the environmental discourse in Indonesia, which is found in *Sunan-at-Tirmidhi*, states that:

> [...] The messenger of Allah, peace and blessings be upon him, said, "Those who are merciful will be shown mercy by the Most Merciful (*Al-Rahman*). Be merciful to those on the earth (*man fi'l-ard*) and the One in the heavens will have mercy (*yarhamkum*, verbal form of *rahmah*) upon you".
>
> *(Gade 2019: 218)*

Here, the example of the Prophet Muhammad is invoked, correlating the ultimate mercy of Allah in the afterlife with the mercy of an individual person or community expressing mercy towards creation. In another hadith, frequently referenced in eco-Islamic discourse, Muhammad instructs believers to plant trees as an act of worship, to "hold on to the sapling in your hand", even "as Judgement Day erupts" (Gade 2019: 113). This argument connects the eschatological reality and apocalyptic motif of Judgement Day to environmental worship in an intriguing way. Presumably the sapling, along with everything else, will be destroyed. Thus, this teaching is sensible insofar as the rationale to plant the tree, one would expect, is judged on the intention and act, and counted as a good deed. Thus, the deed (reward for planting) would endure, although the tree itself would not. In this way, Gade contends that the embrace of apocalypticism does not inevitably lead to the escape from either 'hope' or from 'hopelessness', to use the preferred sentimental terms in the mainstream ethics of climate crisis (Gade 2019: 114). One of Härmä-lä's informants supports this view, considering the same hadith:

> You can plant a tree or plant as many as you can. That is what we are told to do. If you think you are going to die today, then one of the good things you can do is plant a tree. And you get extra credits for those trees that perform functions for any other living thing.
>
> *(Härmälä 2014: 48)*

This refers to another widespread hadith, found in the collection of Al-Bukhari, which says that if one plants a tree or sows crops that a bird, an animal, or a person later eats from, it is considered an act with the rewarded status of a charitable gift (Gade 2019: 230). This situation further fuels the idea that showing mercy towards creation is rewarded.

One figure from Gade's research, K.H. Thonthawi, claims that Allah rewards sincere environmental care more than other religious action because it is giving mercy, which God rewards with mercy in return (Gade 2019: 230–232). In Thonthawi's opinion, *religion* itself as conventionally understood is not enough to respond to the moral magnitude of environmental problems or even enough for religious success in terms of ultimate accounting. In this sense, merely repeating everyday religious actions by rote, like prayer, is not enough to guarantee entry to heaven. Muslim environmentalism, for him, represents a more sincere search for reward in the life to come and an even deeper commitment and meritorious action. The final determination belongs to Allah alone, and it comes through His mercy (*rahmat*). The way to obtain this mercy is through showing loving-kindness to his creatures precisely through Muslim environmentalism, which he calls the 'ticket to paradise'.

In her work, Gade illustrates a pious and religiously oriented Muslim environmentalism that is built on the ethical foundation of engaging crisis as natural consequence of existence, which includes apocalyptic futures while simultaneously affirming care for the Earth as a criterion for the success of the state of the world to come (Gade 2019: 116). In this way, there is a connectivity between the state of the current world and that of the next that highlights relations, merciful acts, and commitments towards creation. It also draws attention to the final accounting across space, time, and generations. The presence of eschatology is at the intersection of Islamic sources and Muslim practice, connecting the underlying worldview of the afterlife as the ultimate focus in this life. However, the accountability of our actions and inclinations in this present life serves as the foundation for such focus. Further, practical norms are embedded within a broader ethical framework in a way where Islam is highly relevant and dynamic. This methodology, approaching Islam through an environmental lens, seeks to understand the Qur'an's role in the context of the practice, performance, and daily life of Muslims, rather than trying to extract environmental teachings that fit into an environmental discourse, as is the case in many eco-theologies.

Conclusion

This chapter has attempted to illustrate that the apocalyptic worldview of the Qur'an and the Islamic tradition, as expressed through Muslim environmentalisms, can be seen as a realization of environmental justice as well as Islamic piety. In this sense, the significance of this life as a 'test' and humanity's relationship with other creatures are to be known when 'the environment' is radically transformed out of human control. In this way, creation (i.e., Earth) is seen as part of an eschatological drama, in which the human figure emerges as highly controversial, exercising an ability to destroy nature beyond what has previously been conceivable in theological discourse. It brings into question the human role and participation in the divine eschatological scheme. This radically transformed natural world is the direct (or indirect) consequences of human agency, in which God's judgement may mirror His mercy in relation to their own commitment to and care for creation.

Ultimately, this message is compatible with the notion of the Anthropocene emphasizing humanity's planet-shaping abilities. The way the Anthropocene is framed, at least in the two accounts examined here, leaves room for human agency and places notions around free will and accountability in a new light. By emphasizing the human role in an unknown ecological future, it also raises questions of submissive attitudes towards the end times and religious fatalism in a way that could possibly draw attention to age-old theological discussions related to free will and predestination (*qadr*). This free will and accountability, however, in the greater scheme of things is ultimately linked to the apocalyptic message of the final judging. Furthermore, it attests to the potential of eco-theology to bring new perspectives to Islamic eschatology and theology, and points to the relevance of eschatology to provide valuable contributions to eco-theology. Interwoven with accountability and seen through the lens of contemporary environmental issues, apocalypticism can point to valuable teachings on the ways human beings are embedded in and interconnected with creation, and therefore anchors human action and interaction in this world.

Notes

1 In its original Greek, apocalypse means 'revelation' as in 'uncovering' or most literally 'lifting the veil'. The current understanding of apocalypse, in more contemporary research, is that of a genre of revelatory literature with a narrative framework, "in which a revelation is mediated by an otherworldly being to a human recipient, disclosing a transcendent reality which is both temporal, insofar as it envisages eschatological salvation, and spatial, insofar as it involves another, supernatural world" (Lawson 2008: 24). There is a scholarly consensus that apocalypse as a genre should not be taken to simply mean destruction. It should be understood as "a supernatural revelation, which reveals secrets of the heavenly world, on the one hand, and of eschatological judgement on the other" (Lawson 2017: 99).
2 The Qur'an is argued to be completely unique in style. The difficulty of identifying genre or a specific literary style has resulted in formulations classifying the Qur'an as a repository of several genres and literary forms (Eggen 2011).
3 The emphasis on individual and collective responsibilities has been formulated in juri'dical doctrine where individual duties (*fard ayn*) refer to, for example, prayer, fasting, paying alms, etc. and collective duties (*fard kifaya*), are defined a communal obligations discharged by the Muslim community. This terminology is often used to discuss social responsibility, such as feeding the hungry, commanding good, and forbidding evil (Fadl 2014: 140).
4 This point is relevant in relation to the current discourse in the mainstream media pleading for future generations inheriting the Earth. The fact that the Qur'an's judgement scenarios point to an intergenerational responsibility, as well as community- and individually based responsibility, makes this parallel to environmental discourse intriguing.
5 For a more elaborate discussion on *ayat* in the eco-Islamic discourse, see the fifth chapter of my 2020 master's thesis, which is entitled "Islam in the Age of Ecological Apocalypse" (2020). It is available at https://www.duo.uio.no/bitstream/handle/10852/84355/Masterthesis.pdf.

References

Abdel Haleem, Muhammad. (2017). "Quranic Paradise: How to Get to Paradise and What to Expect There" in *Roads to Paradise: Eschatology and Concepts of the Hereafter in Islam*, edited by Sebastian Günther and Todd Lawson, pp. 49–66. Brill. Leiden.

Abdul-Matin, Ibrahim. (2010). *Green Deen: What Islam Teaches about Protecting the Planet.* Berrett-Koehler Publishers, Inc.San Francisco.

Breton, Hugh Ortega and Phil Hammond. (2016). "Eco-Apocalypse: Environmentalism, Political Alienation and Therapeutic Agency" in *The Apocalypse in Film*, edited by Angela Krewani and Karen Ritzenhoff, pp. 105–116. Rowman and Littlefield. Lanham, MD.

Crawford, Kate. (2009). "Emergency Environmentalism: On Fear, Lifestyle Politics and Subjecticity". *Angelaki Journal of Theoretical Humanities*, vol. 14:2, pp. 29–35.

Donner, Fred M. (2017). "A Typology of Eschatological Concepts" in *Roads to Paradise: Eschatology and Concepts of the Hereafter in Islam*, edited by Sebastian Günther and Todd Lawson, pp. 758–772. Brill. Leiden.

Eggen, Nora S. (2011). "Koranen – Islams hellige skrift" in *Hellige Tekster i Verdensreligionene*, edited by Jens Braarvig and Justnes Årstein, pp. 87–109. Høyskoleforlag. Kristiansand.

Fadl, Khaled Abou El. (2014). *Reasoning with God: Reclaiming Shari'ah in the Modern Age.* Rowman & Littlefield. Lanham, MD.

Fagan, Madeleine. (2017). "Who's Afraid of the Ecological Apocalypse? Climate Change and the Production of the Ethical Subject". *The British Journal of Politics and International Relations*, vol. 19:2, pp. 225–244.

Gade, Anna M. (2019). *Muslim Environmentalisms: Religious and Social Foundations.* Columbia University Press. New York.

Grim, John and Mary Evelyn Tucker. (2014). *Religion and Ecology.* Island Press. Washington, DC.

Guth, James L., John C. Green, Lyman A. Kellstedt, and Corwin E. Smidt. (1995). 'Faith and the Environment: Religious Beliefs and Attitudes'. *American Journal of Political Science*, vol.39:2, pp. 364–384.

Härmälä, Inga. (2014). "Transformative Islamic Ecology: Beliefs and Practices of Muslims for Sustainable Agriculture and Permaculture". Master's thesis. Lund University.

Hasson, Isaac. (2005). "Last Judgement" in *Encyclopedia of the Qur'an*, edited by Jane Dammen McAucliffe. Washington, DC. Brill Online. [Accessed August 26, 2020].

Kearns, Laurel. (2017). "Climate Change" in *Grounding Religion: A Field Guide to the Study of Religion and Ecology*, edited by Whitney A. Bauman, Richard R.BohannonII, and Kevin J. O'Brien, pp. 137–157. Routledge. New York.

Khalid, Fazlun M. (2019). *Signs in the Earth: Islam, Modernity and the Climate Crisis.* Kube Publishing Ltd.Markfield, UK. Lawson, Todd. (2008). 'Duality, Opposition and Typology in the Qur'an: The Apocalyptic Substrate'. *Journal of Qur'anic Studies*, vol. 10:2, pp. 23–49.

Lawson, Todd. (2017). "Paradise in the Quran and the Music of Apocalypse" in *Roads to Paradise: Eschatology and Concepts of the Hereafter in Islam*, edited by Sebastian Günther and Todd Lawson, pp. 93–135. Brill. Leiden.

National Geographic. (2019). "Anthropocene". https://www.nationalgeographic.org/ency clopedia/anthropocene/. [Accessed June 7, 2019].

O'Hare, Patrick. (2019). "Holocene vs. Anthropocene Debate". *Earth.com*. https://www.earth. com/earthpedia-articles/holocene-vs-anthropocene-debate/. [Accessed June 16, 2019].

Pavid, Katie. (2020). "What Is the Anthropocene and Why Does It Matter?". *Natural History Museum*. https://www.nhm.ac.uk/discover/what-is-the-anthropocene.html. [Accessed June 7, 2019].

84 Marita Furehaug

Rothe, Delf. (2020). "Governing the End Times? Planet Politics and the Secular Eschatology of the Anthropocene". *Millennium: Journal of International Studies*, *vol.* 48:2, pp. 143–164.

Thomassen, Einar. (2009). "Islamic Hell". *Numen*, vol. 56, pp. 401–441. https://www.jstor.org/stable/27793798.

Zakzouk, Mahmoud. (2017). "The Path to Paradise from an Islamic Viewpoint" in *Roads to Paradise: Eschatology and Concepts of the Hereafter in Islam*, edited by Sebastian Günther and Todd Lawson, pp. 39–46. Brill. Leiden.

PART 2

Representing the Environmental Apocalypse

6

THE DISAPPOINTING APOCALYPSE

Climate Collapse and Visual Art since 1960

Andrew Patrizio

Introduction

Apocalypses manifest themselves to the eye. The Greek *apokalyptein* fuses *apo* ('from') and *kalyptein* ('to cover or conceal'), and so speaks to a revelation (a revealing or uncovering) which is, in a sense, revealed again in art. In the context of ecological apocalypse, a depleted future world first emerges from environmental texts and data, and is then re-revealed culturally in an attempt to envisage, through formal, imaginative, and material means, what the coming apocalypse might look like. This chapter, however, examines the incompleteness and limits of this claim, looking at the inability of art, or any imaginative form, to capture an apocalypse to come. There are inherent disappointments deep within the intellectual project of ecological and environmental apocalypse in art, but, as a resul, we are pointed towards some of its striking fascinations too.

Maurice Blanchot's (1997) chapter in *Friendship*, called 'The Disappointing Apocalypse,' has steered some of my thinking on the eco-apocalypse and visual art. Blanchot's text, in retrospect, may seem disappointingly vague in places, refusing particulars beyond shaping a response to nuclear annihilation that was Blanchot's context when he wrote it in 1971. Reviewing many of the art practices discussed above with this framing idea of 'disappointment' might seem to imply acceptance of art's inadequate scale and intensity acknowledged in the very practices that sought to capture environmental collapse. Yet Blanchot's text also posits something necessary in art in the face of this particular end of times:

> art […] plays, in relation to history, the role that, for Hegel, history plays in relation to nature: it gives it a meaning, it assures, beyond the perishable and across the death of duration, life and the eternity of meaning. Art is no longer the anxiety of time, the destructive power of pure change.
>
> *(Blanchot 1997: 29)*

DOI: 10.4324/9781003189190-8

Blanchot's proposition only makes sense, though, in the present, in the creative instance of specific artworks, such as the ones we will discuss. The disappointment in art derives, in part, from its claim to give meaning to unknown audiences today and, perhaps, to future audiences, and to offer these meanings in new ways, which is to some extent unimaginable or unpredictable now. Art, then, is the opposite of acquisitive prospecting and technological reach.

I want to draw attention to the varieties of disappointment that lie at the centre of artistic projects on ecological and environmental apocalypticism. I do not mean 'disappointment' as a negative judgement, a failure, on the part of the artists or their artworks. Rather, I believe that there are a variety of expectations, revelations, horrors and violence deeply embedded in apocalyptic texts that, when they come to be expressed or re-imagined in visual arts, further complicate our relationship with 'the end of the world' (itself a concept full of its own disappointments).

Art once revealed through images the coming truth of what was merely descri-bed in text.[1] Art now nests within a complex ecosystem of visualisation as a form of image production that merges seamlessly with, and often draws on, wider broadcast media, internet-based news, NGO and social activism, experimental film and video broadcast through financial, institutional, discursive, and technological infrastructures. Art's agency is set against, in Thacker's phrase, an 'unthinkable world [...] of planetary disasters, emerging pandemics, tectonic shifts, strange weather, oil-drenched seascapes, and the furtive, always-looming threat of extinction' (cited in Apter 2013: 139). Collectively, these worldly threats are so devastatingly transformative that art or any form of cultural representation falls short. Art historian T.J. Demos sees images themselves as part of the problem – 'a salvage paradigm, compensatory, fetishistic, taxidermical, a last-ditch effort to deny the undeniable, to restore hope in hopelessness' (Demos 2019: 5). Images are complicit, inadequate, compensatory.

Any review of the religious and environmental apocalypse in art reveals a per-sistent thread of ambivalence towards the end of times as both dramatically cata-strophic and rather ordinary. It is surprising, surely, that the human mind copes through such strategies – normalising, as Susan Sontag put it, 'what is psychologi-cally unbearable, thereby inuring us to it.' She counterpointed apocalyptic fantasy as fearful yet opposed destinies of 'unremitting banality and inconceivable terror' (Sontag 1965: 42). For art and creative culture more generally, the consequences of thinking of ordinary and disappointing apocalypses is that it becomes possible to live through and around them. Indeed, contemporary artistic practices that explore apocalyptic ideas might be said to be the first that do so from inside the dimensions of the end times. If the Sixth Mass Extinction is already on us, if the Anthropocene may come to be the last geological era in which humans actually live, then there is a present realism rather than a prospective imaging that sits at the centre of eco-apocalyptic art.

Here, I want to consider three internal subgroups of the artistic engagement with climate apocalypse – the *marine*, the *nuclear*, and the *mineral*. Each can be prefaced by another meta-category that shapes them all: *extinction*. This reflects the fact that eco-apocalyptic immanence is global, varied and differential. It occurs intersectionally and unevenly, so any account of cultural understandings of eco-apocalyptic presence should reflect this non-singularity. As Elizabeth Povinelli writes: 'changes to the earth system are heterogeneous and diachronous, diffused and differential geographies that only appear as instantaneous earth events when viewed from the perspective of millions of years of stratigraphic compression' (Povinelli 2016: 11). There is no united idea of the apocalypse here but a fragmented, unevenly spread set of environmental spaces where conditions of survival are grim, even if the potentialities for eloquent artistic expression are great.

There is a fundamentally self-reflexive element to eco-apocalypse both in how it can be conceptualised as a threat to all biotic life and to how it is culturally and artistically expressed, or imagined, by a species – humans – who are at the malignant causal root of the very concept. Art brings a speculative and critical function – as cultural probe – that has long moved the topical foci towards humanity's destructive presences on the planet. We are drawing on a page that we are simultaneously erasing. This is either folly, as some of the more sceptical historians of the visual have, or, as I suggest below in the conclusion, it is the source of energy and the power to catalyse change in the midst of this very folly.

Some interpretative creations on our theme come from positions of anger and deep pessimism; and anger at structural forms of greed, acquisition and colonisation seems entirely justified. There is also an increasing number of activist and engaged practitioners who, whilst inspired by the same circumstances as the doomsayers, look to work against the grain and seek material change now. The emancipatory and expansive artistic practices of the 1960s and 1970s provide some of the options, allowing practitioners to draw in data, research, performance, collaboration, intervention and installation through text, film, photography, computer visualisation and sound – all to supplement and expand resilient forms of expression in painting and sculpture. There are myriad affirmative practices across activist and community art in the West and in Indigenous contexts globally. Aligned with this expansiveness, belated attention is being paid to Black artists and those of colour. The brutal extensiveness of colonial and capitalist violence ignites global creative responses, like multiple inverted mirrors set to interrogate the aggressor.

We now look at artists working within the meta-category of extinction, before refining our focus on artists who have sought the eco-apocalypse in the mineral, marine and nuclear worlds.

The Extinction Apocalypse

Extinction, whilst a normal evolutionary process, has, since the rise of techno-modernity and anthropogenic pollution, led to the vastly increased eradication of many species. We have self-initiated a Sixth Mass Extinction with visual art as a kind of melancholic handmaiden, making icons of the extinct. This is not a one-directional slide to disaster, as Neel Ahuja has argued:

speculations of climate-driven extinction […] in contemporary literature and visual culture operate as postcolonial fantasies of a universal human precarity, fantasies that are coming under increasing duress as transborder risk migration, indigenous and Southern environmental activisms, and persistent forms of ecological resilience challenge totalizing environmental visions of the human.

(Ahuja 2017: 44)

For all other life forms, the persistence of humanity is not only disappointing but dangerous.

The painter Alexis Rockman has made a substantial career of imagining 'the world without us', to employ Alan Weisman's telling phrase. In Rockman's *Manifest Destiny* (2004), unimaginable sea level rises in New York reverse pastoral idylls of expansion and bounty, and give the United States' Eastern Seaboard to pelicans, jellyfish, and cetaceans. His *Great Lakes Cycle* (2016–2017) is a large series of paintings unpopulated by humans and liminally focussed on the precarious water line. In many of the works, for example *Spheres of Influence* (2016), broken relics of industry and transport encounter marine and bird life that circulate confused and bemused around the wreckage. The incomplete and frustratingly unproductive aspects of 'the disappointing apocalypse' are spread out in vivid painterly detail before us (Rockman 2017).

Indian artist-activist Ravi Agarwal focusses on another kind of apocalyptic extinction that resonates across Indian societal boundaries. *Extinct?* (2008) was located at two separate sites in New Delhi and comprised an installation of objects and repurposed taxidermy displays in the Natural History Museum, and large photographs and wing-like structures set in a nearby busy roundabout. *Extinct?* denotes the catastrophic decline of vultures across India brought about by their ingestion of Diclofenac, a cheap painkiller given to domestic cattle. Hindu religious

FIGURE 6.1 Alexis Rockman, *Spheres of Influence* (2016) from *The Great Lakes Cycle*. Oil and alkyd on wood panel. 72×144 inches. (Collection of Jonathan O'Hara) © ARS, NY, and DACS, London, 2022.

FIGURE 6.2 Ravi Agarwal, *Extinct?* (2008), installation image from 'Extinct' – 48 deg C, Public Eco Art Project, outside the Natural History Museum, New Delhi, © Ravi Agarwal.

practices honouring the cow mean that when they fall ill they are neither killed nor consumed, so scavenging vultures eat and are fatally poisoned by the cow carcasses. The vultures are appropriated from stuffed museum specimens displayed in the museum. As Russell Stephens notes: 'taxidermy's inherent, simultaneous proximity to both corporeal life and that life's physical negation stands as a trope for the concept of extinction itself' (Stephens 2018: 33). He further observes 'a grotesque double negative', as the fleshless birds are 'paradoxically depicted greedily eating the very living material that they themselves no longer possess' (Stephens 2018: 34). Cows, vultures and humans are enfolded in a toxic embrace which is entirely the result of misapplied agro-medical technology (Agarwal n.d.).

If Agarwal reanimates the dead for eco-political purposes, US-based Danish artist Jakob Kudsk Steensen reanimates death differently through virtual-reality computer animation based on analogue fieldwork transformed to the digital. Steensen is fascinated with niche ecosystems under stress. In *Re-Animated* (2018–2019), hypnotically designed three-dimensional calligraphies are derived from the bird call of the extinct Hawaiian Kaua'I 'ō'ō, and so the artwork offers a visual and aural spectacle of de-extinction and ecological restitution.

The flora and fauna in the artist's work are also algorithmically programmed to grow and flourish in the digital realm (Steenson 2019). One message from

FIGURE 6.3 Jakob Kudsk Steensen, *Re-Animated* (2018–2019), image courtesy of the artist.

Steenson's work is that the environmental apocalypse for one species is a place of flourishing for another species. Disappointing and thwarted for one, a life-source for the other.

Critical Art Ensemble (formed in 1987) has an influential history of art and interventions in which art, technology, political activism and disobedience intersect that, over recent years and like many other artists, has become increasingly environmentally focussed. CAE's *Graveyard of Lost Species* (2016), done in collaboration with YoHa in Leigh-on-Sea, England, is one such project (Critical Art Ensemble 2016). It elegiacally commemorated around the walls of an abandoned fishing boat not just extinct wildlife and marine creatures lost in the Thames Estuary region, but also people, livelihoods, fishing methods, landmarks, mythologies and local dialects. Certain phrases link across this apocalyptic landscape, so freighted as it is with disappointment: 'The sea is a place to dispose of things one never wishes to see again: a dead body, a rotting hulk, illicit goods, cadmium batteries, arsenic, diesel, exhausted gear box oil, shopping trolleys, chip rappers, tin cans, and shoes.' The marine, then, takes on the disasters of land life in addition to suffering its own apocalypse underwater.

John Akomfrah's major immersive six-channel video installation *Purple* (2017) represents another elegiac statement on extinction, memories of ice, human suffering, colonial violence and shifting landscapes. The motif of interrelated, biospheric apocalypses once more comes to the fore, as the footage is assembled from archives and specially shot landscapes in Alaska, Greenland, the Tahitian Peninsula and the Marquesas Islands of the South Pacific. The title, *Purple*, refers in part to the colour's relation to mourning in West African culture, and in this context to planetary mourning caught in a monstrous traffic of multiple, evocative images as they pass around the observer. It draws heavily on the slow violence of colonialism, seafaring and what Akomfrah calls the 'finitude and the encroaching closure' of environmental collapse.

The Marine Apocalypse

It is the nature of apocalypses to be invisible unless and until they are rendered in art, text or data. The collapse that is happening across oceanic ecosystems belongs to an extensive set of spaces that largely unfold unseen by humans. The declining fortunes of marine life and those riparian communities that live along its shores have been subject, increasingly, to artistic analysis and capture in recent times.

The marine apocalypse is a set of nested disasters that include rises in temperature and sea level, salination and acidification, and marine areas are increasingly becoming dumping grounds for pollutants, plastics and chemicals whose infiltration into the hydrosphere is reaching its limit. The effects are as various as they are global. The dependency on water by all living forms means that marine decline gets our attention. There is too much water in some places; too little in others. The associations between extreme dryness and extinction offer artistic resonances with other apocalypses discussed here, such as desertification, the nuclear and the extinct. Yet again, we see eco-apocalypse as fragmentary, distributed and overlapping.

Artists have responded to the marine apocalypse globally, capturing their own disasters alongside projects further afield, which have the character of expressive fieldwork. The Argos Collective comprises French writers and photographers who created a series based on the Maldives tsunami (2006) which fed into a larger series on *Climate Refugees* (2010). Rising sea levels depicted as a biblical flood hitting consumption culture is at the centre of Superflex's *Flooded McDonalds* (2009). Sarah Cameron Sunde has created a simple performance mode – *36.5 / A Durational Performance with the Sea* (2013–) – with collaborators who have extended and repeated the event that uses the human body as a oceanic tidal marker of sea-level rise in places like the Bodo Inlet, Kenya (Sunde 2013; see also Dobraszczyk 2017).

Significant pioneers in ecological art with complex political intent are Helen Mayer Harrison and Newton Harrison, whose research-based installations set in motion series that have lasted decades, often under the working principle that, rather than simply provide an aesthetic expression of climate and environment, their work should 'do no harm'. *The Lagoon Cycle* (1974–1986) features projects in Tibet, across peninsular Europe, and, more recently, the Sacramento River.

Large-scale ecosystems are examined, in particular the interactions between food production and watersheds: crab-filled lagoons in Sri Lanka and Southern California. Some of the lagoons had apocalypse immanent as part of the process where herbicides flooded down from nearby valleys and killed aquatic life. Indigenous methods were contrasted with intensive farming. And in *The Seventh Lagoon*, the relationship between Pacific volcanoes and melting polar ice as forcing up water levels in oceans, rivers and lagoons was mapped to a rise of 300 feet. The effects of course were apocalyptic: sometimes averted, other times not. The foundational

FIGURE 6.4 Helen Mayer Harrison and Newton Harrison, from *The Lagoon Cycle* (1974–1986), image courtesy of The Harrison Studio.

idea is that if, from our present perspective, we fully imagine the disaster to come, then we are moved to take aversive action. Art becomes a warning strategy.

A disaster that has already come, and that is subject to an extraordinary form of visual and community repair of sorts, is *The Crochet Coral Reef* (ongoing since 2007) by Margaret Wertheim (a science writer and cultural historian) and Christine Wertheim (a cultural studies academic). Coral collapse, bleaching and desertification – a disastrous fact in the real marine world - is symbolically reversed by the Wertheims through their remaking of colourful coral structures in textiles and recycled plastics. *The Crochet Coral Reef* uses a technique derived from mathematical modelling, namely, crochet stitched in an increasingly large number of stitches in each row so as to replicate colourful zooxanthellae living healthily within coral.

The crenelated frills normally found in sea slugs, corals, kelp and sponges are crocheted back into a colourful flourishing. Handicraft's domestic practices, retooled with discarded plastic rather than wool, has produced sub-projects such as the *Ladies Silurian Atoll* and the *Blue Grove*, each created by dedicated associates of the Wertheims. Indeed, in total, 7,000 people across 25 global locations are involved in this community effort of reparation. The result is 'a vast and growing act of witness' built around the model of the community workshop that mixes

FIGURE 6.5 Margaret Wertheim in the Fohr Satellite Reef at the Museum Kunst der Westkuste, photo © Institute For Figuring.

practical instruction with 'mathematical background, biological insight, feminist implication, and ecological alarm' (Weschler 2011: 134). As Donna Haraway notes:

> This is not a project of melancholy and mourning. Theirs are figures of response-ability [...] By proposing fundamental questions about extinction and survival and response through material figuring, both the crocheting and the installations create publics that learn to care, to make a difference.
>
> *(Haraway 2015: 264)*

The project simultaneously marks inevitable coral loss whilst also seeking prevention and rehabilitation. Here, they share the Harrisons' approach in making art that is agential and materially interventionist. Yet, Margaret Wertheim reflects that

> it's been disappointing to me, personally, because [...] I really thought that I was beginning a project that would lead to the enfranchisement of a certain population that the official rhetoric of the science community claims to want to include and that they would embrace the project. That hasn't happened so far.
>
> *(Buszek 2011: 286)*

Swiss artist Ursula Biemann has done many works drawing together the geographical, ecological and political set in troubled global locations such as the Ecuadorian forests (*Forest Law*, 2014) and the Caspian Sea (*Black Sea Files*, 2005).

Deep Weather (2013) pivots on the motif of oil and water (as archetypically unmixable materials). The oil element is represented in 'Carbon Geologies', which is set in the tar sands of the Boreal forests of Northern Canada, Indigenous lands now scarred and wasted beyond measure in the name of oil extraction. The water element, 'Hydro Geographies', focusses on Bangladesh and how its land is under threat from melting ice fields in the Himalayan ranges and on the extreme weather events that rock that region. Attempts to frustrate apocalyptic water rises are mainly done through the building of protective mud embankments by thousands of local citizens. As Beimann writes on her website, here 'water is declared the territory of citizenship' (Beimann 2013). In *Deep Weather*, both Canadian and Bangladeshi communities experience cataclysm but through different means: aggressive extraction (oil) and incremental incursion (water) that represent a slow apocalypse, which is in marked contrast to the dramatic flood imagery of religious orthodoxies.

The long tradition of apocalypse caused by flooding is evoked in Yun-Fei Ji's work, as it articulates contemporary eco-political themes through traditional Chinese techniques. In works like *The Empty City: East Wind* (2003), detailed and seemingly benign images of Chinese life reveal, on closer inspection, troubled landscapes and ecological imbalances. *Three Gorges Dam Migration* (2009–2010), now in the collection of New York's Museum of Modern Art, is a woodcut that so clearly calls back to earlier art examples of apocalyptic representation.

As a scroll, it also registers a filmic unfolding of environmental catastrophe. It questions the impact of rapid industrialisation in Yun-Fei Ji's native China and the ecological and social upheaval that has been meted out to many parts of its territory. (Whilst his work has received positive attention in the West, the devastation is parallel to that of the Alberta oil tar fields explored by Biemann.) In a subtle detournement, by using traditional Chinese painting techniques the artist conveys a message of return – specifically to the Chinese philosophy of *tian ren he yi*

FIGURE 6.6 Yun-Fei Ji (b. 1963): *Three Gorges Dam Migration* (2009–2010), woodcut scroll, composition (image): 13 3/8 × 120 11/16" (34 × 306.5 cm). Publisher: Library Council of The Museum of Modern Art, NY. Printer: Rongbaozai, Beijing. Edition: 20. Gift of the Library Council of The Museum of Modern Art. Acc. no: 936.2010. New York, Museum of Modern Art (MoMA). 2022. Digital image, The Museum of Modern Art, New York / Scala, Florence.

('harmony and unity between humankind and nature') that has been ruptured by contemporary Chinese water strategies as the central government seeks to hydrate its burgeoning cities.

There is a haunting quality to Yun-Fei Ji's work; explicitly so in relation to the Three Gorges Dam, as it lies nearby to Fengdu, a ghost city. As the artist reflects:

> I wondered where all those ghosts would go when everyone had been relocated, and imagined they would have to move too [...] The ghosts might have been scholars in former days who had lost their position at court and so left for the country and, when they had a little time, they would write stories about what was happening in their lives, put them in jars and bury them. They were creating historical documents, recording facts, but interspersed with ghost stories as if it were the ghosts that were responsible for what had happened.
>
> *(Wei 2016)*

The ghosts here echo the Wertheims' *Crochet Coral Reef* too, which is a ghosting of nature, an artwork that recalls the other-worldly, the reimagined, repopulating the bleached coral reefs with surrogate colourful forms. The marine apocalypse is a haunted space – not only disappointingly abandoned but disappointing too in that these worlds can only be experienced through artistic re-enactment and creation.

The Nuclear Apocalypse

Nuclear and environmental apocalypses are closely entwined, and ongoing research such as Ele Carpenter's *The Nuclear Culture Source Book* (2016) makes it clear that the nuclear is a very large subset of environmental forms of violent revelation. Maurice Blanchot, writing in the 1970s, saw the destructiveness of the atomic bomb as inaugurating 'a beginning in history' based on auto-destructive urges in the human: 'Science has made us masters of annihilation; this can no longer be taken from us' (Blanchot 1997: 101). The nuclear is the pivot between religious forms of apocalypse and anthropogenic ones; moreover 'the environmental apocalypse mode continued and complicated what nuclear apocalypse began' (Buell 2010: 15).

Whilst 'nuclear-war exterminism differs from contemporary eco-catastrophism' in the different weightings of human agency and dispersed causality inherent in each (Demos 2016: 246), both conjure a prosaic and quotidian apocalypse that is at the centre of my theme. Jean Tinguely's two performance pieces, *Study for an End of the World, No. 1* and *No. 2* are key here. The first happened in 1961 in the Louisiana Museum of Modern Art, Copenhagen; the second the following year, on March 21, 1962, on Jean Dry Lake, Nevada (just south of Las Vegas). Here, the beloved mid- and southern US desert sites of the Land Artists were politicised and repurposed. Tinguely was the first artist ever to stage a site-specific performance in an American Southwest playa.

Study for an End of the World, No. 2 was a set of self-destructing sculptures formed from gathered Las Vegas landfill. Most of the audience comprised media

FIGURE 6.7 Jean Tinguely, *Study for an End of the World, No. 2* (1962), installation at Jean Dry Lake, Nevada, March 21, 1962, photo courtesy of Allan Grant / The LIFE Picture Collection / Shutterstock.

reporters. (If, in the popular slogan coined by 1960s Black Power activists, 'the revolution will not be televised', Tinguely certainly made sure the apocalypse was.) The bombs and props were assembled in Tinguely and Niki de Saint Phalle's Las Vegas hotel room and transported to the dry lake. This was, in a sense, an apocalypse for cultural institutions too, with Tinguely musing:

> I've reached the end, you see, for museums in this kind of thing. I need a place where I can build as big as I want, and destroy as violently. The only two settings I can think of as appropriate are the Sahara and the American Desert.
>
> *(Byron 1962: 76)*

The work evoked Cold War narratives as an ironic restaging of the Nevada Test Site explosions (sited only 90 miles away). These nuclear tests, as Emily Scott makes clear, were similarly mediated events in the 'heavily coded' space of the American desert (Scott 2012: 69). The pronounced social and environmental crises in relation to the land (for example, the growing reality and recognition that humans were having an impact on certain areas to the point of rendering them unproductive), and likewise that as increasingly rare 'blank spots on the map', one could deal with supposedly wasted lands and the contemporary condition.

Tinguely pre-empted the visits of the American Land Artists and fused differently (yet just as apocalyptically) the environmental, the nuclear, the parched, and the abject explosions that returned these temporary assemblages to ruination. For Pamela Lee, Tinguely's work was 'cheap symbolism' and 'bad metaphor' which, for my purposes, seems like an interesting version of disappointment (Lee 2004: 147). Matilde Nardelli, in 'No End to the End: The Desert as Eschatology in Late Modernity', also inadvertently speaks to the theme of the 'disappointing', noting Tinguely's

> tongue-in-cheek performance [...] which overtly spelled out the terminal, apocalyptic connotations of the nuclear desert even as it also seemingly mocked them. By contrast with the military conflagrations to which it so obviously made reference, Tinguely's performance was a rather low-tech, DIY affair, involving the artist himself lighting some dynamite and home-made bombs to destroy a precarious and intricate assemblage of found objects, dolls, metal scraps and wires he had built.
>
> *(Nardelli 2014)*

I would argue that the slacker apocalypse is Tinguely's end of the world and seems part of its point.

A much later film by Korean artists Kyungwon Moon and Joonho Jeon shares the same title. *El Fin del Mundo* (2012) has the equally dark umbrella title for the film, book and associated projects of *News from Nowhere* (borrowed from William Morris' eponymous novel of 1890). *News from Nowhere* is a place for artists, architects, designers, scientists, poets and others, imagining a post-apocalyptic future. *El Fin del Mundo* is just over 13 minutes, two-channels, and follows both a woman's journey in a toxic environment and a male character living in an exploded disaster site. The woman works for an imaginary corporate giant Tempus, and is required to both collect material samples and be herself a subject of scientific research. In a clinical laboratory, she washes pieces of botanical specimens, so redolent of Covid cultures we know today, and hangs them on a grid of protruding pins. The narrative envisages a future ecological-disaster world; humans are now endangered species, contaminated and forced to reconsider their lives, values and behaviours.

As Moon and Jeon say:

> An apocalyptic event – whether environmental, military or political – was simply a way for us to reset the stage. By starting over again, beginning our story on an empty page, it forced us and our collaborators to focus on what is truly important.
>
> *(cited in Choi 2013: 109)*

Whilst the artists clearly see their interests paralleled in the conjuncture between beauty, ethics, and interdisciplinarity exemplified by the Arts & Crafts and Bauhaus movements, their utopianism does not seem quite so beholden to socialist and

anarchist models, and they symbolise the split in Korean culture divided north and south today. As they speculate: 'Multinational corporations could be the new form of governmental bodies, with their ability to transcend the nation, race and geography, and to provide a sense of belonging and friendship – we are, after all, social animals' (cited in Choi 2013: 109).

Trevor Paglen's *The Last Pictures* (2012) is relevant here too, continuing both historical links to post-war cultural narratives. They are pictures from the planet (images designed by MIT labs in materials that should last as long as the sun) depicting humanity, progress and post-human melancholia. They will stay in orbit and constitute a view from a world beyond the apocalypse. This is a millennial 'message in a bottle' embodying a deeper time for humans, posthumans and non-humans alike, an updated version of Carl Sagan's famous 'Pioneer Plague' for NASA's *Voyager* space capsules (1972–1973). Paglen has approached the nuclear more directly, for example in *Trinity Cube* (2017) which, whilst looking ostensibly like a minimalist sculpture, around 20 cm^2, is formed from irradiated glass from the Fukushima Exclusion Zone (established following the 2011 explosion in Japan). The cube, made from a form of glass, trinitite, came and derives its name from the 1945 Trinity nuclear bomb test sites in New Mexico. The first version of Paglen's work remains in the Fukushima Exclusion Zone and will only be viewable directly again once the zone is declared safe – anywhere from 3 to 30,000 years from now. Modern sculpture and nuclear histories are fused in oceanic green glass and in decoded notions of deep time.

Susan Schuppli (n.d.) is an artist-researcher working as part of Forensic Architecture. She can be put alongside a number of artists who work on the nuclear–environmental border, such as Jane and Louise Wilson (in their series of photographs, *Atomgrad (Nature Abhors a Vacuum, VII,* 2010) or Stefan Gec (in his glass work *Elephant's Foot,* 2004, which was inspired by the horrors of Chernobyl. Schuppli (n.d.) also sifts through material evidence from war and conflict to environmental disasters. She explores the ways in which toxic ecologies from nuclear accidents and oil spills to the dark snow of the arctic are producing an 'extreme image' archive of material wrongs, always alert to the way that weather, nuclear material and other natural forces have both national and global aspects that play out across geopolitical boundaries.

Atmospheric Feedback Loops (2017) is a 35 mm vertical film lasting 18 minutes that was commissioned for Vertical Cinema & Sonic Acts, Amsterdam. Schuppli's research took her to an area just south of Amsterdam in a rural setting where an open-air laboratory, the Cabauw Experimental Site for Atmospheric Research, tunes into the atmospheric frequencies of nature. Its subject is the complex behaviour of clouds, aerosols, radiation, precipitation, and turbulence interacting with terrestrial events. Since 1970, the lab has been measuring and monitoring the changes taking place in the feedback loops between land surface processes and the airborne dynamics of our planet. *Atmospheric Feedback Loops* uses a vertically stacked video projection to echo the shape of the tower at Cabauw. The video follows two scientists who analyse climate and weather, and creates a soundscape made by the research instruments at the site. As one of the scientists based there said: 'We look at the atmosphere, we listen to the atmosphere. Because we are humans, we

have to interpret our measurements, so we like to make them audible or visible to ourselves' (cited in Schuppli n.d.). Whilst the work starts with the environmental attentiveness cultivated by both scientists and artists, it also stands as a metaphor for a cultural probe, picking up information from the atmosphere that is excessive, dissonant or obstructive, and making it visible culturally.

The Mineral Apocalypse

Mineral and material apocalypses arise from extractivist settings or, we might say, places where the logic of industrial-scale removal has been imposed. At its core, our ecological disaster is founded, metaphorically and physically, on the aggressive removal of material, for profit, without regard for the ethics of ownership and locality, and on financialising the materials that lie underground (coal, oil, silica, water and precious metals like gold and diamonds) and that can also be grown and harvested at large scales above the ground (tobacco, cotton, sugar, soya and palm oil). The art practices have, again, emerged locally and through travel from outsider artists, all bearing visual witness to the collapse of local economies and ethical forms of exchange and markets, which have been distorted by global capital and its appetites. As Achille Mbembe observes in *Critique of Black Reason*:

> The progression from *man-of-ore* to *man-of-metal* to *man-of-money* was a structuring dimension of the early phase of capitalism. Extraction was first and foremost the tearing or separating of human beings from their origins and birthplaces. The next step involved removal or extirpation, the condition that makes possible the act of pressing and without which extraction remains incomplete.
>
> *(Mbembe 2017: 40)*

The slave trade and mineral extraction alike were based on this logic, whose 'materials' could be gathered, used and discarded once exhausted.

A number of art projects have explored the linked themes of extractivism and devastation. Sandra Lahire (1950–2001), an important feminist experimental film-maker, produced *Uranium Hex* (1987, 11 minutes) as a means to explore Canadian uranium mining and its destructive effects on the environment, and the women working in the mines, doing so through a kaleidoscope of experimental filmic techniques. Images range from the women at work, cancerous bodies, industrial sounds and textual research on the effects of uranium mining. In a way that was to become increasingly popular in art filmmaking, the celluloid was doubled as a metaphor for skin and surface. As Marina Grinz writes: 'The radiation of the body is transferred to the radiation of the picture. The radon 222 that disintegrates the skin seems here to over-expose the film image. […] Radioactivity is deployed as a radioactivity of the film image in itself' (cited in Carpenter 2013).

Coal mining is also the subject of the Slovakian group APART Collective's video *The Most Beautiful Catastrophe* (2018), which, using a quiet, informational voice-over approach, explores the effects of coal mining in the country's Nitra region, highlighting land destruction, species extinction, cancerous chemical ingress and social degradation all circulating around a wetland, mixing apocalyptic romanticism with toxic, post-industrial landscapes. The voice-over draws on T.S. Eliot's *The Waste Land* but lacing the poetry with factual description. The abandoned mines are entropically reclaiming the site, and dystopian distractions (fairgrounds and cinemas) are a form of interference, a techno-fantasy driving extractive industries and space exploration (that is to say, the deep and the lofty in cataclysmic embrace).

There are many examples of projects from Africa, where extractivism has reached biblical proportions, first through the removal of human populations and their subsequent placement into slavery, and second through the extraction of palm oil and, since the 1950s, crude oil. As a consequence, the titles of such bodies of work draw on the biblical tropes of hell and paradise, for examples in Nyaba Leon Ouedraogo's *The Hell of Copper* (2008) and George Osodi's *Delta Nigeria: The Rape of Paradise* (2011). The Delta region of Nigeria is the central concern in Otobong Nkanga's (n.d.) *Delta Stories* (2005–2006), where the oil-rich region has been devastated ecologically, politically and socially through war and resource colonisation. In 18 drawings, *Delta Stories* captures how a landscape has been radically intruded upon, mutated and made hostile to local people and itself. *Flow Up North*, in simple form, shows a lightly sketched landscape and the pipework that extracts its wealth.

Delta Stories as a whole juxtaposes visual modes, from mapping, aerial views and poster-style activism to naturalistic landscape, with the surface of many works

FIGURE 6.8 Otobong Nkanga, *Delta Stories – Flow Up North* (2005–2006), 36.5×26 cm, acrylic on paper, image courtesy of the artist.

dripping with blood or oil. Jerry Buhari, another Nigerian, works through abstract painting but with eco-activist intent. His *Fall and Spill History* (2005) uses colour-fully coded oil paint to invoke disaster and apocalyptic assault on nature. Here 'oil' and 'paint' are fused and freighted with unfamiliar associations.

Aerial aesthetics come up in a number of eco-apocalyptic projects. The estab-lished collective Forensic Architecture (n.d.) have been at the forefront of this field, in particular *Ecocide in Indonesia* (2015), a project reflecting on making 'ecocide' a crime, which is reference to the Indonesian wildfires of 2015, which emitted enormous clouds of carbon, methane, ammonium and cyanide into the air. Their investigation raises the question of Indigenous and animal rights (orangutan rights in particular, which have since been more formally established in law, after the project), as the wildfires have been directly linked to the expansion of palm oil plantations that, with help from the country's army, international firms took away from Indigenous people. What is significant in this complex project is the interplay between local disaster and the toxic effects of such extremity that global temperature rises are possible as a direct result. Indo-nesian catastrophes, by being viewed through the lens of environmental law and criminality, show how visual research can embrace intersectional tensions around eco-disaster.

Mineral extraction is a form of slow violence, alongside botanical, species and population extraction and is fuelled, at its base, by a cruel set of appetites – largely colonial and financial. Disappointment threads through in a number of ways. Material and resource colonisation has an insatiable appetite, in both its imperial and con-temporary globalised forms; its injustice and environmental destruction are brought into the open through artistic visualisation in the projects and practices discussed above, but however much modern capital and colonialism may fear the particular mobilisation of artistic affect, this seems never enough to satiate the greed behind the extractive reflex. Art operates with the immaterial and affective as much as the mate-rials in which it is embodied, but artists are always and already compromised by the conditions of their own creation – the very material of their craft is often testament to the intractability of this violence.

Conclusion

In drawing attention to the non-consensual relationship that the capitalised and industrialised West has violently imposed, since around the sixteenth century and beyond, on other parts of the world, and in recognising that eco-apocalypse is distributed patchily and with varieties of intensity, art becomes an expression of different standpoints from which we look upon this disastrous assemblage that in the late twentieth and early twenty-first centuries has become undeniably bleak. This is not to set Western artists (and the privileges of infrastructure and access they enjoy) against Indigenous, marginalised and neglected practitioners, but to point, perhaps in hopeful and less disappointed spirit, to allegiances and common purpose in these global artists' mission, creativity and sense of agency.

Wendell Berry uses the word 'love' in a way that freights it with a commitment to the particular, the placed and the tangible, collectively something that can 'turn us from this deserted future, back into the sphere of our being, the great dance that joins us to our home, to each other and to other creatures, to the dead and the unborn' (Berry 1980: 520). The recent turn towards more affirmative and transgressive art practices (say from around the year 2000) and the critical commentaries that have followed is a reflection of the human mind's unease about always staying with anxiety, negativity and dread, as well as its tendency to seek out more fruitful and creative options for itself. Artists reflect their wider connection with human proclivities here. Haraway's influential injunction to 'stay with the trouble' of ecological collapse, 'not as a vanishing pivot between awful or edenic parts and apocalyptic or salvific futures', creates an imaginative space for artworks as knots of past histories and future meanings, yet as ones that are caught in the entangled present, which is where they do their work (Haraway 2016: 1). Indeed, the ever-present is the *only* place art ever works. In a sense, as I have implied throughout this chapter, making art as a response to ecological and environmental apocalypse is disappointingly inadequate, and properly so. It is to offer the immediate prospect of community with, and insight on, our perilous condition whilst alluding modestly to the loneliness of the historical past and the uninhabited wasteland of the unknowable future.

Acknowledgements

I would like to thank Sarah Bezan, Ele Carpenter, Edward Christie, Eszter Erdosi and Mark Hallett for their support and advice during the writing of this chapter

Note

1 We should note the shared imagery between religious apocalypse and environmental apocalypse, the former being mostly expressed through the devastation of trees, animals and humans in natural disasters like earthquakes, volcanic eruptions, or floods, or through famine and disease. Pre-Enlightenment European culture believed in the link between the weather and human transgression, which is captured by the term 'ecological sin' or 'eco-sin'. Whether it be the Norse Twilight of the Gods or the Sixth Assessment Report of the UN Intergovernmental Panel on Climate Change (IPCC 2022), humans are found both wanting and guilty (see Buell 2010: 18; and Behringer, 2010: 133).

References

Ahuja, Neel. 2017. 'Posthuman New York: Ground Zero of the Anthropocene'. In *Animalities: Literary and Cultural Studies beyond the Human*, edited by Michael Lundblad. Edinburgh: Edinburgh University Press, 43–57.

Agarwal, Ravi. n.d. *In the Shadow of the Vulture: An Extinction of Memory*. Delhi: 48° C Public Art Ecology.

Apter, Emily. 2013. 'Planetary Dysphoria'. *Third Text* 27(1): 131–140. https://doi.org/10.1080/09528822.2013.752197.

Behringer, Wolfgang. 2010. *A Cultural History of Climate*. Cambridge: Polity Press.

Berry, Wendell. 1980. 'Standing by Words'. *The Hudson Review* 33(4): 489–521. https://doi.org/10.2307/3851419.Biemann, Ursula. 2013. 'Deep Weather'. https://www.geobodies.org/art-and-videos/deep-weather. [Accessed February 27, 2022].

Blanchot, Maurice. 1997. *Friendship*, translated by Elizabeth Rottenberg. Stanford, CA: Stanford University Press.

Buell, Frederick. 2010. 'A Short History of Environmental Apocalypse'. In *Future Ethics: Climate Change and Apocalyptic Imagination*, edited by Stefan Skrimshire. London: Continuum Books, 13–36.

Buszek, Maria Elena. 2011. 'Crochet and the Cosmos: An Interview with Margaret Wertheim'. In *Extra/Ordinary: Craft and Contemporary Art*, edited by Maria Elena Buszek. Durham, NC: Duke University Press, 276–290.

Byron, William R. 1962. 'Wacky Artist of Destruction'. *Saturday Evening Post*, April 21, 76–78.

Carpenter, Ele. 2013. 'Nuclear Culture on Film'. https://nuclear.artscatalyst.org/content/nuclear-culture-film. [Accessed March 23, 2022].

Carpenter, Ele. 2016. *The Nuclear Culture Source Book*. London: Black Dog Publishing.

Choi, Jayoon. 2013. 'Beauty and the Apocalypse: Moon Kyungwon and Jeon Joonho'. *ArtAsiaPacific* 85: 104–113. https://doi.org/10.3316/ielapa.738875972222732.

Critical Art Ensemble. 2016. 'Graveyard of Lost Species'. http://critical-art.net/graveyard-of-lost-species-leigh-on-sea-uk-2016/.

Demos, T.J. 2016. *Decolonizing Nature: Contemporary Art and the Politics of Ecology*. Berlin: Sternberg Press.

Demos, T.J. 2019. 'The Agency of Fire: Burning Aesthetics'. *E-Flux* 98. https://www.e-flux.com/journal/98/256882/the-agency-of-fire-burning-aesthetics/.

Dobraszczyk, Paul. 2017. 'Sunken Cities: Climate Change, Urban Futures and the Imagination of Submergence'. *International Journal of Urban and Regional Research* 41(6): 868–887. https://doi.org/10.1111/1468-2427.12510.

Forensic Architecture. n.d. 'Ecocide in Indonesia'. https://forensic-architecture.org/investigation/ecocide-in-indonesia. [Accessed March 2, 2022].

Haraway, Donna. 2015. 'Anthropocene, Capitalocene, Chthulhucene. Donna Haraway in Conversation with Martha Kenney'. In *Art in the Anthropocene: Encounters among Aesthetics, Politics, Environments and Epistemologies*, edited by Heather Davis and Etienne Turpin. London: Open Humanities Press, 255–270.

Haraway, Donna. 2016. *Staying with the trouble: making kin in the Chthulucene*. Durham, NC: Duke University Press.

IPCC. 2022. *Climate Change 2022: Impacts, Adaptation, and Vulnerability*. Contribution of Working Group II to the Sixth Assessment Report of the Intergovernmental Panel on Climate Change, edited by H.-O. Pörtner, D.C. Roberts, M. Tignor, E.S. Poloczanska, K. Mintenbeck, A. Alegría, M. Craig, S. Langsdorf, S. Löschke, V. Möller, A. Okem, and B. Rama. Cambridge: Cambridge University Press. https://www.ipcc.ch/.

Lee, Pamela M. 2004. *Chronophobia: On Time in the Art of the 1960s*. Cambridge, MA: MIT Press.

Mbembe, Achille. 2017. *Critique of Black Reason*. Durham, NC: Duke University Press.

Moon, Kyungwon, and Joonho Jeon. 2012. *News from Nowhere: A Platform for the Future and Introspection on the Present*. Seoul: Workroom Press.

Nardelli, Matilde. 2014. 'No End to the End: The Desert as Eschatology in Late Modernity'. *Tate Papers* 22, Autumn. https://www.tate.org.uk/research/tate-papers/22/no-end-to-the-end-the-desert-as-eschatology-in-late-modernity.

Nkanga, Otobong. n.d. https://www.otobong-nkanga.com.

Povinelli, Elizabeth A. 2016. *Geontologies: A Requiem to Late Liberalism*. Durham, NC: Duke University Press.

Rockman, Alexis. 2017. http://alexisrockman.net/great-lakes/.

Schuppli, Susan. n.d. www.susanschuppli.com.

Scott, Emily Eliza. 2012. 'Elsewhere: Desert Ends'. In *Ends of the Earth: Land Art to 1974*, edited by Philipp Kaiser and Michelle Piranio. Munich: Prestel, 66–85.

Sontag, Susan. 1965. 'The Imagination of Disaster'. *Commentary* 10(42). https://www.commentary.org/articles/susan-sontag/the-imagination-of-disaster/.

Steenson, Jakob Kudsk. 2019. http://www.jakobsteensen.com/#/re-animated/.

Stephens, Russell. 2018. 'Extinct? – An Art Intervention by Ravi Agarwal in Delhi'. *Art Journal*, 77(2): 28–51. https://doi.org/10.1080/00043249.2018.1495531.

Sunde, Sarah Cameron. 2013–. *36.5 / A Durational Performance with the Sea*. https://www.36pt5.org/.

Wei, Lilly. 2016. 'Jun-Fei Ji. I Am Pessimistic about China'. https://www.studiointernational.com/index.php/yun-fei-ji-interview. [Accessed on February 27, 2022].

Weschler, Lawrence. 2011. 'The Hyperbolic Crochet Coral Reef'. *The Virginia Quarterly Review* 87(3): 124–139. https://www.jstor.org/stable/44714499.

7

AVOIDING THE APOCALYPSE

The How-To Guide as a Method

Francesca Laura Cavallo

In the Cold War, how-to guides preparing people for the nuclear fallout gave a new shape to the apocalypse of Judaeo-Christian tradition. As the end of the world was no longer just a myth but a 'man-made' risk to be managed – to borrow Ulrich Beck's (1992) terminology – the visual representations of 'the End' became institutional, illustrative, didactic, accessible and close to home. The apocalypse came dressed up in a 'safety suit' in the form of a guidebook.

Survival manuals and toolkits have permeated popular culture for many decades through self-help literature and DIY movements, but they have become entirely mainstream in the age of climate change. Bill Gates' recent book *How to Avoid a Climate Disaster* (2021), for example, lays down a plan for a carbon-free future by 2050 in what a reviewer has called a 'cross between a planetary instruction manual and a global warming for dummies guide' (Watts 2021). Extinction Rebellion's *This Is Not a Drill* (2019), conceived and designed as a call for action and a manual for activists in the climate emergency, combines apocalyptic rhetoric and practical tips. At the same time, a quantity of behavioural guides and toolkits for saving the planet appeared in the British newspapers only in the last year: *50 Simple Ways to Make Your Life Greener* (Berrill et al. 2020), *How to Reduce Your Carbon Footprint* (Albeck-Ripka et al. 2019), *How to Be More Eco-Friendly in Everyday Life* (Rahman-Jones 2021), and *What You Can Do for Your World* (Natural Resources Defense Council and Um 2020). As tips in bullet points suggest changes in how customers eat, travel, live, dress and partake in civil society, they also propose strategies to cope with climate change distress. These guides turn preoccupations for the future into action plans and simple illustrations through the mixed repertoire of achievable goals and apocalyptic warnings.

This chapter presents a critical interrogation of these how-to guides by (a) framing them in the contexts of the survival guide manuals from the Cold War up to very recently; and (b) exploring their function and relevance for climate change

DOI: 10.4324/9781003189190-9

activism by explaining the different strategies they propose for *avoiding* the apocalypse. It dwells in the grey area that separates how-to guides from the reality they aim to transform, analysing them as tools for social amplification and immanence, and as the by-products of our risk cultures, which are often used to conceal the fact that we frequently do not know how to deal with change. Offering tools for the critical interrogation and production of climate change guides, it lingers on the structure, language, and visual features of survival guides from the past to understand their role in both governance and militant action today.

Surviving the Nuclear Apocalypse

In the Cold War, enduring apocalyptic myths and fantasies in art, cinema and literature were overtaken by a new, very real existential threat: the nuclear bomb. Produced initially for war-time civil protection campaigns, the genre of survival guides gained popularity in the post-war period in the UK and the US, where the possibility of a thermonuclear war seemed to have haunted humanity's future. While science fiction provided the ideal space for traumatic anxieties to be exercised and *exorcised*, the emphasis on preparedness attempted to tackle the practicalities of this perceived end of the world (Rose 2001).

Between the 1940s and the 1960s, the British and the American Civil Defence Corps distributed thousands of booklets containing necessary how-to steps for surviving nuclear fallout. In a process Tom Vanderbilt (2002) defined as the 'domestication of doomsday', civilians were asked to build shelters, stockpile food, and learn a repertoire of gestures to perform in case of emergencies (Davis 2007). The design of those booklets was instrumental to these didactic purposes. It needed to normalise even the most terrifying situation, in so far as maintaining morale was concerned.

Between You and Disaster, a 1960s American leaflet featuring a stereotypical scene of a family at the supermarket, was produced as part of a preparedness programme that invited the American consumer to stockpile food and medicines to prepare for the fallout (USA, Office of Civil and Defense Mobilization 1958). Leaflets such as this were intended to simultaneously acquaint the population with the possibility of a disaster and reassure them with more or less clumsy how-to survival techniques. In reality, the nuclear bomb was a relatively new technology, and it was impossible to estimate its long-term effects, let alone *offer* the appropriate survival advice to the public. Fear and panic, however, were considered just as dangerous as an actual conflict, so rather than focus on accurate estimations of damages, these booklets targeted the morale (and sense of safety) of the population (Boyer 1985; Scheibach 2015). The design of nuclear fallout booklets was instrumental to these didactic purposes. The famous *Bert the Turtle Says Duck and Cover* (USA, Federal Civil Defense Administration 1951), for example, was tasked with normalising even the most terrifying of situations to help adults and children navigate the wide range of emergency measures set in place for them (see also USA, Federal Civil Defense Administration 1946). In the booklet *Advising the Householder on Protection against*

Nuclear Attack (Great Britain Home Office 1963), British Civil Defence instructed readers on how to build a nuclear fallout shelter and the actions required following a siren going off. The layout was particularly telling. Small headlines and illustrations framed each situation (or scenario), with miniature human figures resembling demonstrative and emotionless pictograms. The red colour in these illustrations matched the red bricks used for building in the UK, giving a sense of familiarity to scenes. The language was also 'to the point', conveying seemingly factual information with no dramatic headlines and no alarmist tones. The overall effect must have attempted to be quite encouraging: presenting survival as a likely possibility and suggesting readers welcome pets and neighbours in need of shelter.

The design of step-by-step illustrated guidelines in these nuclear fallout booklets conveys something like an instructional aesthetic, where an impartial, anti-emotive tone is used to avoid panic. This instructional style effectively inherited the anti-alarmist spirit of earlier guides for civilians, including those produced as a part of the Safety First movement. *How-to's* were a popular genre at the beginning of the 20th century: tips about avoiding accidents in the street, at home or at work, featured in governmental initiatives of public information, but also in those of private companies and associations (Mohun 2013; Rennie 2005, 2015). Even the insurance sector and cigarette companies were keen producers of survival instructions (Cavallo 2020). Nuclear fallout booklets appropriated the same style and adapted it to the nuclear threat, substituting seasoned techniques with what authorities believed to be the best strategies. The structure was always relatively standard in these booklets. Beginning with a description of the risk and a statement of intent, they went through step-by-step guides of emergency procedures divided into 'before', 'during' and 'after' scenarios, ending with a conclusive checklist. From a design perspective, this formula was practical and understandable, serving as a model for many booklets to follow, including the notorious *Protect and Survive* (Great Britain Home Office 1976). However, as a political operation, these were controversial.

Fallout manuals had little evidence supporting their advice. The population was recalcitrant to execute the instructions: many did not believe that anyone could survive a nuclear attack, and only a few actually built a shelter. Moreover, the reassuring tone also carried an inconsistent message: in case of a nuclear attack, one must expect to be self-reliant (Lichtman 2006). Consequently, after the intense iterations of the danger of nuclear war, this emphasis on preparedness attempted to tackle, but consequentially also contributed to, the creation of proverbial end-of-the-world paranoia. Apocalypticism began to permeate both official propaganda and cultural production. Perhaps unsurprisingly, several alternative manuals were also produced, criticising authorities for being the actual cause of the risks against which they attempted to shield the people. These guides incorporated instructions and adopted the same *how-to* formula to subvert their meaning.

In *The Intelligent Woman's Guide to Atomic Radiation* (Bennett 1964), author and screenwriter Margot Bennett explained in 'plain English' the dangers and advantages of nuclear energy, touching on issues such as pollution and health. Bennett states in the introduction: 'the only real protection is to understand what is happening, to form your own judgement, to decide if you are satisfied with the position. If you are not, you are entitled to say so. Informed public opinion is infectious, even to government' (Bennett 1964: 14). In a similar spirit, activists in the UK a few years later produced *Protest and Survive* (Palmer Thompson and Campaign for Nuclear Disarmament 1980), whose title mocked the above-mentioned controversial Civil Defence guide *Protect and Survive*. Rather than 'pointless' survival instructions, this alternative guide offered 'scientific facts' about radiation.

To contemporary eyes, the instructions contained in civil defence booklets have become the *topoi* of Cold War preparedness: the rehearsals for an event that did not happen but that manifested itself as an immanent presence – that is, a *presence* that materialised in the perceptions and practices of preparation that it determined. The booklets were a tool for such immanence to become manifest, framing apocalyptic thinking with a risk management outlook. Moreover, the idea of using *how-to's* to prepare for the worst has had a long-lasting legacy. Transcending the specific field of civil defence preparedness, survival guides became, in the years that followed, much more involved in the practices, habits and choices of everyday life: they became tools for behavioural change. If official pamphlets addressed specific issues such as the nuclear fallout, it was not the content but the method that persisted in the following years.

From Eco-Warriors to Preppers

The publication of books such as *Silent Spring* in 1962 (Carson 2002), followed by the development of ecological consciousness and the awareness of the ecocide perpetrated by the Americans in Vietnam, generated several manuals for 'retreaters', advocating a life off-the-grid in America. These ranged from eco-friendly books such as *Survival with Style* (Bradford 1974) to the survivalist guides suggesting relocating to the countryside to escape the corruption and violence of society.[1] Ignited by the DIY movement in America, these eco-guides in the 1970s shared all sorts of tips, products and techniques for self-reliance and sustainable living, in the name of 'access to tools' – undeniably inspired by the most famous of them all: *The Whole Earth Catalog* (Brand 1968).

Although not a survival manual in the strict sense, this highly illustrated publication featured articles appropriated (or stolen?) from different sources on various disparate subjects, which were organised through a principle of connectivity and free association – instructions, poems, products and book reviews where the reader could 'find his own inspiration, shape his own environment, and share his adventure with whoever is interested' (Brand 1968: 2). What was extraordinary about *The Whole Earth Catalog* (WEC) was its agile repurposing of different sources of knowledge into *tools* for living: there was no separation between theory and practice, at least in the intent. Conceived as a toolkit, the WEC became an extraordinarily non-hierarchical experiment in multi-disciplinarity at the service of the 'community' and communities across the 1970s US. Shaken by the prophecies of eco- or nuclear apocalypse, these communities believed in new world-making possibilities through solidarity, creativity, and innovation. In this context, eco-guides represented a way to respond positively and pro-actively to the looming prospect of overpopulation, pollution and biosphere degradation.

Similarly to contemporary guides to reducing climate change, eco-guides in the 1970s had an almost ecumenical, civilising function – enrolling an army of individuals in the collective battle against pollution, overpopulation and injustice. In this sense, they *mediated* (or at least attempted to) individual actions and the collective good through mobilising populations against the environmental damages that scientists have been increasingly monitoring since. The terms of such mediation between individual action and mobilisation were the emphasis on tools, guidelines, and action plans that enhanced the collective sense of belonging to something greater than individual interest (something that has become incredibly visible since the outset of Covid-19).[2]

Effectively, this ideology of togetherness directly responded to what was perceived as an existential threat to humans and 'their' planet. In *Operating Manual for Spaceship Earth*, Buckminster Fuller (1969) had described the Earth as a spaceship propelled into the universe with a finite number of resources that could not be resupplied. It is up to the 'ship's captain', he wrote, to create technologies and techniques that can harness, maintain, nurture and manage its resources. The optimism (also enlarged by the recent moon landing) in Fuller led him to believe that once the Earth's fragility had been understood, humanity had in itself the power and ability to maintain this precarious balance and even innovate it for the best. He confided in innovation, design and technology to form a new 'regime of efficiency' that would get *all* humans out of trouble (Höhler 2015). Inspired by this vision and featuring the first image of Earth from the moon, Steward Brand's WEC understood the planet as both a living organism and a spaceship, connecting humans to nature and nature to technology – the interconnectedness of all things on the planet required humans to form more connections. As Sabine Höhler (2015) has noted, Fueller's ideology was, whilst presciently imbued with ecological sensibility, an anthropocentric and universalistic one. 'We are like God', Brand famously wrote in the WEC purpose statement, 'so we may as well get good at it' (Brand 1968).

By the end of the 1970s, however, the survival guide had become a method for militant action way beyond ecological concerns and the nuclear threat to humanity, cultivating survival as an individual enterprise. In 1975, for example, right-wing sympathiser Kurt Saxon started self-publishing the magazine *The Survivor*, 'The first publication preparing its readers to the total collapse of our economy and social system' (Saxon 1976). Saxon was disillusioned with American politics and, like Steward Brand, also advocated a DIY attitude. His pamphlets contain reprints of old guidebooks and technical manuals as much as his writing and advice, from money-saving to surviving a nuclear winter, from growing earthworms to burglar-proofing one's house. Saxon's publications also included information about making explosives and other dangerous activities, all in the name of self-protection (Saxon 1976).

The self-proclaimed 'inventor' of the word 'survivalist', Saxon also insisted on the urgency to survive as heroes and not just 'sit and watch' as his contemporary pacifists would do:

> My definition of a Survivalist is a self-reliant person who trusts himself and his abilities more than he trusts the Establishment. Insofar as the Establishment is deteriorating, the Survivalist prepares to leave it.
>
> *(Saxon 1980)*

The Survivor represents an almost uncanny doppelgänger of the WEC: if the first was, perhaps a bit naïvely, all about sharing tools to make a better world, that latter was focussed on self-defence from the corrupted society in the blink of an apocalypse. Saxon had responded to the Cold War prospect of human annihilation with a transgressive attitude that combined the instructional style with the recycling of old-fashioned skills *à la* Robinson Crusoe. This strange mix of practical tools and apocalyptic rhetoric tapped into the fascination for weapons and heroism of disillusioned Vietnam veterans, and cultivated individualistic, libertarian ideas of self-sufficiency through divisive rhetoric. As in the historical apocalyptic traditions (not just the Judaeo-Christian, but Hindu as well), only the good (or the brave) deserved to survive.

This very ideology, fostered by the efficacy of the how-to guide *as a method*, turned out to be the formula for the numerous survivalist guides produced later. Some of them circulated mainly in underground circles and right-wing counterculture, but others reached mainstream attention by preparing readers for apocalyptic visions of different kinds and scales. In the 1980s, John A. Pugsley's (1981) book *The Alpha Strategy: The Ultimate Plan of Financial Self-Defense* became a best-seller by suggesting stocking up on food and household goods as a hedge against inflation. In 1999, several guides prepared readers for Y2K (also known as the 'millennium bug'), a 'millenary' disaster that worried high-flying managers and fringe religious groups who saw it as the fulfilment of end-of-the-world prophecies.[3] After 9/11, as could be expected, several books quickly appeared addressing terrorism, such as *Life after Terrorism* by Bruce D. Clayton (2002), a former bodyguard trained in Israel, published a mere six months after the event. And the list continues: *Pocket Guide to the Apocalypse: The Official Field Manual for the End of the World* (Boyett 2005); *Emergency: This Book Will Save Your Life* (Strauss 2009); *Arming for the Apocalypse: Assembling Your Survival Arsenal While You Still Can* (Ballou 2012).[4]

The popularity and diversity of survival manuals discussed above demonstrate that their 'instructional style' has succeeded as a formula for knowledge transfer regardless of their content. Despite their undeniable differences, such as the shape and nature of the worst-case scenarios that they claim to prepare people for, survival guides capitalise on the same conceptual presupposition: the awareness that bad-new-days are coming and the need to devise of strategies to face them. Their success, I have argued on another occasion, may rest on their ability to frame apocalyptic thinking within a risk management outlook, presenting even the most daunting of threats as something that can and should be managed (Cavallo 2020).

In return, they feed and breed on apocalypticism as a marketing strategy. Most importantly, many of these guides prospect survival as a possibility for only a few against what is perceived as a hostile, corrupted society. In a way, survivalist guides perpetuate an *us vs them* narrative resting on a very different ethical ground from the eco-inspired DIY movement. Let us now briefly consider this ethical ground and how it shifts to demonstrate how survival manuals are fundamental tools for transforming ideology into practice.

Ideologies of the How-To Guide

By tapping into anxieties regarding the future and attraction to adventure, weapons, or heroism, many survival guides that have emerged in past decades have contributed to shaping an individualistic and future-oriented approach to survival that has very little to do with what survival truly means. When worldviews and lifestyle choices are based on the imagined imminence of the worst-case scenario, life is also modelled around real and fictional tales of human endurance and their protagonists. Far away from the actual adaptation, ability to improvise and luck of those that have survived a trauma, this style of survival is perfectly encapsulated by the figure of the trained *prepper*. Highly informed about what comes from the future and equipped with the tools necessary to deal with it, the prepper lives in a constant sense of threat from the outside world, yet she is stranded by the inability to truly control it. She does not embrace improvisation like the fictional models that have inspired her (see Rambo or Robinson Crusoe) but fights against the unpredictable by training and preparing for any disaster; the apocalypse is, for the prepper, an immanent narrative to rehearse and exercise every day.

Is this all thanks to the preppers' how-to guides? Or is it the case that the survival guide is a perfect medium for propagating their ideology, letting it infiltrate into every single thing they do in their lives? To quote Frank Kermode, 'No longer imminent, the End is *immanent*', meaning that the apocalypse is not in their future, but it exists and remains within their life (Kermode 1968: 25).[5] By attempting to avoid the collapse of society, survivalists are performing their apocalypse every day by ending what connects them to the society they reject. Their how-to-guides are tools for embracing the apocalypse and not avoiding it.

What is, then, the role of the how-to guide in the age of climate change (or the eco-apocalypse)? Does it really help us avoid the 'climate apocalypse', or does it instead *perform it* for better or for worse?[6]

These are arguably apocalyptic times. And not because of the widespread use of apocalyptic language in the media and popular culture. The contemporary climate apocalypse – as of the revelation to humans that the end of their world is imminent and as an occurrent, ongoing phenomenon of extinguishment – is a palpable circumstance. What in the eyes of many is a sudden revelation that their world is ending is just a fact in the past for many populations, not to mention animal species and other forms of existence and landscapes. On the one hand, the consistent monitoring in recent years of the irreparable damages that humans are causing for other forms of existence and the prospect of the escalating consequences that this is producing constitute a sense of endangerment and threat for contemporary Western societies. On the other, the present symptoms of climate change are shaping the environmental disasters that are already happening in and beyond the Global South. Clearly, the apocalypse has already happened many times, way beyond the realm of myths. In their radically different understanding of time, Indigenous cultures have persistently painted themselves as post-apocalyptic (Dillon 2012; Powys Whyte 2017). Genocide, death, and the encounter with the coloniser have signified the end of their world and the beginning of a new one, and have defined their existence ever since.

In this sense, the apocalypse is unavoidable because it has already happened as both a *presence* and *immanence*: both a fact and a revelation that one world needs to end so another one can be born. Is not, for example, the discovery of the Anthropocene a *revelation*, a promise for the end of a worldview and the beginning of a new one? Does not discovering what Bruno Latour (2017) has called the 'terrestrial' imply the end of anthropocentrism and myths of humans' control over nature), perhaps even the end of linear notions of time in the name of simultaneity and interdependence?

So how can we make sense of the current plans and guides to *avoid* the climate apocalypse? More than 50 years after Buckminster Fuller (1969), Bill Gates's *How to Avoid a Climate Disaster* lays down plans for 'each of us' to save Spaceship Earth. He calculates that to preserve the planet's temperature, we need an integrated deployment of green technology, policy, and investment in renewable energies and carbon capture techniques. Like Fuller, Gates (2021) wants to invest in the future, but some of his technological proposals seem worrying: creating thermal energy from hot rocks underground; nuclear fusion; de-acidifying the oceans; air-capturing; and storing emissions deep underground. 'Gates' most important proposals involve new technologies', writes Gordon Brown (2021) in his review of the book. 'His principal interest is in a technological breakthrough, the environmental equivalent of the Manhattan Project or the moon landing'. History repeats itself: eco-modernist solutions and men taking upon themselves the task of saving the world. The hope still invested in extractivist methods that may cause more damage than good is troubling when it fails to engage with present battles and histories of injustice. It continues to position the apocalypse in the future, even if the revelation has already happened. This 'future-oriented-ness' fails to recognise (and thus *avoids*) what Elizabeth Povinelli, based on the work of Native writers such as Vine Deloria Jr., calls 'the ancestral catastrophe of late capitalism' (Povinelli 2021): the slow extinguishment of resources many now call the 'Anthropocene'.

As several Native and Indigenous writers have pointed out, this preoccupation with their future 'End' is just the capitalist inability to accept the apocalypses of their own making over cultures that have already faced the end of their world (Dillon 2012; Mitchell and Chaudhury 2020). Furthermore, Indigenous post-apocalyptic strategies consist of ancestral practices of everyday life such as maintaining supportive community relations with all forms of existence, including rocks, plants, rivers and animals, as well as other humans, in order to maintain very precarious balances.

As two recent books (Elizabeth Kolbert's *Under the White Sky* and Michael Mann's *The New Climate War*) published in the same year as Gates' (2021) volume have pointed out, techno-capitalist responses to climate change propose reducing the damages done by technology and industry with more technology and industry – to the point that some attempt to 'avoid the apocalypse' by planning ways through which humanity could colonise another planet. As intriguing, absurd or promising plans to avoid might sound, they must never be dissociated from the ideologies that produce them. By wanting to avoid the apocalypse, guides enable us to perform it as an ideology.

The same is also true for the other extreme in the spectrum of the climate-change how-to guides: Extinction Rebellion's (2019) total surrender to the end of capitalism and strategic use of apocalypticism to mobilise activism in *This Is Not a Drill*. As Gates' plans of action speak to business people's common sense, Extinction Rebellion's embracing of the apocalypse is a tactic designed to enrol an army of eco-warriors to defend the planet and provoke change. Will any of these ideologies and tactics ever prevail over the other? However persuasive their argument might be, the point is how 'performative' they will be, that this, how much of what they say will be effectively implemented as policy or as collective or individual action for change.

In all of these disquisitions, the humble how-to guide for a green life that comes in the Saturday newspaper might be of good use. As activists scream that 'this is our darkest hour', 'the house is burning down' and 'extinction is imminent', concentrating on the daily achievable task of reducing waste, cycling when one can and eating less meat can be both reassuring and powerful. Nothing like the pandemic has shown how silent, individual actions such as staying at home can be selfless and radical. Not by chance, Latour (2021) defined the pandemic as the 'big rehearsal'. Welcome, then, to the *how-to's* as far as they turn the awareness of climate change into militant action for the greater good. Different instructions work for different people and contexts, and their plurality is an asset rather than an issue, but the *how-to's* must also be critically interrogated. Survival guides play an essential part in how individuals negotiate a 'collective new world'.

In this regard, it becomes crucial to examine survival guides as objects and methods for ideologies that (or ask one to) perform the apocalypse. Looking back to the preppers' guides, for example, with their rehearsals and daily tasks based on fear and distrust, the figure of the prepper might be an extreme – and worrying – version of other, more benevolent models of everyday preparedness that accept the apocalypse as a revelation (and live with its immanence as a premise for transformation). Similarly, one should be wary of those guides that want readers to avoid the apocalypse by transposing it into the future without acknowledging its 'ongoingness' for other forms of existence (to borrow Donna Haraway's [2016] terminology) and the responsibilities it demands.[7] In conclusion, with preppers showing how manuals that *prepare* people for the apocalypse can degenerate into radical ideological directions, plans to *avoid the future apocalypse* are, in fact, bluntly denying its ongoing reality.

Conclusion

The handbooks discussed in this chapter show how institutions, individuals and groups continue to rely on instructions to promote survival. Over the years, people who perceived risk differently have challenged them as unreliable sources. Alternative guides have been produced, adopting the very same instructional approach and promoting different kinds of strategies. Neither fictional, nor properly factual, the survival manual is a hybrid genre, but it has something unique in the modalities through which knowledge is produced and transferred.

First, the manual presents itself as the recipient of expert knowledge (that is, knowledge that comes from direct experience and embodied practice), yet it is a vehicle for ideology. By framing realities that one needs to respond to, by codifying plans and rules that people take on board, it is a potent tool for social amplification and immanence. *How-to's* can freely multiply according to whichever situation one perceives as the worst-case scenario and become the ideal vehicle for any narrative, whether adventurous or apocalyptic. Even if presented as the tools necessary for overcoming challenging futures, they are instead lifestyle tips for the present.

Second, the relation between how-to guides for the apocalypse and what *really* happens is neither mechanical nor predictable. Users guides, for instance, clearly explain the mechanics of how a product works; travel guides give clear evidence about locations; medical manuals give information about the body. However, can survival guides prepare somebody for the unpredictable? Perhaps they just express the desire to control the unexpected and then simultaneously become seduced by it. In this sense, they are tools for managing (or taking) risks. They turn 'real facts' (or 'experiences') into *exempla* for the future, combining info pasted from all sorts of sources.[8] Unlike user guides, however, survival manuals cannot guarantee that what they say will work. In this sense, they are comparable to magic books: saying the formula will not necessarily do the trick. There is a crucial difference between reading about techniques and acquiring the ability to reproduce them. Thus, survival instructions may just be coping strategies that further increase this difference, rather than tools for pre-emption and preparation. In the absence of knowledge about the future, survival instructions give us a plan, but they do not necessarily guarantee what they promise.

Third, sometimes how-to guides are used to conceal the fact that we simply do not know what to do. The nuclear fallout booklets, discussed at the beginning of this chapter, offered how-to instructions that partially reassured the reader by breaking down the big unknown of the bomb into a set of reproducible actions. However, did having those action plans imply that people were prepared? How could that have been possible if even the authorities did not know a nuclear explosion's real and long-term effects? The booklet functioned as a palliative for the absence of essential knowledge about the bomb, which remained ultimately unknown or 'occult' (from the Latin *occultus*, meaning 'concealed').[9]

In a similar vein, books such as *How to Avoid a Climate Catastrophe* guide us through possible scenarios over which we have no control. Suppose the future worst-case scenario haunts us with almost supernatural traction. In that case, the survival manual substitutes anxieties about the future with histories of self-efficacy and endurance (think about Gates' success). Today's self-help books also promise us life-changing results if we follow their simple advice, but between the hope of success and its realisation remains a gap that authors desperately aim to conceal. It is even more extreme regarding survivalist literature that prepares the reader for the apocalypse, which is so intensely perceived as a real possibility so as to become a lived experience of pseudo-apocalyptic situations.

In conclusion, how-to guides have an awkward relationship with reality. They may help readers deal with the uncertain future; reassure them; or foster transformation; or they may enable them to rehearse worst-case scenarios mentally. However, as Maurice Blanchot (1980) so poignantly wrote, it is impossible to 'write' the disaster.

How-to guides are a by-product of our risk culture – or of the present condition of uncertainty. It is not surprising that more recently, with climate change and extinction featuring prominently in political and public arenas, a whole new generation of survival manuals is coming into circulation. Where there is no factual information about possible dangers, instructions step in by giving plans for action to manage them as risks. However, having a plan does not mean being prepared, and we make even more plans when we do not know what will happen. Instructions do not necessarily generate reliable knowledge but mostly a set of actions to compensate for its absence. Action plans reiterate a literary negotiation and struggle with the unknown via speculations about the future and our attempts to prepare for it. In a sense, they compensate for the absence of knowledge that we inevitably experience when thinking about our future. How-to manuals guide us (or pretend to) through the uncertainty that is at the core of situations laden with risk.

How can one negotiate individual and collective responsibility? Born from the awareness of climate change and its implications for the planet, how-to guides on the climate apocalypse offer plans for action by linking the quotidian with the universal, the immediate with the long term, the personal with the collective. If the viral apocalypse has taught people something, it is about individuals taking responsibility for the collective (this is why the pandemic is a rehearsal). The climate-apocalyptic visions are necessary for this awareness: survivals of genocides still living in a post-apocalyptic present, the site of sea-level rises, social unrest, loss of comforts, the face of history disappearing. These apocalypses are all revelations of the need to look beyond self-sustained life and preservation – to question one's place in human, animal, vegetal and inanimate societies *of being*. The ancestral catastrophe of populations worldwide, the writings of Lynn Margulis (Margulis and Sagan 1995), the virus itself – so small and powerful – can open paths for different ways of sensing and being. Perhaps a metamorphosis is needed to destroy individualism in the awareness that each individual is a colony or an aggregate of infinitesimal being making love with (and within) each other (Barad 2007, 2012). Without this awareness, plans of action can only get us as far as our noses. The instructions only show how things *could be*.

Notes

1 See, for example, Fadiman and White (1971); Stephens and Stephens (1976); and Benson (1983).
2 From the point of view of international law, some progress was made (e.g. Clean Air Act 1970; Clean Water Act 1972; Endangered Species Act 1973; European Environment Action Programme 1972) but not enough to actually address the acceleration of extractionist capitalism that continued indiscriminately to destroy environmental resources and the livelihoods dependent on them.
3 With the arrival of the new millennium, the bug, which would have supposedly crashed all computers, never materialised. However, it had already claimed an abundance of scare-mongering media coverage and conspicuous investment in contingency planning across several private and public institutions.
4 See also *How to Bury Your Goods: The Complete Manual of Long-Term Underground Storage* (The Wire 1987); *Maritime Terror: Protecting Yourself, Your Vessel, and Your Crew against Piracy* (Gray et al. 2011); *Survive the Economic Collapse: A Practical Guide* (San Giorgio 2013); *How to Prosper During the Coming Bad Years in the 21st Century* (Ruff 2014); and *How to Survive a Robot Uprising: Tips on Defending Yourself Against the Coming Rebellion* (Wilson 2018).
5 Writing in 1967, the literary theorist argued that the persistence of apocalyptic myths reflects a need to structure our existence: beginnings, intervals, and ends give meaning to the interval we are in, which is our life. For further discussion on immanence, see Deleuze (1997). Deleuze reappraised Baruch Spinoza's concept of *Deus sive Natura* ('God or Nature'), asserting that there is no transcendent principle or external cause to the world, and that the process of the creation of life is contained in life itself.
6 Please note that I do not give performativity here a diminished value as opposed to reality. The guides are performative, as they prescribe models for actions that readers should emulate or avoid it they want to manage or cope with the coming apocalypse. For more about this understanding of the performative, see Butler (1997).
7 'To make kin in lines of inventive connection as a practice of learning to live and die well with each other', Haraway famously wrote (Haraway 2016: 1).
8 Survival manuals are rooted in travelogues and encyclopaedias, where 'real facts' (or experiences) are turned into exempla: household encyclopaedias, accounts of disasters, medical manuals, the Boy Scouts' books (the motto of the Boy Scouts being 'be prepared!'), travel guides, housekeeping manuals, military tactics, even manuals for magicians and tricksters. To discuss each one of these, unfortunately, is beyond the scope of this research, and I shall just note a few examples. *Vitalogy* (1899), for example, was a quirky health encyclopaedia for the household comprising warnings and instructions of how to deal with practical accidents and unexpected illnesses via a mix of quack medicine, herbal remedies and spiritual beliefs (Wood and Ruddock 1923); *The Household Cyclopaedia* from 1856 is a 'must have' according to the website modensurvival.com, and includes a variety of topics such as manure, pottery, animal diseases, and metallurgy (The Household Cyclopaedia 1881). Accounts of disasters, such as *The San Francisco Calamity by Earthquake and Fire* by Charles Morris (1906), were among the first attempts to estimate the damages and the survival strategies for future calamities in the US, establishing the basis for what would later feed into in civil defence preparedness booklets and strategies. A similar example is *The Complete Story of the Italian Earthquake Horror, The World's Greatest Disaster, Death and Ruin by Earthquake, Tidal Wave and Fire* by Martin Miller (1909).

9 I use this word here not so much as a reference to any so-called 'occult arts', but to emphasise the level of incommensurability of radiation, as in Robert Bendiner's definition in the Merriam Webster dictionary: 'occult matters like nuclear physics, radiation effects and the designing of rockets'. See https://www.merriam-webster.com/dictionary/occult [accessed February 21, 2021].

References

Albeck-Ripka, L., Simpson, A., and Croall, F. 2019. 'How to Reduce Your Carbon Footprint'. The Times, January 31. https://www.nytimes.com/guides/year-of-living-better/how-to-reduce-your-carbon-footprint.

Ballou, J. 2012. *Arming for the Apocalypse: Assembling Your Survival Arsenal While You Still Can.* Boulder, CO: Paladin Press. Barad, K. 2007. *Meeting the Universe Halfway: Quantum Physics and the Entanglement of Matter and Meaning.* Durham, NC: Duke University Press.

Barad, K. 2012. 'On Touching: The Inhuman that Therefore I am'. *differences* 23(3), 206–223. https://doi.org/10.1215/10407391-1892943. Beck, U. 1992. *Risk Society: Towards a New Modernity.* Trans. M. Ritter. New Delhi: Sage.

Bedford, L and Jones, B. 2014. *Happy Healthy and Prepared: Top Tips from the Host of Survival Mom Radio Network.* Amazon Digital Services LLC.

Bennett, M. 1964. *The Intelligent Woman's Guide to Atomic Radiation.* Baltimore: Penguin.

Benson, R. 1983. *The Survival Retreat: A Total Plan for Retreat Defense.* Boulder, Colorado: Paladin Press.

Berrill, A., Card, N., Cable, J., Harper, L., Blanchard, L., Hughes, S., and Ferguson, D. (2020) '50 Simple Ways to Make Your Life Greener', *The Guardian*, 29 February.https://www.theguardian.com/environment/2020/feb/29/50-ways-to-green-up-your-life-save-the-planet?

Blanchot, M. 1980. *L'écriture du désastre.* Paris: Gallimard.

Boyer, P.S. 1985. *By the Bomb's Early Light: American Thought and Culture at the Dawn of the Atomic Age.* New York: Pantheon.

Boyett, J. 2005. *Pocket Guide to the Apocalypse: The Official Field Manual for the End of the World.* Orlando: Relevant Books.

Bradford, A. 1974. *Survival with Style: In Trouble or in Fun … How to Keep Body and Soul Together in the Wilderness.* New York: Vintage Books.

Brand, S. 1968. *Whole Earth Catalog.* Menlo Park, CA: Portola Institute.

Brown, G. 2021. *'How to Avoid a Climate Disaster* by Bill Gates Review – Why Science Isn't Enough'. *The Guardian*, 17 February. https://www.theguardian.com/books/2021/feb/17/how-to-avoid-a-climate-disaster-by-bill-gates-review-why-science-isnt-enough. Butler, J. 1997. *Excitable Speech: A Politics of the Performative.* New York: Routledge.

Carson, R.L. 2002. *Silent Spring.* Boston: Houghton Mifflin.

Cavallo, F.L. 2020. 'Sensing It Coming: Regarding the Aesthetics of Risk'. PhD diss., University of Kent.

Clayton, B.D. 2002. *Life after Terrorism: What You Need to Know to Survive in Today's World.* Boulder, CO: Paladin Press.

Davis, T. 2007. *Stages of Emergency: Cold War Nuclear Civil Defense.* Durham, NC: Duke University Press.

Deleuze, G. 1997. 'Immanence: A Life…'. *Theory, Culture & Society* 14(2), 3–7. https://doi.org/10.1177/026327697014002002.

Dillon, G. 2012. 'Imagining Indigenous Futurisms', in Dillon, Grace (ed.), *Walking in the Clouds: An Anthology of Indigenous Science Fiction.* Tucson: University of Arizona Press, 1–12.

Extinction Rebellion. 2019. *This Is Not a Drill: The Extinction Rebellion Handbook*. London: Penguin Books, Limited.

Fadiman, C. and White, J. 1971. *Ecocide: And Thoughts Toward Survival*. New York: Center for the Study of Democratic Institutions.

Fuller, B. 1969. *Operating Manual for Spaceship Earth*. Carbondale, IL: Southern Illinois University Press.

Gates, B. 2020. *How to Avoid a Climate Disaster: The Solutions We Have and the Breakthroughs We Need*. London: Allen Lane.

Gray, J., Monday, M., and Stubblefield, G. 2011. *Maritime Terror: Protecting Yourself, Your Vessel, and Your Crew against Piracy*. Boulder, CO: Paladin Press.

Great Britain Home Office. 1963. *Advising the Householder on Protection against Nuclear Attack*. London: Great Britain Home Office.

Great Britain Home Office. 1976. *Protect and Survive*. London: Great Britain Home Office.

Haraway, D.J. 2016. *Staying with the Trouble*. Durham, NC: Duke University Press.

Höhler, S. 2015. *Spaceship Earth in the Environmental Age, 1960–1990*. London: Routledge.

Kermode, F. 1968. *The Sense of an Ending: Studies in the Theory of Fiction*. Oxford: Oxford University Press.

Kolbert, E. 2021. *Under the White Sky: Can We Save the Natural World in Time?* London: Vintage.

Latour, B. 2017. *Facing Gaia: Eight Lectures on the New Climate Regime*. Cambridge: Polity.

Latour, B. 2021. 'Is This a Dress Rehearsal?'. *Critical Inquiry* 47(S2), S25–S27. https://doi.org/10.1086/711428.

Lichtman, S.A. 2006. 'Do-It-Yourself Security: Safety, Gender, and the Home Fallout Shelter in Cold War America'. *Journal of Design History* 19(1), 39–55. https://doi.org/10.1093/jdh/epk004.

Mann, M.E. 2021. *The New Climate War: The Fight to Take Back Our Planet*. London: Hachette.

Margulis, L. and Sagan, D. 1995. *What Is Life?*Berkeley: University of California Press.

Miller, M. 1909. *The Complete Story of the Italian Earthquake Horror, The World's Greatest Disaster, Death and Ruin by Earthquake, Tidal Wave and Fire*. Chicago.

Mitchell, A. and Chaudhury, A. 2020. 'Worlding beyond 'the' 'End' of 'the World': White Apocalyptic Visions and BIPOC Futurisms'. *International Relations* 34(3), 309–332. https://doi.org/10.1177/0047117820948936.

Mohun, A. 2013. *Risk: Negotiating Safety in American Society*. Baltimore: Johns Hopkins University Press.

Morris, C. 1906. *The San Francisco Calamity by Earthquake and Fire*. Philadelphia: J.C. Winston Co.

Natural Resources Defense Council and Um, T. 2020. 'What You Can Do for Your World'. *National Geographic*, April, 24–33.

Palmer Thompson, E. and Campaign for Nuclear Disarmament. 1980. *Protest and Survive*. London: Campaign for Nuclear Disarmament.

Povinelli, E.A. 2021. *Between Gaia and Ground: Four Axioms of Existence and the Ancestral Catastrophe of Late Liberalism*. Durham, NC: Duke University Press.

Powys Whyte, K. 2017. 'Our Ancestors' Dystopia Now: Indigenous Conservation and the Anthropocene', in Heise, U.K., Christensen, J., and Niemann, M. (eds), *The Routledge Companion to the Environmental Humanities*. New York: Routledge, 271–273.

Pugsley, J.A. 1981. *The Alpha Strategy: The Ultimate Plan of Financial Self-Defense*. Los Angeles: Stratford Press.

Rahman-Jones, I. 2021. 'How to Be More Eco-Friendly in Everyday Life'. *BBC News*, August 9.https://www.bbc.co.uk/news/newsbeat-47990742.

Rennie, P. 2005. 'An Investigation into the Design, Production and Display Contexts of Industrial Safety Posters Produced by the Royal Society for the Prevention of Accidents during WW2 and a Catalogue of Posters'. PhD diss., University of the Arts London.

Rennie, P. 2015. *Safety First: Vintage Posters from RoSPA's Archive*. Glasgow: Saraband.

Rose, K. 2001. *One Nation Underground: The Fallout Shelter in American Culture*. New York: NYU Press.

Ruff, H. 2014. *How to Prosper during the Coming Bad Years in the 21st Century*. New York: Berkley Books.

San Giorgio, P. 2013. *Survive the Economic Collapse: A Practical Guide*. Whitefish, MT: Washington Summit Publishers.

Saxon, K. 1976. *The Survivor*. Harrison, AR: Atlan Formularies.

Saxon, K. 1980. 'What Is a Survivalist?' *Cdn.preterhuman.net*. https://cdn.preterhuman.net/texts/survival/whatsurv. [AccessedJanuary 20, 2020].

Scheibach, M. 2015. *Atomics in the Classroom: Teaching the Bomb in the Early Postwar Era*. Jefferson, NC: McFarland & Company, Inc.

Stephens, D. and Stephens, B. 1976. *The Survivor's Primer & Up-Dated Retreater's Bibliography*. Spokane, WA: Stephens.

Strauss, N. 2009. *Emergency: This Book Will Save Your Life*. Toronto: HarperCollins.

The Household Cyclopaedia of General Information containing over Then Thousand Receipts in all the Useful and Domestic Arts. 1881. New York: Thomas Kelly.

The Wire, E. 1999. *How to Bury Your Goods: The Complete Manual of Long-Term Underground Storage*. Port Townsend, WA: Breakout Productions.

USA, Federal Civil Defense Administration. 1946. *What Can I Do: The Citizen's Hand Book of War*. Washington, DC: Federal Civil Defense Administration.

USA, Federal Civil Defense Administration. 1951. *Bert the Turtle Says Duck and Cover*. Washington, DC: Federal Civil Defense Administration.

USA, Office of Civil and Defense Mobilization. 1958. November. *Between You and Disaster: For Your Survival—A Civil Defense Home Food Storage Program*. Washington, DC: US Government Printing Office.

Vanderbilt, T. 2002. *Survival City: Adventures among the Ruins of Atomic America*. Princeton, NJ: Architectural Press.

Watts, J. 2021. '"It Is the Question of the Century": Will Tech Solve the Climate Crisis – or Make It Worse?'. *The Guardian*, 6 March.https://www.theguardian.com/books/2021/mar/06/it-is-the-question-of-the-century-will-tech-solve-the-climate-crisis-or-make-it-worse.

Wilson, D. 2018. *How to Survive a Robot Uprising: Tips on Defending Yourself against the Coming Rebellion*. New York: Bloomsbury Publishing USA.

Wood, G.P. and Ruddock, E.H. 1923. *Vitalogy; Or Encyclopedia of Health and Home, adapted for Home and Family Use*. Chicago: Vitalogy Association.

8

WAITING FOR THE END

Narrating and Grieving Extinction

Sarah France

Adam McKay's (2021a) recent apocalyptic satire *Don't Look Up* follows astronomy graduate student Kate Dibiasky and professor Dr Randall Mindy as they discover a comet set to hit the planet in six months' time – a comet with the capacity to be a "planet-killer". The film enacts and alludes to many of the expected genre conventions seen in similar American disaster movies – the discovery of a planetary threat, the scientists' attempts to communicate the threat to those in political positions of power, the attempts at a technological save[1] – yet the film enacts these conventions in order to query them, ultimately refusing the promise of human survival and instead ending with the annihilation of the planet. The film can clearly be interpreted as an analogy to climate change, and much of the political and public response to the knowledge of the inbound comet can be read in terms of climate denialism – climate scientist Peter Kalmus notes that it is "the most accurate film about society's terrifying non-response to climate breakdown I've seen", and that the "panic and desperation [the scientists] feel mirror the panic and desperation that many climate scientists feel" (Kalmus 2021). Despite the still-growing threat of irreparable climate catastrophe, there remains a widespread sense of denial surrounding the extent of environmental risk, whether it be explicit climate scepticism (denial of the existence of climate change), interpretive denial (interpreting the change to be a positive or non-destructive one), or implicatory denial (denial of its socio-political and moral implications).[2]

Refusal to accept the reality of ongoing climate change can partially be linked to the difficulty posed in acknowledging, articulating, and conceptualising climate change and its potentially devastating consequences. Eva Horn argues that this stems in part from climate change being a "catastrophe without event" (Horn 2018: 55) – a catastrophe that lacks a single geographical and temporal location, and that has global, long-lasting, and destructive consequences. Unlike events with clear perpetrators and logical, immediate consequences, "climate change has no

DOI: 10.4324/9781003189190-10

identifiable source or responsible agents – rather [...] a multitude of sources and agents" (Horn 2018: 56–57). The process of climate change is incremental and abstract, with no evident antagonists, protagonists', or even location – no short-term narrative structure to follow, no distinct beginning, middle or end that can be ascertained. This difficulty in identifying a clear origin, or even a clear outcome, is part of what contributes to climate change denial or scepticism, and it also contributes to the difficulty in forging a narrative around environmental catastrophe – a concern that Rob Nixon raises when considering the "slow violence" of climate change:

> How can we convert into image and narrative the disasters that are slow moving and long in the making, disasters that are anonymous and that star nobody [...] How can we turn the long emergencies of slow violence into stories dramatic enough to rouse public sentiment and warrant political intervention?
>
> *(Nixon 2013: 3)*

Nixon raises several vital points: climate change requires action that might be stirred via use of narrative; and by its very definition climate change often resists satisfactory narrativization. With no clear antagonists or protagonists, no clear "moment" of disaster, and the absence of an evident move towards a resolution, how can we articulate a narrative that evidences the potential consequences of climate change in a way that is serious enough to provoke action?

Undoubtedly, the past few decades have seen an influx of narratives which envision the potential planetary consequences of climate change, yet the large majority have situated it as devastating but often survivable. Post-apocalyptic narratives have become particularly dominant: fictions where anxieties surrounding climate catastrophe or other mass destruction manifest in the form of a catastrophic but "survivable" apocalyptic moment. These fictions are generally secular, but their structures are rooted in theological apocalyptic traditions which position "apocalypse" as a preordained and revelatory event centred on a deterministic view of existence: "apocalypse" in this sense can offer comfort in its promise of salvation and continuation in a world beyond our own. The notion of "apocalypse" has transcended this definition, becoming instead synonymous in contemporary culture with a massive global catastrophe, a point of momentary crisis resulting in socio-political upheaval. This view of apocalypse as a rupture or transition appears to have lost its theological drive, instead positioning the threat as secular and often human-driven, with apocalyptic narratives increasingly emerging to express fears surrounding nuclear disaster, terrorism, viral outbreak, otherworldly threat, socio-economic collapse, and numerous other cataclysms. Yet despite the contemporary shift to secular apocalyptic events, the very suggestion of a "post"-apocalypse retains the sense of assurance and revelatory elements found in apocalypse's theological history: the rupture of secular apocalyptic fiction provides an opportunity to break from the imperfection of the previous world and begin anew, having learnt from past mistakes. Themes of renewal and redemption persist, and visions of catastrophic apocalypticism combine with utopian and millennial fantasies, emphasising the capacity of humanity to overcome apocalypse and ensure the persistence of our legacy.

When considering the possibility of catastrophic climate change and imagining these "stories dramatic enough to rouse public sentiment" (Nixon 2013: 3), narratives that invoke the reassurance of a teleological apocalypse risk asserting the guarantee of human endurance in face of calamity – despite the calamity being our own construction. It risks reliance on the assurance of individual "redemption" in the post-apocalyptic world (whether secular or theological), which in turn stands in contrast to the collective action required to prevent the escalation of catastrophic climate change. It also risks asserting the binary logic of apocalypticism that positions the apocalyptic moment as a singular event which neatly splits time into a "before" and "after", and often has a clear perpetrator or cause, which opposes the abstract and intangible nature of climate catastrophe: as Antonia Mehnert argues, "apocalyptic imageries generally based on presenting clear antagonists, may no longer be fitting to grasp the diffuse relation between victim and culprit in contemporary risk scenarios" (Mehnert 2016: 33), demonstrating the complexity inherent in narrating the potential long-term consequences of environmental slow violence.

As a comedy, *Don't Look Up* responds to aspects of these concerns via satirical means. The desire for a protagonist is articulated by President Orlean, who states that despite remote technology being the easiest way to knock the comet off-course, they "need a hero [...] a pilot [with] real guns", as Washington "always gotta have a hero" (McKay 2021a: 51:56–52:04). The film delves into the complexities of articulating environmental breakdown through the analogy of the asteroid: Dr Mindy begins the film reciting facts and figures of its trajectory, yet she is brushed aside. The turning point of the film occurs when the asteroid becomes close enough to see with the naked eye – when consequences literally become "visible". The metaphor is not a perfect one – the asteroid itself is not human-made as climate change is – but in moving away from apocalyptic movies in which the film ends with a technocratic save, McKay's film shifts towards a conceptualisation of planetary threat which refuses to end with a satisfying form of closure that offers assurance of human prevalence and mastery. Instead of offering assurance in a new world post-asteroid, *Don't Look Up* implodes the expectations for the genre by ending the film with attempts to prevent the asteroid failing due to capitalist greed and a desire to profit resulting in an attempt to fracture the asteroid and harvest the valuable minerals it contains – a move which recalls corporate refusal to shift to more sustainable but less profitable environmental practices. The film closes with a harrowing scene where the scientists and their friends and family eat a home meal as they wait for the asteroid to hit, a final attempt at normalcy as the camera cuts become increasingly more jarring, leading to tight shots of the characters as the walls of the house behind them begin to fracture and the water from the ensuing tsunami encroaches on the group once and for all. Scenes of the burning planet are spliced with clips of animals and the environment, emphasising the non-human response to the impact. Finally, the Earth ruptures. A mid-credits scene does show the fate of a small "generation ship" containing the richest and most powerful of the US, alluding to the discrepancies of how climate change can affect people in different positions, and how often the least vulnerable and most culpable are the least affected – yet rather than positioning this survival as a possibility for continuation, the scene reveals that half did not survive the trip, and that all seem older than feasible to for continuation of a species. President Orlean's death at the teeth of an unfamiliar creature shows they are unprepared for a new alien world, and surely won't last long.[3] A final post-credits scene appears to directly mock the notion of post-apocalyptic survival – the president's son crawls out from under the annihilated White House, pulling his phone out to stream to his no-longer-living followers.

By refusing an ending which assures human continuation, the film rejects the narrative of reassurance and redemption that has been so prevalent in both filmic and textual articulations of planetary risk. Texts which take this route and engage with the possibility of human extinction are much rarer than those which offer hope for continuation – they can be seen as too fatalistic, risking a result of climate pessimism, a sense that it is too late to act and reverse near irreversible changes. Yet could the anxiety and grief they invoke alternatively rouse the need for collective action that is so essential in combatting climate catastrophe? Could the invocation of extinction, and the rejection of the revelation, renewal, or reassurances of apocalypticism, incite complex existential anxieties that are specifically aligned with the act of considering anthropocentric environmental breakdown?

In the following sections of this chapter, I examine two texts which, similarly to McKay's film, construct a narrative which refuses a conclusion of redemption and renewal, instead exploring the possibilities of extinction. These types of narratives, which I will refer to as "pre-extinction fiction", differentiate themselves from apocalyptic and post-apocalyptic fictions by positioning the narrative in the anticipatory period prior to the event, moving towards a more quotidian and realistic vision of eco-catastrophe, and ending with a refusal of the possibility of continuation. Karen Thompson Walker's (2012) *The Age of Miracles* and Michelle Tea's (2015) *Black Wave* both position the narrative in the space leading up to a wholly secular "apocalyptic" event, with the implication being that the event is unsurvivable. I examine how the novels invoke certain tropes in order to disrupt them, frustrating expectations for survival and for traditional closure. In doing so, they pose a critique of certain techno-utopian or anthropocentric responses to climate change. I examine the consequences of this future extinction event on the characters of the texts, on the narrative itself, and finally, on the reader. I argue that rather than being purely nihilistic, these kinds of narratives can instead be seen as productive – the grief these narratives invoke is particularly troubled, and I explore how we might understand the absence of satisfactory closure and the experience of anticipatory planetary grief as something that can be productively utilised in coming to terms with the ongoing environmental catastrophe.

Narrating Extinction

Extinction poses unique problems both conceptually and narratologically. Anxieties surrounding mortality are often assuaged in ways that require the assumption of continuous human existence following individual death. This in turn assures the preservation of individual impact: whether via intergenerational continuation through family, or through the continued memory of our actions or influence, there is a requirement for a continuation of the collective. Extinction removes this comfort, eliminating the reassurance of symbolic continuation and inciting fears surrounding the idea of archival erasure. Beyond the difficulty in conceptualising a future with no continuation of humanity, there is the inherent paradox in thinking extinction: to truly think or experience extinction, all must become extinct, yet the moment this were to happen there would be no spectator to acknowledge or comprehend the extinction. Extinction can only really be theorised or anticipated, eluding accurate representation and conceptualisation. Eugene Thacker describes this paradox as being a kind of "speculative annihilation", where in thinking about extinction "[e]ither one successfully thinks extinction (but then the thought ceases to be adequate to the concept of extinction), or, quite simply, one stops thinking (thereby successfully thinking extinction)" (Thacker 2011: 4). Resultingly, texts which explore extinction must be structured around or towards an ending that is, by its very definition, inaccessible – they anticipate an absolute absence that cannot be fully experienced, let alone narrated. Both *The Age of Miracles* and *Black Wave* explore the consequences of an imminent extinction event, positioning the catastrophe as something that builds incrementally in the background, a slow violence that escalates to an unidentifiable point of no return, a point which cannot be clearly ascertained, so as to juxtapose the "before" from the "after". There is no clear, singular antagonist, no evident moment at which the catastrophe is said to begin, and, ultimately, no simple solution to reverse or outlast the catastrophe: the texts end with the assumption that the narrative space beyond the final pages will soon become extinct.

In Karen Thompson Walker's *The Age of Miracles*, the cause of the breakdown is the slowing of the rotation of the Earth; nights and days begin to lengthen until, at the end of the novel, they ultimately last for weeks at a time. This "slowing" has physical, psychological, and economic consequences, and although the reason for the slowing is never ascertained, many of the consequences are notably environmentally linked. The slowing causes damage to the Earth's magnetosphere, resulting in an excess of radiation which prevents people from staying outside for extended periods of time; it triggers solar storms, impacts the ability to grow crops to make food, and disrupts the rhythm of the tides, causing tens of thousands of whales to beach themselves (Walker 2012: 225, 348, 306, 256). The text ends with no form of solution or answer: Julia notes that neither cause nor preventative action for the slowing was ever discovered, that "[i]t's only a matter of time before the fuel that keeps us alive runs out", and that "we had come to suspect that we were dying" (Walker 2012: 367–368).

Michelle Tea's *Black Wave* is set in an alternative 1999 California where the world is succumbing to accelerated environmental catastrophe: many species have gone extinct, there are extreme weather events, the "glaciers [...] long ago melted into floods" (Tea 2015: 37), and much farmland is no longer sustainable, "[t]he water [...] too ruined for effective farming and the animals [...] out of whack" (Tea 2015: 133). Around half-way through the novel, Michelle learns that "[t]he problems, the oceans, we've passed some point where it's going to accelerate" (Tea 2015: 197), and result-ingly, the world has a year remaining before it becomes fully uninhabitable and all of humanity is anticipated to die. There is no solution found in *Black Wave*: Michelle notes that the "*scientists can't reverse anything*" (Tea 2015: 197, emphasis in original), and much of the latter half of the text explores people's attempts to come to terms with and respond to humanity's imminent end. The final pages depict Michelle writing her story as she waits for the world to end, "telling it until the words before her vanished and the very world around her was gone" (Tea 2015: 325).

The texts explore the difficulty in considering this absence of a collective future. Julia acknowledges the struggle of coming to terms with this loss, stating that humanity was "too convinced of our own permanence" (Walker 2012: 58). She often refers to the comfort found in assuming permanence, noting that she "liked the idea, how the past could be preserved, fossilized, in the stars" (Walker 2012: 124). Without the assurance of posterity, many of the meaning-making activities people do lose their meaning – Michelle thinks about writing a novel but is advised to "forget about publishing", as "[i]t might not be worth it to go through the trouble of putting out a book if we're all going to die the day after it comes out" (Tea 2015: 208). Without the knowledge that there is a connection to the future, the characters struggle to know what to do in the time they have left, and struggle with their place in a world with no future.

In addition to impacting the characters, the presence of extinction also impacts the texts structurally, staging an interaction with genre that disrupts and frustrates the generic expectations of the apocalyptic genre and the non-speculative genre they belong to. Although the texts detail the experience of an apocalypse, they can both be seen as hybrid genres, with the apocalyptic event often taking a backseat to the pri-mary narrative. Although the novel explores the consequences of the slowing, *The Age of Miracles* focuses primarily on the protagonist, Julia, as she navigates the difficulties of high school, her budding relationship with her friend Seth, and her father's infidelity. The text sits between genres – in one way, the novel is a coming-of-age novel, yet by setting the narrative against the backdrop of environmental breakdown, the presence of extinction counters the expectations of both the coming-of-age novel and the (post-)apocalyptic or disaster genre. Time is key to the coming-of-age genre in that novels in this genre detail the passing of time: yet here, time is distorted, stretched out, and disrupting the traditional structures of temporality. Julia struggles to distinguish between the impact of the slowing and the experience of growing up:

[I]t seems to me now that the slowing triggered certain other changes too [...] the tracks of friendships [...] the paths toward and away from love. But who am I to say that the course of my childhood was not already set long before the slowing? Perhaps my adolescence was only an average adolescence, the stinging a quite unremarkable stinging.

(Walker, 2012: 44)

Aware of the psychological consequences of the slowing, Julia looks back to her youth and considers the possibility that the slowing impacted her experience of growing up, yet she cannot assert this – instead she links the consequences of the slowing to the everyday experiences of growing up that are detailed in genres such as the *Bildungsroman: The Age of Miracles* is a hybrid coming-of-age novel which refuses to allow the main character to "come of age", as her future is cut short as a result of the extinction event.

Black Wave details Michelle's move from San Francisco to Los Angeles, exploring her relationships with friends, family and partners, and her experiences overcoming her drug and alcohol dependency. Yet the text is also autofiction – part fiction, part memoir. Tea constructs a fragmented narrative and uses meta-fictional techniques to engage with the prospect of extinction and environmental collapse. In the second part of the novel, Tea splits the narrative, with the primary 1999 storyline running alongside the experience of a "second" Michelle: one who lived in a world with no apocalypse and is writing the story of "apocalypse" Michelle. There is uncertainty as to who the "true" Michelle within the fictional world of the novel – whether 2016-Michelle is writing 1999-Michelle, or whether 1999-Michelle is imagining an alternative world where she lives on to become 2016-Michelle. The narrative streams run alongside one another, blending and blurring and leaving the reader with confusion as to which is the "real" narrative strand within the text. The novel closes with Michelle finishing her story as the world comes to an end around her. There is a complex encounter with the genre expectations of the memoir: the blurring between the frame narratives creates an unease that the final frame of the text (that of reality) may collapse too, allowing the eco-apocalypse to infect our own past and present – yet in the invocation of this unease, we understand that it already has.

By invoking various tropes and expectations, the texts frustrate the structure of the genres of the coming-of-age novel and the memoir, respectively, by juxtaposing them with the stark reality of environmental apocalypse, and in doing so emphasise the impact of extinction on narrative and story. Michelle recalls how "[i]n every apocalypse movie she had ever seen, people needed guns" (Tea 2015: 262), using the framework of an apocalyptic movie as guidance, a script to follow that will enable her to survive. Later, she refers to herself as "cold as a gun-toting avatar in a video game" (Tea 2015: 270), continuing this framing of her experiences through a form of media that allows her to distance herself from reality and provide her with the strength and penchant for survival that a video game character might have. Other experiences are described through the language and metaphor of story and film – towards the end of the text, she runs into the actor Matt Dillon and has sex with him in the book shop, as "the day [takes] on the sharp focus of an apocalypse dream" (Tea 2015: 273), offering a sense of unreality to the experience. Michelle describes feeling "like she was a dying little girl granted one last wish by a benevolent organization [...] to be in a movie with Matt Dillon and here they were in a sex scene, his prop pistol against her head" (Tea 2015: 273). She describes Matt "[beginning] his denouement", "[lifting] the gun above Michelle's head and [firing] it into a wall of books" as he finishes, then declaring "*There [...] Act Two*" (Tea 2015: 274). The language of narrative and film – the prop gun, denouement, the "climax" of shooting the gun, and the declaration of the end of an act – frames the experience as a scene from a movie. For Michelle, it becomes easier to experience the events of apocalypse through a narrativized understanding of it, framing it as a structured narrative which might offer clear protagonists and antagonists, demarcated climaxes, closure, and resolution. By attempting to narrativize her experiences, she seeks to assign a secure storyline to her experience which offers comfort from its story-like nature.

However, it is the invocation of these tropes that draws attention to their inadequacy in this pre-extinction state. The texts focus primarily on the quotidian experience of the end of the world rather than on the heroic – instead of constructing post-apocalyptic hero narratives which "reveal the undiscovered heroic potential in the most ordinary of us" (Renner 2012: 206–207), emphasis is directed towards the mundanity of the end of the world and the everyday experiences of the people who experience it. *The Age of Miracles* distinguishes itself from more common spectacle-driven apocalyptic/disaster narratives, in that when the slowing began there was "no footage to show on television, no burning buildings or broken bridges, no twisted metal or scorched earth, no houses sliding off slabs" (Walker 2012: 13), thus distancing the event from expectations of apocalypse, and aligning it more with the initial "slow violence" of climate change. Michelle describes a phone call with her brother who works in LA as a filmmaker's assistant: as he informs her of the imminent apocalypse, she queries the validity of his assertions that the world is coming to an end. Kyle similarly uses filmic language, stating that they have "passed some point where it's going to accelerate and become like some sort of horrible like sci-fi movie where we all start eating each other and bands of crazed rapists roam around murdering each other" (Tea 2015: 197), drawing links to expected tropes of post-apocalyptic/sci-fi narratives. Michelle questions whether Kyle is telling the truth, asking: "This isn't a Treatment for a Film You're Casting?" (Tea 2015: 197), affirming the notion that apocalypse belongs solely in the world of fiction, only for Kyle to respond: "*No, I wish. Bruce Willis is not coming to save us*" (Tea 2015: 197). This phrase asserts how these narratives distinguish themselves from expected tropes – rather than a post-apocalyptic hero narrative with a clear-cut protagonist who prevails, this apocalypse denies the narrative a satisfactory solution that follows the "steps" of a spectacle-driven action narrative. Expectations and invocations of these narratives are raised so as to juxtapose this notion with the stark reality of environmental apocalypse. By invoking these tropes and expectations, the texts incite a frustration of generic expectations, refusing the redemption, renewal, and closure that may be anticipated.

The loss of a future also stages an uncertainty when considering children and future generations. The figure of the child as a symbol of hope and posterity is another common trope of apocalyptic fiction: Rebekah Sheldon has questioned "why, when we reach out to grasp the future of the planet, do we find ourselves instead clutching the child?" (Sheldon 2016: vii), referencing numerous examples of apocalyptic and dystopic narratives which centre the figure of the child as a beacon of hope that will assure the continuation of the human race.[4] The image of the child as a symbol of hope and a signifier for posterity is itself linked to patriarchal and heteronormative constructions of the future: Lee Edelman (2004) has famously explored this in his work on reproductive futurism, the idea that our political desire for progress is motivated by a desire for a better future for our children. Both the pregnant body and the child are resources to secure a continued future for the present, a construction of futurity that renders any alternative imaginings of the future as implausible and unintelligible. Sheldon writes that this understanding of "child as resource" is "freighted with expectations and anxieties about the future", yet when this symbolic child is linked to anxieties relating to environmental catastrophe, the child becomes:

> tethered to a future that can no longer be taken for granted [...] much of the horror of ecological disaster comes from the projected harm to the future [the consequences of climate change] portend. And the future is the provenance of the child.
>
> *(Sheldon 2016: 3)*

The image of the child as a source of hope and an assurance of posterity becomes threatened when facing environmental collapse. As a representation of future continuation of humanity, the child is intrinsically linked to the concept of the future; and when the certainty of the future becomes insecure, so too does the symbol of the child. In *Black Wave*, imagery of children and reproduction is used to demonstrate how the image of the child has lost its ability to represent posterity: with the certainty that complete annihilation is imminent, there is no opportunity for the lone survivors who will bring humanity into a new world. As Michelle states: "No babies, no planet, no future" (Tea 2015: 269), signifying the link between child and future that Sheldon describes. Michelle explains that "everyone who had become pregnant was having an abortion and those who weren't looked disturbed [...] too far along, committed to the things inside their giant bellies. They looked like animals at the pound, stuffed into too small-cages" (Tea 2015: 270). Instead of being seen as evidence of posterity and human continuation, pregnancy becomes a source of additional suffering, a trapping that subverts the trope of the child as resource, instead positioning it as a form of containment. In *The Age of Miracles*, Julia notes that children are more at risk in this slowing world, that "the radiation was more hazardous to children than to adults. Our bodies were smaller, incomplete" (Walker 2012), implying that the next generation is not an era of hope and potential, but one of higher risk. Seth ultimately succumbs to extreme sickness and is taken to South America for treatment, after which Julia never hears from him again. In these worlds of pre-extinction, the child is not a symbol for the future – it is a symbol for the lack of one.

Grieving Extinction

The absence of a future within these texts similarly causes a major disruption to the experience of grief and satisfactory mourning processes. How can characters grieve an extinction event that is not only massive in scope, but is literally, by definition, inaccessible? Many theories have been developed in order to define and discuss the difficulty in grieving both present and future environmental losses: terms such as 'ecological grief', 'climate trauma', 'Solastalgia', 'climate pre-TSD', and 'Tierra-trauma' have emerged so as to attempt to conceptualise – and subsequently make attempts to work through – grief that is directed towards the planet.[5] There are multiple facets as to why environmental loss is difficult to come to terms with, many of which align with its difficulty in moulding to a narrative. The first is its massive impact and scale – the loss is not directed towards one individual, but to sprawling numbers of both human and nonhuman beings both present and future, and towards more abstract concepts in the loss of a secure future. The second is the complexity of grieving for the nonhuman world: environmental spaces and non-human bodies have historically been positioned as inferior and of lesser value, and, as such, grief directed towards the environment and nonhuman bodies is complex: as Catriona Mortimer-Sandilands asks in a consideration of queer environmentalism and melancholia: "how does one grieve in a context in which the significance, the density, and even the existence of loss is unrecognized?" (Mortimer-Sandilands 2010: 339). The third is the loss experienced as ongoing, continual and often future-oriented: it troubles the temporality generally required of mourning pro-cesses. Extinction's disruption of both temporal structure and teleological under-standings of existence has consequences on understandings of mourning practices and processes. Satisfactory mourning has a temporal requirement, a movement from the inciting moment of initial grief through the processes of mourning, resulting (ideally) in satisfactory mourning processes. Yet with extinction, the inciting moment is not accessible – it is, by definition, an inaccessible future loss that can never be moved past, thus troubling the timeline of grief. Paul K. Saint Amour argues that part of the reason why ecological grief troubles temporal frames of mourning is because there is "no privileged vantage outside or above ecological loss from which to narrate it", and as such, "[f]or the foreseeable future, our stories of ecological grief will be related from the midst of ecological grief" (Saint Amour 2020: 141). It is hard to process loss until it has passed – there is no moving through a loss until we can access and know its terminus.

Being in the midst of climate change additionally means that our surroundings assure us that the loss has not yet fully occurred. Martin Jay notes that, despite the loss implied in apocalyptic thinking, the inability to mourn is a result of the "continued presence in what we might call the real world of the object whose apparent loss we cannot mourn" (Jay 1994: 42). He notes that, in the case of individual loss, "the passage of time is enough to allow the realization of genuine absence to achieve its work of consolation", yet with the loss of the Earth this cannot be the case:

> For the earth, however wounded by our depredations, is still around to nurture us. There is no reality testing that permits us to let go of the libidinal investment we seem to have in an object that has not fully disappeared.
>
> *(Jay 1994: 42)*

This concept can be extended to both climate catastrophe and human extinction: whilst both humans and the planet persist, even despite being threatened, it is impossible to realise the absence of the lost object and come to terms with the related grief. The presence of our surroundings reassures us that total environmental degradation has not yet fully occurred, despite knowledge to the contrary. Paradoxically, this inability to mourn environmental catastrophe can also result in its perpetuation. Owain Jones, Kate Rigby and Linda Williams argue that failure to come to terms with and "fully and consciously enter into the grief of extinction" results in the continuation of actions that threaten the environment and human and planetary continuation, thus "[contributing] to the perpetuation of ecocidal practices" (Jones et al. 2020: 398).

As opposed to individual loss, where the passage of time may be sufficient in satisfactory grieving processes, for environmental catastrophe the passage of time may in fact be detrimental and result in increased loss, so long as we as a species do nothing to remedy it.

Both narratives engage with the grief felt as the surrounding environment succumbs to breakdown: Julia talks of how she and Seth preserved items that would soon die off, creating an environmental museum in their homes, "[lining] the shelves with relics from our time": "the last living eucalyptus [...] the neighborhood's last blades of grass [...] the final flowerings of daisies, of marigolds, of honeysuckle" (Walker 2012: 308). Together, they preserve ephemera they anticipate will no longer be alive in their future: "Look here, we pictured saying someday, this one we called maple, this one magnolia, this aspen, this oak", noting that "on dark days, Seth drew maps of the constellations as if those bodies, too, might soon fall away" (Walker 2012: 308). In anticipating the future losses of her surroundings, Julia emphasises the presence and the prescience of experiencing ecological grief – its contemporariness and impact on the present, but also its orientation towards the future. The absence of a future haunts the present of the text, the knowledge impacting the everyday lives of the characters, with Julia noting towards the latter part of the novel that students on the last day of school shared a feeling that "we might never return to those halls", that although "September loomed just three months away" they had "stopped predicting the future" (Walker 2012: 347). When choosing what to write in the wet cement that is recalled in the final lines, Julia notes that she "felt a vague sadness then, the premonition of a future feeling" (Walker 2012: 357), describing a sense of grief located in the future that is reflected onto her present.

Black Wave directly acknowledges the impact of environmental grief: Michelle notes that "[o]n the television Kim learned how the world was making people sick", listing a number of suspected reasons for this "blend of fatigue, ache, anxiety and depression":

> It was time, which passed faster and, therefore, more abusively than it once had. It was the death of God. It was how meaningless everything was. It was the lack of trees and foliage, it was the animals made extinct and the sludge of the sea. […] It was Compound Environmental Malaise. No one knew how to treat it.
>
> *(Tea 2015: 19)*

This "Compound Environmental Malaise" is seen as a consequence of grieving the damage to the environment and its inhabitants, yet it also stems from a loss of the future, an awareness of the lack of time to remedy the situation, and a lack of meaning and sense-making to be found in this non-teleological apocalypse. The presence of environmental decay persists throughout the narrative; Michelle describes "that smell in the air all the time, the tinny stink of environmental collapse […] [t]he fog clung to Michelle's glasses and wouldn't come off, her view of life perpetually smeared" (Tea 2015: 98), emphasising how the environmental malaise (or ecological grief) impacts the way Michelle and the other characters move through life. Michelle notes the difficulty she feels in accepting what is happening, aware that "[s]he was supposed to be feeling something a few layers down, something authentic and meaningful", yet she fears that "she was not having an authentic experience of the beginning of the end of the world. She was having a deeply authentic experience of inauthenticity" (Tea 2015: 213–214), acknowledging the difficulty in experiencing "authentic" feelings towards this ungrievable future loss.

In addition to impacting the characters' experiences of grief, the presence of the future extinction of the narrative world can impact the experience of reading the text itself. In these narratives, it is possible to consider the potential annihilation that cannot yet be accessed whilst the current Earth offers assurance of its persistence, observing the personal difficulties the characters feel in regards to coping with the future loss of extinction, but also feeling a similar sense of loss in the ending. This is in part from the lack of resolution and reassurance, but also from the inaccessibility of the projected final loss of extinction itself. The absence of narrative closure is tied to the absence of closure in mourning, reframing constructions of environmental apocalypse in a way which refuses satisfactory closure or any form of reassurance, instead leaving the reader sitting with the anticipated loss that resides in the extra-temporal space of the narrative, invoking a grief that remains beyond the final pages. The texts thus both explore and simultaneously invoke the existential toll of climate change and potential environmental extinction.

Conclusion

McKay's film sparked much debate, with some critics and viewers arguing that the satire was too heavy-handed, the ending unnecessarily fatalistic. The film was described as "frantic, strident, and obvious. McKay's touch here is considerably blunter and less productive than it has been in a while" (Dargis 2021). The film is said to imply that humanity is "a dumb, doomed species, too perpetually distracted and misinformed and gullible to endure" (Fear 2021), that could have been "great fun if it had been executed with some respect for our intelligence [...] rather than glib nihilism" (Morgenstern 2021). Despite the criticisms, the film broke the Netflix record for most hours of viewing in a week, and incited a huge amount of discussion and, for some people, a provocation of climate anxiety. McKay tweeted: "Loving all the heated debate about our movie. But if you don't have at least a small ember of anxiety about the climate collapsing [...] I'm not sure *Don't Look Up* makes any sense" (McKay 2021b), indicating McKay's intention to incite climate anxiety. If one aspect of the issue with climate and/or extinction grief is that we cannot fully experience and access it, can narratives such as *The Age of Miracles, Black Wave*, and *Don't Look Up* provide us the narrative space in which to explore these emotions? Can these emotions in turn be manifested into something productive – into the personal responsibility, political action, and civic engagement required to prevent the climate apocalypse? Thinking about the difficulty in affective acknowledgement of climate catastrophe, Sylvan Goldberg has argued that "discussions of environmental affect should consider the elongated temporality of climate change's geological timescales" and of Nixon's "slow violence", suggesting that feelings best suited to map onto our current condition are "feelings that *sustain*, that stretch out beyond the flash of fear or of anger" (Goldberg 2015: 56). The unsubstantiated, unprocessed and anticipatory grief that the characters in these texts feel, and that is in turn displaced onto the readers, is a sustainable feeling that will last longer than the security of quick closure and assured continuation – or, at the very least, incite discussion and awareness as it has with *Don't Look Up*. There is political potential in sitting with this anticipatory grief, and by considering the impossibility of ever fully moving through it by thinking extinction it seems possible to productively utilise these emotions so as to fully comprehend the gravity of environmental catastrophe.

Notes

1 Examples of Hollywood films that contain some of these tropes include *Independence Day* (1996), *Armageddon* (1998), *Deep Impact* (1998), *The Core* (2003), *The Day after Tomorrow* (2004), and *2012* (2009).
2 See Norgaard (2011).
3 When asked in an interview whether "President Orlean's violent death [presages] that life on the new planet is doomed", McKay responded: "Does it mean everyone on every one of the ships gets eaten by brontarocs? [...] Actually, yeah. I think it does" (McKay 2021b).
4 Examples referred to by Sheldon include Bong Joon-ho's film *Snowpiercer* (2013), Margaret Atwood's *MadAddam* trilogy (2003, 2009, 2013), and Alfonso Cuaron's *Children of*

Men (2006). Further examples include Stephen King's *The Stand* (1978), Cormac McCarthy's *The Road* (2006), *The Walking Dead* series (2010–), *2012* (2009), and *Knowing* (2009).

5 For some (non-exhaustive) examples, see Albrecht (2019); Cunsolo and Landman (2017); Woodbury (2019); and the Spring 2020 special edition of *American Imago*, which responds to the theme of ecological loss.

References

Albrecht, G. (2019) *Earth Emotions*. Ithaca, NY: Cornell University Press.

Cunsolo, A., Landman, K. (2017) *Mourning Nature: Hope at the Heart of Ecological Loss*. Montreal: McGill–Queen's University Press.

Dargis, M. (2021) "'Don't Look Up' Review: Tick, Tick, Kablooey". *The New York Times*, December 23. https://www.nytimes.com/2021/12/23/movies/dont-look-up-review.html.

Edelman, L. (2004) *No Future: Queer Theory and the Death Drive*. Durham, NC: Duke University Press.

Fear, D. (2021) "'Don't Look Up' … or You Might See One Bomb of a Movie Hurtling Right Toward You', *Rolling Stone*, December 24. https://www.rollingstone.com/movies/movie-reviews/dont-look-up-review-leonardo-dicaprio-jennifer-lawrence-1268779/.

Goldberg, S. (2015) "'What Is It about You … That So Irritates Me?': Northern Exposure's Sustainable Feeling." In: Oppermann, S. *New International Voices in Ecocriticism*. London: Lexington Books, pp. 55–70.

Horn, E. (2018) *The Future as Catastrophe: Imagining Disaster in the Modern Age*. New York: Columbia University Press.

Jay, M. (1994) "The Apocalyptic Imagination and the Inability to Mourn". In: Robinson, G., Rundell, J. *Rethinking Imagination: Culture and Creativity*. New York: Routledge, pp. 30–47.

Jones, O., Rigby, K., Williams, L. (2020) "Everyday Ecocide, Toxic Dwelling, and the Inability to Mourn: A Response to Geographies of Extinction", *Environmental Humanities*, 12(1), pp. 388–405. https://doi.org/10.1215/22011919-8142418.

Kalmus, P. (2021) "I'm a Climate Scientist. Don't Look Up Captures the Madness I See Every day". *The Guardian*, December 29. https://www.theguardian.com/commentisfree/2021/dec/29/climate-scientist-dont-look-up-madness.

McKay, A. (2021a) *Don't Look Up*. Film. USA: Hyperobject Industries.

McKay, A. (2021b) 'Adam McKay on the Ending(s) of "Don't Look Up": DiCaprio's Last-Minute Line, Streep's Improv and Brontarocs'. Interviewed by K. Aurthur, *Variety*, December 27. https://variety.com/2021/film/news/adam-mckay-dont-look-up-ending-spoilers-1235142363/.

Mehnert, A. (2016) *Climate Change Fictions: Representations of Global Warming in American Literature*. London: Palgrave.

Morgenstern, J. (2021) "'Don't Look Up' Review: A Cosmic Disaster". *The Wall Street Journal*, December 23, https://www.wsj.com/articles/dont-look-up-review-a-cosmic-disaster-dont-look-up-leonardo-di-caprio-jennifer-lawrence-meryl-streep-kate-blanchett-timothee-chalamet-11640294136.

Mortimer-Sandilands, C. (2010) "Melancholy Nature, Queer Ecologies". In: Mortimer-Sandilands, C., Erickson, B. *Queer Ecologies: Sex, Nature, Politics, Desire*. Bloomington: Indiana University Press, pp. 331–358.

Nixon, R. (2013) *Slow Violence and the Environmentalism of the Poor*. Cambridge, MA: Harvard University Press.

Norgaard, K. (2011) *Living in Denial: Climate Change, Emotions, and Everyday Life*. Cambridge, MA: MIT Press.

Renner, K.J. (2012) "The Appeal of the Apocalypse", *LIT: Literature Interpretation Theory*, 23(3), pp. 203–211. https://doi.org/10.1080/10436928.2012.703599.

Saint Amour, P.K. (2020) "There Is Grief of a Tree", *American Imago*, 77(1), pp. 137–155. https://doi.org/10.1353/aim.2020.0007.

Sheldon, R. (2016) *The Child to Come: Life after the Human Catastrophe*. Minneapolis: University of Minnesota Press.

Tea, M. (2016) *Black Wave*. New York: Feminist Press.

Thacker, E. (2011) *In the Dust of This Planet*. London: Zero Books.

Walker, K.T. (2012) *The Age of Miracles*. New York: Random House.

Woodbury, Z. (2019) 'Climate Trauma: Toward a New Taxonomy of Trauma', *Ecopsychology*, 11(1), pp. 1–8. http://doi.org/10.1089/eco.2018.0021.

9

'THE EVENING(S) OF OUR DAY'

Melville, McCarthy, and the Anthropocene's Double Apocalypse

Lindsay Atnip

In September 2019, as wildfires raged across California, supplying a vivid image of catastrophic global warming, the *New Yorker* published an article by Jonathan Franzen with the tagline "The climate apocalypse is coming. To prepare for it we need to admit that we can't prevent it" (Franzen 2019).

Franzen's article reflects a broad tendency in recent years to refer to the current situation of global warming and environmental degradation as "apocalyptic." Most who use this term seem to intend it in the secular colloquial sense of a large-scale future catastrophe potentially inclusive of "the end of the world," the extinction of human life. "Climate apocalypse" now has its own Wikipedia page which defines the phrase as a "permanent global catastrophe, ecological collapse and the los[s] of most species, general misery in a new environment which is hostile to human life, and all the indirect chaos and problems which result [from] these changes." (The discourse of environmental apocalypse may also reflect a sense of inevitability, in contrast to similar terms such as "climate emergency" or "climate crisis," though many also speak of "averting," "preventing," or "avoiding" the climate apocalypse.) The predominant emphasis is therefore on the anticipated destruction of the material conditions of human (and non-human) life.

However, much of the current discourse suggests a catastrophe that is more than material, retaining something of the sensibility of apocalypse's origins and history within the religious imagination. Apocalypse, on these accounts, seems not only to describe an empirical (possible) state of affairs but also to serve as a characterization of an already-present "spiritual" reality. In what follows, I consider what it means to conceive of our present situation as "apocalyptic" in a different and more immediate sense, and argue that this situation is uniquely and necessarily mediated to us by imaginative literature.

DOI: 10.4324/9781003189190-11

Apocalypse Then and Now

Apocalyptic visions are historically associated with crisis, a crisis which is beyond the power of the individual or the collective to rectify, at least *de facto*.[1] Traditional apocalypticism emerges when it seems (to some) that reform is impossible and that only some utter disruption of the current state of things could create the conditions for the restoration or establishment of right order. The imagining of the end of the current world and the ushering in of a new one is thus also a matter of apprehending the current state of the world as unsustainable, incoherent, and evil.

The apocalyptic vision therefore also implies a wholly different ground of judgment and sense than that of the present world, a different standard of "good" and "bad" than that of the reigning powers on Earth. Indeed, the recognition and inhabiting of this different ground may be the most important aspect of the apocalyptic vision. As Frank Kermode (2000) notes, from early on, with the first disappointment of the prediction of the coming of God's kingdom believers had to adjust their understanding of the meaning of the revelations and expectation of the End. In his well-known formulation, Kermode writes that Christians as early as the first century began to conceive the End as "immanent" rather than "imminent"[2]; that is, rather than a one-time event expected in the (near) future, "eschatology is stretched over the whole of history, the End [...] present at every moment," (Kermode 2000: 26) a way of making sense of and judging the present: collectively, contemporary events are viewed in terms of apocalyptic types (e.g. Hitler perceived as the Antichrist); individually, one's own life is viewed in terms of a battle between good and evil culminating in a Last Judgment.

In popular allusions to the "climate apocalypse," the alternative ground of judgment—the "good" in light of which the present state of things is seen as "evil"—could simply be conceived as material sustainability, the limits of the planet's capacity to support a flourishing human civilization into the future, limits which we would seem to be greatly exceeding. But the current public discourse on "climate apocalypse" suggests that the problem goes deeper; our "failure to live sustainably" (Manjoo 2019) is conceived not only as an error or a practical problem to be solved but as a deep moral failure, both individual and collective. Implicit and often explicit in this discourse is the charge that we are destroying our own habitat because we are greedy and rapacious; we take more than we need, and we take it from others—from poorer citizens and nations, from Indigenous peoples, from nonhuman living beings, from future generations. The degradation of our environment is, on this view, ultimately both a product and a reflection of our failure to care about the right things—not just "the planet" but each other. "Climate apocalypse" would therefore seem to pertain not only to the material conditions of human life but also to ideal and normative conditions—to what makes human beings (properly) human.

I propose that these conditions are, therefore, distinctively—and in certain respects uniquely—mediated through reflection upon what one might call the "modern apocalyptic tradition"—that is, works of art and literature that take up apocalyptic imagery and themes in the context of a more general modern crisis or malaise, including literature that would appear to be resolutely secular (if not positively profane) and not primarily concerned with climate or environmental issues. I will here primarily consider two works of American literature—Herman Melville's *Moby-Dick* and Cormac McCarthy's *Blood Meridian*—that reflect the peculiarly American extremity of the "progress" that has culminated in this unsustainability. The consideration of such literary works is, I contend, essential to comprehensively theorizing the "environmental" crisis which is, in the end (so to speak), a crisis of our humanity—not only because environmental degradation undermines the material conditions of human life, nor only even because our failure to care for the environment is a moral failure, true as these both may be, but also because the climate crisis reflects our more general failure to apprehend, and still less adhere to, certain imperatives of and constraints upon our humanity—namely, those imperatives and constraints that belong to what, following Charles Elder, I would call "the grammar of humanity" (Elder 2021): imperatives and constraints which are irreducible to any calculus of utility and outside of our existing social, political, and cultural institutions, and so which might therefore be called transcendent (though in a sense which remains to be specified).

More specifically, I will show that Melville's and McCarthy's works reveal two seemingly disparate but intertwined destructive tendencies that can be seen as both causes and manifestations of this disregard or ignorance of transcendent constraints—the creeping rationalization of capitalism, on the one hand, and an exorbitant violence that emerges in part in reaction against this rationalization—and both of them, as I will try to show, most appropriately seen in light of a sublimity which, against our habitual vanity and indifference, manifests itself through the nonhuman world.

These works further suggest that while our apocalyptic situation may not lead to the establishment of a just order, as in traditional apocalypse, it does retain the aspect of revelation or uncovering. This revelation is not only a "negative" revelation of our sins and of their fruits, but a "positive" revelation of the true conditions of human life—fundamentally of the *fact* of our radical conditionedness and that these conditions are not only material but ideal and, in some sense, transcendent: humanity as we know it depends not only on food, water, a stable and temperate climate, and so on, but on a proper relation to *immensity*, to the vastness of a world not made to human measure.

Moby-Dick: Fast Apocalypse, Slow Apocalypse

The sense of an ending has long shaped the American imaginary. In *Toward a New Earth*, John R. May identifies an apocalyptic strand in American literature which emerges in the nineteenth century,

> reflect[ing] a strong reaction against the easy optimism of the late nineteenth and early twentieth centuries and later a poignant response to the succession of

global hot and cold wars [...] as well [as] the process of secularization that began in the nineteenth century and blossomed into the anomie of the century of unrestrained technology.

(May 1972: 32–33)

As Warren Wagar summarizes, "May defines a vision as apocalyptic if it possesses at least two of the specific symbolisms of the traditional apocalypse: catastrophic change and a norm of judgment." Wagar notes that "[a]lmost all serious contemporary American fiction qualifies as apocalyptic in this radically immanentized sense" (Wagar 1982: 8–9). (The third "symbolism" of the traditional apocalypse is some form of renewal, which May notes is not a characteristic feature of the modern works he is considering, leading him to characterize them as "apocalypses of despair" [May 1972: 37], a point to which I will return.)

Amongst those works that both May and Wagar identify as paradigmatically apocalyptic by these criteria is Herman Melville's *Moby-Dick*. Melville's work, published in 1850, predates the awareness of climate change, but it exemplifies this broader apocalyptic sensibility I wish to unfold, which is already bound up with a sense of transgression against nature.

In *Moby-Dick*, the "catastrophe" is, of course, the sinking of the Pequod with all souls aboard lost save that of the narrator, Ishmael. This catastrophe appears as apocalyptic because the Pequod seems somehow representative or typical, a microcosm for America, and its fate therefore taken to have broader implications. But what of the norm or ground of judgment which is expressed and revealed through this catastrophe?

The Pequod is driven to her fate by her mad captain, Ahab, who would seem to be the primary transgressor, embodying the sins that provoke the judgment of the cosmos (*nemesis*, in the terms of classical tragedy). In recent years, this transgression has been interpreted environmentally, for instance by Justin Slaughter, who, in an essay relating *Moby-Dick* to climate change denialism, characterizes Ahab as "the chief executive who wields centuries of accumulated knowledge and labor for his own gain, but who—not unlike Donald Trump and his circle—would blindly throw all of it into the abyss" (Slaughter 2017). This would be continuous with the moralistic element of climate crisis rhetoric that frames the environmental crisis as a product of greed and ignorance. Whereas the environmentalist interpretations of *Moby-Dick* focus on the destruction of the natural world (most obviously the commercial killing of whales), I would emphasize the ways in which Melville shows how our humanity is at stake in this destruction. I propose, furthermore, that the catastrophe-provoking transgression represented in *Moby-Dick* is more complex than Slaughter's characterization—that Ahab is, in some degree, reacting *against* the forces of capitalism and the utilitarian ethos that are more directly responsible for the destruction of the natural world, in Melville's time and in our own, and that this indicates the double-sided character of our contemporary apocalypse: we are at risk not only of a "fast" apocalypse that might come in a blaze of spectacular violence, but also of a "slow" apocalypse that is concomitant with the very spread of (modern, capitalist) civilization.

Though the world of *Moby-Dick* is not yet ravaged by global warming, the text does express the anxiety that human beings are encroaching upon nature and might ultimately consume it. In one chapter entitled "Does the Whale's Magnitude Diminish? Will He Perish?" Ishmael entertains an apocalyptic vision of the extinction of the whale, wondering

> [w]hether owing to the almost omniscient look-outs at the mast-heads of the whale-ships [...] whether Leviathan [...] must not at last be exterminated from the waters, and the last whale, like the last man, smoke his last pipe, and then himself evaporate in the final puff.
>
> *(Melville 2017: 337–338)*

Ishmael answers his own hypothetical in the negative, enumerating at length the reasons that the whale will not be wiped out like the buffalo (the relative hostility of their habitat, the difficulty of killing them, etc.), and affirming in the end that the whale is "immortal in his species, however perishable in his individuality" (Melville 2017: 339). The narrator arguably "doth protest too much" here, sounding as if he is trying to convince himself, and even if his ultimate conclusion is meant in earnest, we are left also with the image of the extinguished buffalo, which "shook their iron manes and scowled with their thunder-clotted brows upon the sites of populous river-capitals, where now the polite broker sells you land at a dollar an inch" (Melville 2017: 338). What the extinction of the buffalo represents, what the annihilation of the whale would represent, is not only the end of those species but the diminution of the human—from the intrepid men who tried themselves against wildness (like Bulkington, who would "better [...] perish in that howling infinite, than be ingloriously dashed upon the lee, even if that were safety" [Melville 2017: 91]) to men domesticated, calculating, "mean" in the old sense of "low, inferior." The persistence of the whale, and by extension wild nature, is an essential condition for a properly human existence, not only providing a worthy adversary for the hunt, but embodying a magnitude and power without which the world we inhabit would be smaller, meaner, and too wholly domesticated.

The rapine of the natural world—which is by extension, then, the destruction of certain conditions of a full humanity—is represented as the product of a small-minded business sense, of capitalism and commercialism—the broker measuring out inches of land to sell by the dollar, the magnificent whale killed in bulk "to fill men's lamp-feeders" rather than, as Ishmael remarks somewhat tongue-in-cheek, as in "the knightly days of our profession, when we only bore arms to succor the distressed" (Melville 2017: 272), comparing the present commercial enterprise of whaling unfavorably with its mythical predecessors, which he finds, for instance, in Perseus's slaying of the sea monster.

What complicates the picture is that throughout *Moby-Dick* Ahab tends to stand in *opposition* to this money-mindedness, while the flag of capitalism is staunchly waved by Starbuck, one of the novel's most sympathetic characters. While Starbuck's first objection to Ahab's vendetta against the White Whale is that it is "blasphemous" to take "[v]engeance on a dumb brute [...] that simply smote thee from blindest instinct", his second is that "it will not fetch thee much in our Nantucket market" to obsessively pursue just *one* whale. Ahab replies contemptuously that "If money's to be the measurer, man, and the accountants have computed their great counting-house the globe, by girdling it with guineas, one to every three parts of an inch; then, let me tell thee, that my vengeance will fetch a great premium *here*," pounding on his chest—his heart (Melville 2017: 132). Ahab would in fact abandon the pursuit of other, lesser whales except that he recognizes that the crew needs the prospect of cash to motivate them: "The permanent constitutional condition of the manufactured man is sordidness," he reflects (Melville 2017: 169), again expressing his contempt and the sense that industrialization and commerce have colonized and diminished the human soul by transforming all goods into fungible commodities and, in Max Weber's terms, supplanting "value rationality" with instrumental or utilitarian reasoning.

C.L.R. James proposes that Ahab is himself a product of capitalism (James 2001:6 ff), which has taken his life and his leg from him and left him with no way of making sense of his deprivations and suffering, of the "forty years of privation, and peril, and stormtime" (Melville 2017: 389)[3]; Ahab's quest for Moby Dick would then be an insistence on finding the sense that the commercial world has failed to provide, which he does by conceiving the whale as representative of "all evil" (Melville 2017: 148) and taking up arms against that evil. As he puts it to Starbuck:

> All visible objects, man, are but as pasteboard masks. But in each event – in the living act, the undoubted deed – there, some unknown but still reasoning thing puts forth the mouldings of its features from behind the unreasoning mask. If man will strike, strike through the mask! How can the prisoner reach outside except by thrusting through the wall? To me, the white whale is that wall, shoved near to me.
>
> *(Melville 2017: 133)*

Indeed, one could see this "unknown but still reasoning thing" lurking behind all objects, events, and deeds as a description of capitalism's "invisible hand" in its less benevolent manifestations and effects. Yet the only way Ahab is able to strike back is by reifying the causes of his (and humankind's) suffering into a singular malevolent intelligence and projecting this intelligence onto the albino whale, with the obvious disastrous consequences.

Arguably, then, two apocalyptic possibilities, one "fast," and one "slow," are presented in *Moby-Dick*. There is the spectacular disaster that concludes the narrative, which could be seen as representing the culmination of hubristic and egotistical (anthropocentric) pursuits, but there is also the sense of a more diffuse, quieter End, one which could be said to anticipate and universalize the closure of the frontier: the final taming of the wild and with it all of the human possibilities that depend upon it, the buffalo and the whale vanished and the globe girdled with the accountants' guineas. (Melville's Bartleby inhabits this latter "post-apocalyptic" world, an urban desert, wholly rationalized, in which there is no longer anything to desire.) Ahab's animus—leading to the first apocalypse—is not just against the hostile natural world but against this latter social ethos that would lead to the second apocalypse; his is a rebellion against the spread of a calculating prudence which is the colonization of morality by capitalism's amoral laws of profit and democracy's pragmatic rules of toleration—such that Starbuck's Quakerism becomes indistinguishable from good business, and even Ishmael initially rails against Queequeg's fast and "all […] Lents, Ramadans, and prolonged ham-squattings in cold, cheerless rooms" as "opposed […] to the obvious laws of Hygiene and common sense" (Melville 2017: 77), as if there could be no other principle of action than these "obvious laws."

If these were the only two possible visions of the end of history—a wholly rationalized, commercialized world, or disaster brought on by an irrational or anti-rational reaction against that world[4]—this would certainly be an "apocalypse of despair." However, as we all know, there is a survivor of the sinking of the Pequod, and Ishmael's "survivor's narrative"—the book *Moby-Dick*—is arguably a kind of apocalyptic "renewal": not the establishment of God's kingdom on Earth at least in any literal, temporal sense, but the birth or rebirth of a certain consciousness represented by the book's narrative perspective. This consciousness could be said to be, or to intimate, the standard or ground of judgment to which Melville's apocalypse—or apocalypses—educate us.

What kind of consciousness is this? It is difficult to singularly characterize the narrative viewpoint or voice of *Moby-Dick*, and this multivocality is one of its prime attributes. We can say, however, that the narrator (Ishmael or Melville), like Ahab, continually experiences and articulates the world as fraught with elusive and indeterminate meaning, or meaningfulness, but—unlike Ahab—refuses the temptation to fix this meaning. A passage in the first chapter encapsulates this tension. After describing at length how people are drawn to bodies of water, Ishmael says:

> Surely all this is not without meaning. And still deeper the meaning of that story of Narcissus, who because he could not grasp the tormenting, mild image he saw in the fountain, plunged into it and was drowned. But that same image, we ourselves see in all rivers and oceans. It is the image of the ungraspable phantom of life, and this is the key to it all.
>
> *(Melville 2017: 18)*

One might see all of what follows as a quest to grasp that "ungraspable phantom" which nonetheless, like the White Whale, will not finally be reeled in. Over and over again—to a degree that becomes comic and even tedious—Ishmael describes a natural or human phenomenon and then interprets it symbolically or allegorically. The very proliferation of meanings calls into question whether there is any final basis for them, or whether rather the meaning-making mind is like "Nature […] paint[ing] like the harlot" (Melville 2017: 57), obscuring with color the terrifying whiteness and void at the heart of reality. Yet this irony attends a compelling humaneness as well as a deep appreciation for the natural world. Ishmael too is on a quest for sense against the melancholy and even violent or suicidal despair which drive him to sea. But he finds, or creates, the sense that he seeks not by piling it onto the hump of the White Whale but in his telling of his (or Ahab's) tale—and telling it against a background of immensity or sublimity: that which exceeds the fullness of human comprehension.

The tremendous liveliness and often light-heartedness of the narrative voice marks an important difference between Ishmael's consciousness and Ahab's. But it is essential that the former includes, among other things, a certain admiration for Ahab and a recognition of those conditions that would produce the man and his dire quest. It is also a consciousness of the compelling power of a vision such as Ahab's and of the inefficaciousness of the ironic consciousness in disarming such a vision. Ishmael's narrative ultimately implies that all we can do is, in Alfred Kazin's words, "take the span of [the] magnitude" (Melville and Kazin 1956: xiii) that overwhelms. This magnitude consists on the one hand of the magnitude of the inhuman conditions of the world—"the heartless voids and immensities of the universe" represented by the whiteness of the whale (Melville 2017: 157), "the universal cannibalism of the sea; all whose creatures prey upon each other, carrying on eternal war since the world began" (Melville 2017: 215), "the unwarped primal world" of the ocean floor that the cabin boy Pip imagines when abandoned on the open ocean (Melville 2017: 308)—and also of the magnitude of the human impulse that would insist, impossibly and catastrophically, on fronting them.

This span-taking, the unresting search to understand and to poetically represent this magnitude, is itself a kind of preservation of humanity, in eternity if, perhaps, not temporally. It entails the recognition that at some point the human project will end "and the great shroud of the sea [roll] on, as it rolled five thousand years ago" (Melville 2017: 410), but at the same time affirms those attempts to live and to perceive humanly even while acknowledging their complicity, active or passive, in the disastrous trajectory of human history: Ahab's quixotic quest as well as the "attainable felicity" (Melville 2017: 309) with which that quest stands in tension, that admixture of humor, curiosity, and fellow-feeling that we discern in Ishmael and that makes him such good company on the journey to the disaster he cannot prevent.

The narrative consciousness of *Moby-Dick*—which arguably exceeds that of the greenhorn sailor whose journey it recounts—can thus itself be seen as a modern form of apocalyptic renewal, a consciousness which emerges from the catastrophe and which recognizes the sources of that catastrophe, which one might broadly characterize as the failure to see the non-utilitarian character of the world as intimated by nature in its immensity and terror and beauty and mystery. This failure of recognition then enables, on the one hand, the "slow" catastrophes of capitalism and its conversion of everything into commodities and, on the other, the "fast" catastrophes of violence in conscious or (more often) unconscious reaction against that meaningless utilitarian world.

Consciousness of all of this not only intimates but in a sense *is* a different and non-catastrophic way of perceiving and being. Such a way of being may not be realizable on this Earth, at least not broadly or quickly enough to avert the catastrophe (this catastrophe). But it would at least allow for the recognition of the apocalyptic reality of our situation, and of the background against which it occurs, which is of a magnitude beyond full reckoning and one that is always threatening to whelm us.

The character of this inhuman background—and what it might mean to see and respond to it—is illuminated by a twentieth-century work that takes up both the concrete subject of *Moby-Dick* and its symbolic resonances, Cormac McCarthy's unpublished screenplay "Whales and Men." The drama follows a marine biologist, Guy Schuyler, who gets entangled with two British aristocrats who are attempting to leverage their wealth and influence to prevent the extinction of the whale. As in *Moby-Dick*, McCarthy associates the whale with God. As Guy tells his friend the Lord Peter Gregory:

> When I get a fleeting vision of the pure platonic whale and I have a sense that what we are after, the whaleness of the whale, does exist as an idea, but an idea with which we are inadequate to deal. I think that's why we kill them. I think their value as produce is secondary to their sacrificial value. It's our inadequacy to the overwhelming fact of the whale that will doom the whale. I think [M]elville is right. It's like killing God. *That* whale's existence is the whale I can't deal with. That it is the whale who strings the world together on the vectors of his breath and I won't know in this world where that whale will have gone when the last whale is slaughtered and hauled from the sea.
>
> *(McCarthy n.d.: 25)*

Despite Guy's reference to the "pure platonic whale," he is not referring only to our image of the whale, the whale as symbol, but the reality or "overwhelming fact" (one might say "sublime fact") of the whale—a reality that nonetheless transcends the physical creature. The existence of the whale is essential to consciousness—one wants to say "human consciousness," but the point seems to be that consciousness is itself more-than-human, or that human consciousness is only human if it encompasses or can take the perspective of the extra-human. One might say that the whale and the inhuman sublimity that suffuses it belong to the grammar of our humanity, to those conditions that are determinate for what it means to be human, which include the recognition of and relationship to that immensity which stands beyond human comprehension and control.

Blood Meridian: Apocalypse Barbarous and Civilized

While prescient, Melville could not, of course, wholly foresee how the destructive forces represented in *Moby-Dick* would develop over the following century or beyond. McCarthy takes up some of these forces in "Whales and Men," in which, *pace* Ishmael, it has become clear that the whale is very much in danger of perishing from the Earth. As Peter puts it, grimly:

> When we have slaughtered and poisoned everything in sight and finally incinerated the earth itself then that black and lifeless lump of slag will simply revolve in the void forever. There is a place for it too. A nameless cinder of no consequence even to God. That man can halt this disaster now seems so remote a possibility as to hardly bear consideration.
>
> *(McCarthy n.d.: 59)*

McCarthy's apocalyptic vision is fully if not literally realized in his 1985 historical novel *Blood Meridian, Or the Evening Redness in the West*, which also bears Melville's imprint, if more subtly. The novel alludes to *Moby-Dick* in the form of its story as a whole—a young man ("the kid") joins up with an outfit headed by a mad captain on a doomed quest—as well as in the digressive and episodic narrative style, the sublime and terrible images of natural immensity and mystery, and the apocalyptic sensibility and apocalyptic conclusion. And then there are countless local references, from the prophetic Mennonite who, like Elijah in *Moby-Dick*, warns the kid and his fellows of the fatality of their "voyage," to the descriptions of the villainous Judge Holden—pale, naked, enormous—which uncannily evoke the White Whale.

I propose that we can see *Blood Meridian* as reflecting on and extending the apocalyptic premonitions of *Moby-Dick* in light of the wars and horrors of the intervening century—such that in McCarthy's work we see not only human hubris vis-à-vis the power of nature but also "intra-human" violence and atrocity as apocalyptically self-destructive. And in this light, McCarthy's novel suggests that only a more austere consciousness, even further distanced from our ordinary humanity than the narrative consciousness of *Moby-Dick*, might adequately grasp our situation.

Blood Meridian could be called the apotheosis of the American apocalyptic sublime. It may stand alone in American literature in its insistence, as it would seem, on depravity and inhumanity as the prevalent and defining features of human beings. Not only uncompromisingly unromanticizing but unrelentingly (one might almost say gleefully) *de*romanticizing of the American West, the work offers an image of the end of humanity as we know it, or as we have dreamed it to be. Set mostly in the late 1840s, the novel follows its protagonist, the unnamed "kid," from Texas into Mexico, where he ends up joining up with a gang of outlaws and mercenaries hired by the Mexican government to kill Apaches. The bulk of the book relates the gang's travels (the refrain "They rode on" recurs dozens of times) and their ever more violent and unlawful activity: they are paid by the scalp and become increasingly indiscriminate about where they get those scalps.

The spiritual leader of the gang is the erudite but monstrous "Judge" Holden (his legal credentials are questionable), who wantonly drowns puppies and sodomizes and murders children, and who articulates the philosophy that seems to govern the world as McCarthy shows it: "Moral law is an invention of mankind for the disenfranchisement of the powerful in favor of the weak. Historical law subverts it at every turn" (McCarthy 1985: 250), he pronounces, and more simply "War is god" and war is the "ultimate game" (McCarthy 1985: 249). It is Holden who, standing amongst Anasazi ruins, makes explicit the apocalyptic vision that the novel's action appears to illustrate:

> The way of the world is to bloom and to flower and die but in the affairs of men there is no waning and the noon of his expression signals the onset of night. His spirit is exhausted at the peak of its achievement. His meridian is at once his darkening and the evening of his day. He loves games? Let him play for stakes. This you see here, these ruins wondered at by tribes of savages, do you not think that this will be again? Aye. And again.
>
> *(McCarthy 1985: 146–147)*

Environmental concerns are not forefronted in *Blood Meridian*, and its more obvious contemporary referent would be the Vietnam War and the violence associated with the American civil rights movement (Owens 2000: 19–44). Yet as in *Moby-Dick*, the natural world is the sublime and inhuman background of the action in *Blood Meridian*, and the latter similarly suggests that the destruction of the conditions of human life, including environmental conditions, is bound up with a failure to discern, respect, and respond to this immense background—though again, this is not a straightforward morality tale; the work implies too that there is something fated in this destruction, that it is bound up with the progress of civilization and perhaps even innate to the human constitution and human ways of meaning-making.

Throughout most of *Blood Meridian*, nature seems at no risk from human beings. For the most part, it is described as vast and direly inhospitable, dwarfing the gang as they "ride on" through the desert. Descriptions like the following abound: "All night sheetlightning quaked sourceless to the west beyond the midnight thunderheads, making a bluish day of the distant desert, the mountains on the sudden skyline stark and black and livid like a land of some other order out there whose true geology was not stone but fear" (McCarthy 1985: 47), desert and sky and weather serving as a cosmic backdrop to the novel's human enormities.

At the story's end, however, after the gang has been massacred by the Yuma and the few survivors dispersed, the kid escaping both the "indians" and, for the time being, the murderous judge, the action jumps forward briefly to the early 1860s and then, for the concluding chapter, to 1878. This leap allows the story to reach beyond the lawless violence of the mid-nineteenth century to the rationalization and "civilization" of the late nineteenth century, furthered by the completion of the transcontinental railroad (1869) and the introduction of barbed wire (patented in 1874), and leading toward the domestication of the West and the closing of the frontier (1890). Opening this final chapter in the narrative is the encounter of the kid, now "the man," with a buffalo hunter. Like Melville, McCarthy offers an apocalyptic description of the annihilation of the herds, the hides pegged out over actual square miles of ground and the teams of skinners spelling one another around the clock [...] and the meat rotting on the ground and the air whining with flies and the buzzards and ravens and the night a horror of snarling and feeding with the wolves half crazed and wallowing in the carrion.

(McCarthy 1985: 317)

"I wonder if there's other worlds like this," the hunter muses after concluding his account of going out to kill the last herd. "Or if this is the only one." There is the sense that the end of the buffalo is the end of the world, or a world, not only for the Native Americans who depended on it but for White civilization as well. The kid's *peripeteia* has taken him from his childhood home in Tennessee, bordering "woods [...] that harbor yet a few last wolves" (McCarthy 1985: 3)—or one might say "only" a few last wolves—to a West that harbors not even one last buffalo.

Thus, in *Blood Meridian*, too, there is a kind of "secondary apocalypse," the coming of a time of rationalization, when the buffalo are gone—a time which is, in some degree, already upon us. As in *Moby-Dick*, this threat to the human (and nonhuman) world does not map simply onto the threat represented by the nihilistic violence of the scalphunters. It is associated just as closely, if not more so, with that civilization with which the scalphunters seem to be contrasted. Halfway through the novel, the gang arrives in Chihuahua to present to the governor the scalps of the Apache (etc.) they have slain; Glanton and his men are paid and fêted, lavished with gifts by the people of the city, with a ball thrown in their honor. They consume all they are given and take more and more until the citizenry bar their doors and hide away; the gang then takes another excursion for more scalps, which they end up taking not from the Apache but from Mexican villagers. Returning to Chihuahua a second time, they

entered the city haggard and filthy and reeking with the blood of the citizenry for whose protection they had contracted. [...] Within a week of their quitting the city there would be a price of eight thousand pesos posted for Glanton's head. They rode out on the north road as would parties bound for El Paso but before they were even quite out of sight of the city they had turned their tragic mounts to the west and they rode infatuate and half fond toward the red demise of that day, toward the evening lands and the distant pandemonium of the sun.

(McCarthy 1985: 185)

This would seem to be the "apocalyptic turn"; the gang have certainly already engaged in wanton violence, but now they have turned on the civilized world that called them into being and have driven themselves beyond the Pale. The reader's civilized sensibilities are disgusted when the scalphunters behave so loutishly, and appalled when they terrorize and murder peaceful villagers, both Mexican and Native American. We feel similarly near the end of the novel, when one of the gang, Davy Brown, insists on sawing off the barrels of an exquisitely crafted shotgun, terrorizing a farrier who refuses to mutilate the piece. Vereen Bell notes that Brown is in unconscious rebellion against civilization:

> he seems to know instinctively that the gun is a symbol of an order of being, aesthetic and economic, that his whole existence denies. [...] [W]hat Brown does to it is therefore symbolic as well. He knows this, one could say, by not knowing it.
>
> *(Bell 1988: 117)*

But *Blood Meridian* suggests that the existence of these goods of civilization is not separable from the criminality and violence of the scalphunters, even though the two are worlds away from each other in sensibility. Civilization is not merely a veneer—the beauty of the gun is real, evoked powerfully by McCarthy's description so that, like the farrier, we feel what Brown proposes and carries out to be a desecration. But it is significant that the item in question here is a *gun*, and that the civilization to which it belongs is built upon and by those forces wielded by men like Trias and Captain White (leader of an ill-fated band of filibusters and sponsor of the kid's entry into Mexico) who, Frankenstein-like, produce monsters they cannot control—not only the scalphunters but also the barbarity of the Native Americans, by intruding on their land and violating treaties and uprooting them from their soil and customs, more or less forcing them into actual "savagery." One might even see the gang as scapegoats, expelled from that civilization and bearing its sins of human bloodshed while still-"civilized" men wipe out the buffalo and the people of color who impede White Manifest Destiny. The destructions wrought by the gang and those wrought by the society from whose mores they have become unmoored are, then, another iteration of interdependent "fast" and "slow" apocalypses.

Blood Meridian would seem, even more than *Moby-Dick*, to be an "apocalypse of despair," ending with the judge killing the kid for his mulish resistance to Holden's attempts to seduce him to the ethos of total war; the judge at the very end "dancing, dancing" like the sea rolling on, "say[ing] that he will never die" (McCarthy 1985: 335). An obscure epilogue follows this, in which a man "progress[es]" over a plain making what appear to be post-holes ("sparking the fire out of the rock which god has put there" (337)), but if this is an apocalyptic renewal it is at best ambiguous, as the others on the plain are described as mechanistic and devoid of interiority or vitality.

And yet, as in *Moby-Dick*, there is something that comes "after" and out of the catastrophic story: the telling of that story. The teller is no Ishmael—none survive the "wreck" in *Blood Meridian* except the judge—and even if the laconic, barely literate kid *cum* man were to have survived, the narrative perspective of the novel is surely not his. It seems in fact hardly human, taking scarce interest in the characters' thoughts and feelings, and while it expresses elliptic judgments through its imagery of atavism, darkness, and void, it does so from a tremendous distance and does not seem a participant in the human world upon which it gazes. Its perception of the material world is, by contrast, acute, but the consciousness is not just a consciousness of things. The visible is constantly figured as the manifestation of some astonishing and threatening invisible through McCarthy's hallucinogenic similes. To cite just one of numerous instances:

> That night they were visited with a plague of hail out of a faultless sky and the horses shied and moaned and the men dismounted and sat upon the ground with their saddles over their heads while the hail leaped in the sand like small lucent eggs concocted alchemically out of the desert darkness. When they resaddled and rode on they went for miles through cobbled ice while a polar moon rose like a blind cat's eye up over the rim of the world.
>
> *(McCarthy 1985: 52)*

The vehicle of these similes is generally abstract, if not, practically speaking, inconceivable ("a land [...] whose true geology was not stone but fear" [McCarthy 1985: 47]), while the tenor is something concrete, material, visible, if uncanny (lightning, "gravel reefs," etc.), again and again evoking the other-worldly—not the Judaeo-Christian kingdom of God but a world in which what watches is a "blind cat's eye." The descriptions of the landscape suggest that the alternative ground to which McCarthy's apocalyptic tale refers us is an inhuman sublime: that which dwarfs human beings and surpasses their understanding—a dark sublime, a supernatural or demonic world, dark to human understanding, a gnostic world in which some substantive principle of evil predominates and in which the good is alien and distant. The descriptions are not all dark and terrifying; the landscape has a stark and inhospitable beauty, and it often seems portentous (like the burning tree [McCarthy 1985: 215]). The portents, however, seem not to be aimed at human beings for them to guide their actions, at least not collectively, but to be only intimations of some wholly different order of things, a world in which the human ego is radically displaced from the center.

"Turning the Glass": Toward a Contemporary Apocalyptic Consciousness

There remains the question of what it would mean, what it would look like, for us to recognize and live out this decenteredness. Suggestive in this regard is the response given by another character in "Whales and Men," Kelly, to the question of whether she believes in God. Reflecting on the pure givenness of the world, she says:

> even as a skeptic it's hard for me to ascribe purposelessness to something that exists. Isn't existence a purpose? I think we confuse purpose with utility. It's as if existence itself were somehow not noble enough or sacred enough to justify its own undertaking. I think the failure is ours. The failure in reverence. Even if God were only [...] What is it that Thomas says? The force that through the

green fuse drives the flower? [...] I guess what I would say is that yes I believe in God but I don't know who he is. I don't know what his name is. Why would he have a name? To distinguish him from what? (She pauses) I can't parse him out from all I see and feel. I want a God so large that there's no place to even stand outside of him to say that he's not so.

DAVID: What about life after death?

KELLY: I don't know. I wonder if people who are really holy would even care? I think that our egos force us to look through the wrong end of the glass. It's not that we have a soul. It's that the soul has us.

(McCarthy n.d.: 44–45)

This could be taken as a New-Age-like spiritualism, a kind of pantheism that costs nothing, but it seems that there is an imperative here, an imperative to "turn the glass" and look through from the proper end and to put our egos in the proper, minute perspective. Such a conversion is likely only possible for the individual, which in turn raises the concern of a quietistic or apolitical apocalyptic vision, one that entails giving up on the possibility of averting the catastrophe. In this regard, it is worth noting that, despite his belief that all the current forces work toward the doom of the whale, and to human beings as well, Guy works against these forces.

There is no prescription here for hope of "saving the world." Neither Melville nor McCarthy promises that, even if we were to attain the decentered consciousness that their works evoke, we would be able to alter our dire historical course. Both, however, imply that it would be necessary and that, without cultivating such a "reverence" for the nonhuman conditions of human life, for that incomprehensibly immense natural world and the yet more immense ideality of which it is an image, our destiny will be the vision with which Guy's monologue on the whale ends: "At other times of course I see only the empty silent seas through which he has passed forever" (McCarthy n.d.: 25–26)—an emptiness and silence that portends the absence of the human as well, or at least, at best, a severely diminished humanity. And these works imply that whether or not it is efficacious, it is essential that we recognize this nonhuman immensity and its indispensability to the fullness of our humanity.

Such a consciousness of the conditions of our humanity may be, moreover, the only positive moment of transformation afforded by the apocalyptic Anthropocene—a consciousness made possible by what Amos Wilder (1971) calls the "nakedness to Being" that comes with the breakdown, current or anticipated, of those psychological and social structures on which we have depended.[5] This consciousness is not just a consciousness of human failings, of the fragility of the civilized world or of the possibility of ecological collapse, but of the unfathomable whole of which human beings are a part. Such a consciousness would be not only, possibly, a condition for material sustainability, but would also be itself an essential part of sustaining our humanity, achieving the humanity that is available to us in an unprecedented and catastrophic historical situation.

This spare consolation of consciousness is arguably why we read apocalyptic novels, why we read novels in the face of the apocalypse. Doubtless, we come to a certain consciousness of these nonhuman conditions of the human world through encounters with the natural world and, differently, through what science teaches us, and these modalities are essential. But through its dramatizations and its poetic language, literature distinctively brings to light the human meaning of the nonhuman world and the character of the human threat to that world and thereby to humanity itself, as well as what it means, or would mean, for humanity to be lost. We need the humanities—works such as Melville's and McCarthy's, and humanistic reflection upon them—as well as the sciences to grasp the reality of the crisis—perhaps to avert or mitigate it, but at least to see clearly what is at stake.

Notes

1 As L. Michael White writes, the Judaeo-Christian genre of apocalypse emerges from prophecy, which was originally not a prediction of the future but an indictment of present sin and a summons to God's people to do God's will. But after the destruction of Jerusalem and the Temple, oracles began "calling for people to hold fast, saying that there would be a restoration of the nation and that the enemies would eventually be punished by God. A future-looking sense of history was born" (White n.d.).
2 "No longer imminent, the End is immanent. So that it is not merely the remnant of time that has eschatological import; the whole of history, and the progress of the individual life, have it also, as a benefaction from the End, now immanent. History and eschatology […] are the same thing" (Kermode 2000: 25).
3 Compare the industrial description of the rendering of the whale and Ishmael's comment at the unending labor of the whalers: "this is man-killing!" (Melville 2017: 317).
4 Compare Benjamin Barber's "Jihad vs. McWorld":
    Just beyond the horizon of current events lie two possible political futures—both bleak, neither democratic. The first is a retribalization of large swaths of humankind by war and bloodshed: a threatened Lebanonization of national states in which culture is pitted against culture, people against people, tribe against tribe—a Jihad in the name of a hundred narrowly conceived faiths against every kind of interdependence, every kind of artificial social cooperation and civic mutuality. The second is being borne in on us by the onrush of economic and ecological forces that demand integration and uniformity and that mesmerize the world with fast music, fast computers, and fast food—with MTV, Macintosh, and McDonald's, pressing nations into one commercially homogenous global network: one McWorld tied together by technology, ecology, communications, and commerce. (Barber 1992)
5 "Common to all true apocalyptic is a situation characterized by anomie, a loss of 'world,' or erosion of structures, psychic and cultural, with the consequence of nakedness to Being or immediacy to the Mystery" (Wilder 1971: 440).

References

Barber, Benjamin R. "Jihad vs. McWorld." *The Atlantic*, March1992, https://www.theatlantic.com/magazine/archive/1992/03/jihad-vs-mcworld/303882/.

Bell, Vereen M. *The Achievement of Cormac McCarthy*. Baton Rouge: Louisiana State University Press, 1988.

Elder, Charles. *The Grammar of Humanity: The Sense and Sources of the Imperative to Consciousness*. Manuscript in preparation, 2021.

Franzen, Jonathan. "What If We Stopped Pretending the Climate Apocalypse Can Be Stopped?" *The New Yorker*, September 8, 2019. www.newyorker.com, https://www.newyorker.com/culture/cultural-comment/what-if-we-stopped-pretending.

James, C.L.R. *Mariners, Renegades, and Castaways*. Hanover, NH: Dartmouth College Press, 2001.

Kermode, Frank. *The Sense of an Ending: Studies in the Theory of Fiction*. 2nd ed. Oxford: Oxford University Press, 2000.

Manjoo, Farhad. "Opinion: It's the End of California as We Know It." *The New York Times*, October 30, 2019. https://www.nytimes.com/2019/10/30/opinion/sunday/california-fires.html.

May, John R. *Toward a New Earth: Apocalypse in the American Novel*. Notre Dame, IN: University of Notre Dame Press, 1972.

McCarthy, Cormac. *Blood Meridian, Or the Evening Redness in the West*. New York: Vintage Books, 1985.

McCarthy, Cormac. *"Whales and Men"*. Cormac McCarthy Papers, Southwestern Writers Collection, The Wittliff Collections, Texas State University-San Marcos, n.d.

Melville, Herman. *Moby-Dick*, 3rd ed. Edited by Hershel Parker. New York: W.W. Norton & Company, 2017.

Melville, Herman, and Alfred Kazin. *Moby-Dick*. Boston: Houghton Mifflin, 1956.

Owens, Barcley. *Cormac McCarthy's Western Novels*. Tucson: University of Arizona Press, 2000.

Slaughter, Justin. "C. L. R. James in the Age of Climate Change." *Jacobin Magazine*, June2017. https://jacobinmag.com/2017/06/moby-dick-clr-james-mariners-renegades-castaways.

Wagar, W. Warren. *Terminal Visions: The Literature of Last Things*. Bloomington: Indiana University Press, 1982.

White, L. Michael. "Apocalyptic Literature in Judaism and Early Christianity", n.d. https://www.pbs.org/wgbh/pages/frontline/shows/apocalypse/primary/white.html. [Accessed November 5, 2018].

Wilder, Amos N. "The Rhetoric of Ancient and Modern Apocalyptic." *Interpretation*, 25(4), 1971, pp. 436–453. https://doi.org/10.1177/002096437102500402.

PART 3

The Ethics of the Environmental Apocalypse

10

"GUILTY?"/"NOT GUILTY?"

Kierkegaardian Reflections on *Carbon Ideologies*

Simon Thornton

Introduction

How can we make sense of the guilt that we feel upon recognizing that we are leaving our descendants a ruined world? In this chapter, I will explore this question as it arises in William T. Vollmann's *Carbon Ideologies* (2019). Addressed to future generations, whose nightmarish plight is foretold in jarring fragments throughout the work, *Carbon Ideologies* comprises an excruciating survey of the major carbon ideologies of our day—our more or less apparent justifications of nuclear, coal, fracking, and natural gas—aimed at providing an answer to the simple question: "What was the work for?" (Vollmann 2019a: 35, *passim.*). Accordingly, across its two volumes, *Carbon Ideologies* adopts a methodology which employs "induction to generalize from subjective case studies into analytical categories of the phenomenon under investigation" (Vollmann 2019a: x), as it transports us from the "abandoned homes of nuclear refugees" (Vollmann 2019a: 461) in Tomioka (Vollmann's "metonym for Fukushima" (Vollmann 2019a 471)) to the "poisoned" and "hollowed out" (Vollmann 2019a: 189) coal towns of West Virginia and the Emirati oil encampments for Bangladeshi and Pakistani guest workers (encampments that seemed to Vollmann "like alien fabrications on some low-gravity planet" [Vollmann 2019a: 534])—all with a view to "tell[ing] you [in the future] 'how it was' for us" (Vollmann 2019a: 599).

In my discussion, I shall be less concerned with the dizzying array of factoids that Vollmann gathers in *Carbon Ideologies* than the schizoid mood that pervades his presentation of them. Specifically, I will suggest that Vollmann's ironically manic apologia for carbon ideologies vividly manifests a complex and tension-filled form of guilt—his own—and the guilt he suspects lies in the hearts of others. My main aim in this chapter is to gain a technical grasp of Vollmann's difficult guilt complex (I prescind from assessing its appropriateness). I shall draw on some ideas presented

DOI: 10.4324/9781003189190-13

in Søren Kierkegaard's pseudonymous authorship in order to show that *Carbon Ideologies* is pervaded by a complex form of guilt that includes tragic and apocalyptic elements alongside its more familiar ethical one. My conclusion will be that one thing we can learn from Vollmann's massive work on the climate crisis is that the complex guilt we feel in recognizing that we are leaving our descendants a ruined world cannot be understood in ethical terms alone—it is also tragic and apocalyptic.

First, a note on guilt. "In ordinary usage," Michelle Kosch has remarked, "'guilt' is present only if the individual freely did (or omitted) something morally reprehensible (or morally required). One cannot ordinarily be guilty for something one did not do, or did not intend, or could not have … avoided" (Kosch 2006: 167). In other words, in ordinary usage, guilt is linked, unambiguously, to moral responsibility. By contrast, in his pseudonymous authorship, Kierkegaard carefully considers "attributions of types of wrongness … that are not unambiguously ethical in character" (Watts 2021: 170). In tragic guilt, the extent to which the hero is seen to be morally responsible for the tragic outcome ambiguously commingles with a sense that the outcome was unavoidable, a fate that the hero must suffer. And in apocalyptic guilt, the individual is gripped by an anxious feeling that despite their best efforts their whole way of life may yet be radically mistaken. I will show that *Carbon Ideologies* can be seen to exemplify each of these senses of guilt—moral, tragic, and apocalyptic.

"Appalachian Tragedy"

I shall begin by considering the tragic guilt of the Appalachian coal miners. In his exploration of coal ideology, Vollmann finds that in addition to the harmful contribution that coal mining has made to climate change, it has also ravaged the lives of countless West Virginian miners. There have been deadly disasters such as the Buffalo Creek flood, the Elk River chemical spill, and the Upper Big Branch accident. And these disasters have often produced deleterious long-term consequences, such as acidified tap water, polluted residential wells, and radioactive groundwater contamination (from fracking). Then there is the irreparable aesthetic damage that mining-related practices like mountaintop removal have inflicted on the Appalachian backcountry. And this is not to mention the fact that "more than 100,000 American coal miners were killed on the job in the 20th Century" (Vollmann 2019a: 52). Many others have suffered from mining-related illnesses, such as black lung, and still others struggle with debilitating psychological trauma arising from the brutality of the mines.

Sadly, it comes as no surprise to discover that what Vollmann calls "*the regulated community*," following local usage, "by which was meant the coal lords, chemical cowboys and kindred entities" (Vollmann 2019b: 120), gives little weight to the environmental and social costs of coal mining:

> After all, as a famous muckraker once concluded: *The business of a coal-operator was to buy his labor cheap, to turn out the maximum price to parties whose credit was*

satisfactory. If a concern was doing that, it was a successful concern; for anyone to men-
tion that it was making wrecks of the people who dug coal, was to be guilty of senti-
mentality and impertinence.

(Vollmann 2019b: 63)

This is one reason why the members of this regulated community—shameless
grotesques like Jim Justice and Don Blankenship—make up the select few for
whom Vollmann reserves unmixed moral disapprobation. By contrast, the mixture
of weary resignation and defensiveness—verging on defiance—that Vollmann
encountered in the long-suffering West Virginian miners themselves when invited
to comment on the environmental and social costs of coal is striking. It is shocking
to hear Vollmann report that one local environmentalist usually tells his neighbors
that "I'm a biologist, not an environmental scientist, because immediately people
say *tree hugger*—and can be aggressive" (Vollmann 2019b: 199). For an outsider, the
miners' often strident resistance to so much as acknowledging the harms of coal—
harms that fall most severely on the miners themselves—is deeply disturbing.

This seemingly self-destructive defensiveness becomes more intelligible when we
learn that according to a popular self-image—one encouraged by the *regulated
community*—the miners style themselves as "soldiers of coal" (Vollmann 2019b:
197). The miners are soldiers first of all because they have made a "sacrifice for the
country" (Vollmann 2019b: 60) by providing America with energy at great personal cost. After all, coal has long been "America's best friend" (Vollmann 2019b:
192)—it "kept the lights on" (Vollmann 2019b: 59). But, Vollmann adds, "not
only were the miners soldiers who fought and sometimes died so that we could
keep the lights on, they were also defenders of a people, region, and tradition now
under siege" (Vollmann 2019b: 199). For these West Virginians, coal was seen as
their unique "heritage" (Vollmann 2019b: 293). This is well exemplified by the
photographs of the surreal and extravagant West Virginia coal festival beauty
pageant and the imposing Soviet-style statues commemorating fallen miners included in *Carbon Ideologies*. The community-defining practice of coal mining must be
defended against "cultural devastation" (Vollmann 2019b: 216), even celebrated—
there is nothing else.

Perversely, the hardness of mines seemed only to vindicate the miners' self-
image as soldiers—martyrs—of coal, who suffered environmental devastation and
physical and psychological maiming for some greater good. As Vollmann percep-
tively reports, "it sometimes seemed … that West Virginians' elegiac memories of
coal got conflated with memories of what coal had taken from them" (Vollmann
2019b: 61). To this we may add that the West Virginians sometimes also seemed to
conflate the brutal fact that they had no choice but to enter the mines with a belief
that coal mining was their special vocation or calling. This cruel fact is well captured in the following vignette:

A woman down in Keystone (her name was Jean Battlo) once said to me:
"I'm an environmentalist, so I'm very concerned about the damage, and as a

> matter of fact I used to be all on that side and then I went to a meeting and heard a man say to me he was tryin' to raise his three kids and coal was his only hope." *His only hope!* How could that man have worried about climate change?
>
> *(Vollmann 2019a: 71)*

One can well imagine the resignation of this man, who, as a matter of bare survival, had no choice but to go into the mines. This thwarted man did not have the luxury of worrying about climate change. To him, Vollmann's gentle environmentalism must seem like an unjust attack. He had no choice! And it is small wonder that a man in this position of wretched powerlessness might be pushed to defiantly re-cast his painful fate into a heroic destiny. He is suffering for us! He is a soldier of coal.

In what follows, I want to suggest that the strange mixture of resignation and defiance that we can see in Vollmann's portrayal of the West Virginians can be fruitfully understood in terms of Kierkegaard's theory of tragic drama. Kierkegaard's (1987a, 1987b) most sustained analysis of tragic drama can be found in the pseudonymously authored *Either/Or*, under the auspices of a "fragmentary endeavor" titled "The Tragic in Ancient Drama Reflected in the Tragic in Modern Drama," authored by "A"—the fictional aesthete and dominant voice of the book's first part. According to A, ancient tragic drama essentially imitates action characterized by a specific kind of "collision" (Kierkegaard 1987a: 144) between an individual's assumed autonomy and their subjection to a seemingly inescapable fate in the production of an unwanted outcome. Consider the case of Antigone:

> [W]hen Antigone, in defiance of the King's injunction, decides to bury her brother, we see in this not so much a free act as fateful necessity, which visits the iniquities of the fathers upon the children. There is indeed enough freedom in it to enable us to love Antigone for her sisterly love, but in the inevitability of fate there is also a higher refrain, as it were, that encompasses not only Oedipus's life but also his whole family.
>
> *(Kierkegaard 1987a: 156)*

On A's assessment, Antigone's "defiance" lies somewhere between suffering and action: whilst—given her bad inheritance—a transgression of some kind was unavoidable, she nonetheless, and however sorrowfully, chose *this* transgression ("defiance") rather than sully her brother's honor. Or consider the case of Oedipus himself. Whilst Oedipus did indeed perform the crimes of patricide and incest, these crimes were perpetrated inadvertently. He was ignorant of his true circumstance and parentage when he killed a man and later married a woman, and he strove for his whole life to avoid fulfilling the prophecy that the events of patricide and incest would befall him vis-à-vis his bloodline. In both cases, the tragic hero's agency manifests as "something intermediate between action and suffering" (Kierkegaard 1987a: 144): substantial determinants such as lineage, society, and misfortune ambiguously commingle with autonomy in the tragic hero's action, making it appear both free and not free.

It is worth noting right away that A's characterization of the ambiguity of tragic action resonates somewhat with the situation of the West Virginian miners. As Vollmann emphasizes, the miners have practically no choice but to go into the mines. If for nothing else, they must do it for the sake of their families. But the effects of coal mining are devastating for the miners themselves, for their communities, and for the world. The miners know this; they have felt it. And they have come to terms with this unfortunate fate by thinking of themselves as soldiers who toil in the mines for the good of the country. Emboldened by this self-image, the miners are prepared to fight to defend their own community and its culture—even if it kills them. In this way, for these West Virginians, mining seems at once free and not free, both autonomous and determined, between action and suffering. Although they have practically no choice but to mine—and to suffer the tragic consequences of this—it is nonetheless a life that they have chosen to embrace and defend.

Kierkegaard's aesthete believes that modern theories of tragedy have tended to elide the ambiguity essential to the tragic hero's action. These theories portray the hero as "stand[ing] and fall[ing] on his own deeds," "load[ing] his whole life upon his shoulders, and mak[ing] him accountable for everything" (Kierkegaard 1987a: 144). As Jacob Howland has noted, this way of thinking about tragedy reflects "modernity's emphasis on freedom of choice" which "renders contingent on personal preference fundamental elements of individual identity that were previously understood to be ineluctably given … thereby mak[ing] particular individuals responsible for things that had previously been taken to be beyond individual control" (Howland 2017: 116). In other words, these theories transform the hero's properly "aesthetic [or tragic] guilt into ethical guilt" (Kierkegaard 1987a: 144). According to this harsh view, we would judge the miners unequivocally to have brought devastation upon themselves—and upon others. And we would hold them morally responsible for their own wretchedness and ignorance, without reservation.

Against this, A insists that just as the action imitated in ancient tragedy is "something intermediate between action and the suffering, so also is the guilt" (Kierkegaard 1987a: 144). But how can we make sense of this? Surely one is either guilty or not guilty—there is no "between". Daniel Watts has noted in this context that, generally speaking,

> Tragic guilt involves attributions of types of wrongness (*harmartia*) that are not unambiguously ethical in character. That is reflected in the way such wrongness is apt to be described in terms other than the narrowly moral: in terms of one's having become tainted or polluted, for example, or having fallen short of the mark, or of one's life having become out of joint or dissonant.
>
> *(Watts 2021: 170)*

In a separate context, Northrop Frye has shown that in "low mimetic" or "domestic" tragedy, the tragic hero can take the form of a *pharmakos* or scapegoat. In this role, the scapegoat is neither innocent nor guilty. He is innocent in the sense that what happens to him is far greater than anything he has done provokes, like the mountaineer whose shout brings down an avalanche. He is guilty in the sense that he is a member of a guilty society, or living in a world where such injustices are an inescapable part of existence.

(Frye 1957: 41)

What these descriptions elicit is a sense of one's being in the wrong—and, thus, of being guilty—which is not unambiguously the result of one's own autonomous, intentional action. Notably, we can see our West Virginian reflected in Frye's image of the mountaineer in as much as the environmental devastation visited upon the former also seems "far greater than anything he has done provokes." We can also note that the lives of many West Virginians are tainted or polluted in a literal sense. Consider the West Virginian waitress who would not drink her local water. When asked why, Vollmann reports that "instead of explaining the root cause—acid drainage from coal mines, which perhaps she didn't know—she replied, 'Just everything falling apart one stage at a time.' Was she angry or merely resigned? I'll never know" (Vollmann 2019b: 382). It is striking that in this exchange the waitress does not think that something can and must be done about her increasingly polluted world; it is just the way things are, another thing that she just has to put up with, as though deserved.

Tragic guilt, then, involves attributions of types of wrongness that are not unambiguously ethical in character. Yet, this is not to say that the guilt does not retain any moral or ethical aspect. Rather, on A's view tragic guilt must reflect the genuine ambiguity of tragic action: "If the individual has no guilt whatever," A explains, "the tragic interest is annulled for in that case the tragic collision is enervated. On the other hand, if he has absolute guilt, he no longer interests us tragically" (Kierkegaard 1987a: 144). In other words, the tragic downfall of heroes such as Antigone and Oedipus cannot be attributed to fate alone, for in that case the hero would experience no guilt but only victimhood. Nor, however, can the events that have transpired be attributed solely to the hero's free agency, in which case moral guilt could be ascribed to them. Rather, for A, tragic guilt "involves the contradiction of being guilt and not being guilt" (Kierkegaard 1987a: 150). Whilst the tragic hero cannot be fully exculpated by reference to fate, since they did freely contribute to their own downfall, they can just as little be held fully responsible for the tragic events that have come to pass, as those events were seemingly inevitable and unavoidable. Thus, the tragic hero oscillates interminably between the two judgments "guilty" and "not guilty." This is the "dialectic implicit in the concept: tragic guilt" (Kierkegaard 1987a: 151).[1]

We can readily understand Vollmann when he admits that "I could not help but pity [the coal miners'] situation" (Vollmann 2019b: 204), which was "abused by corporations with shameless ingenuity" (Vollmann 2019b: 189). But it is hard for us to ignore the fact that, at some level, these people freely embraced the coal ideology, and fought for it as its foot-soldiers. Perhaps the most striking symbol of this attitude is the battered pickup truck photographed by Vollmann with a license plate frame that reads "Friends of Coal" and "Coal keeps the lights on" and a bumper sticker that declares "If you don't like coal don't use electricity" (Vollmann 2019b: 209). How can such arrogant defiance not provoke moral revulsion? To call the situation of the West Virginian miners tragic is to view them as, in a sense, morally culpable for their own wretched situation. Yet, as with all tragedy—at least on Kierkegaard's view—the extent to which we are moved to find the miners morally responsible for their complicity in perpetuating coal ideology is matched by an equally forceful feeling of deep sympathy for the miners' plight. The extreme economic and cultural dependence of these West Virginians on the coal industry has been pitilessly exploited by that industry, leaving the miners feeling as though they have no choice but to cling to a dying industry that is destroying them. Tragically, the West Virginians are both guilty and not guilty.

"Back When I Was Alive"

The bulk of the second part of Kierkegaard's *Either/Or* consists of two lengthy letters sent to A by his *soi-disant* mentor, Judge William. These letters excoriate A's aesthetic endeavors as presented in *Either/Or*, Part One, as symptomatic of A's overall emotional detachment, moral evasiveness, and adolescent self-indulgence. The Judge clearly suspects that A's insistence on the aesthetic ambiguity of tragic guilt all too conveniently serves his implied policy of avoiding responsibility for his own life. By contrast, the Judge "assures" A that

> if my life through no fault of my own were so interwoven with sorrows and sufferings that I could call myself the greatest tragic hero, could divert myself with my affliction and shock the world by naming it, my choice is made: I strip myself of the hero's garb and the pathos of tragedy; I am not the tormented one who can be proud of his sufferings; I am the humbled one who feels my offense.
>
> *(Kierkegaard 1987b: 237)*

As we shall see, the ethical counterpoint championed here by the Judge is inspired by a Christian notion of hereditary sin: it rests on the thought that I should take full responsibility not just for what I have done, but for who I am. That is why the Judge passes on to a little homily composed by a (fictional) Jylland pastor tellingly entitled "The Upbuilding That Lies in the Thought That in Relation to God We Are Always in the Wrong." This homily is supposed to give a religious *imprimatur* for the Judge's own moralistic sermonizing. However, most commentators agree that the sermon does not so much reinforce the Judge's hyper-moralized worldview as "explode" it by revealing *inter alia* a qualitative difference between the Judge's wholly moral conception of guilt and (what I shall term) an "apocalyptic sense of guilt" sketched out in the sermon.[2]

The notion of apocalyptic guilt may be introduced by noting that it is oriented by a revelatory "judgment of the world" at the "end of history" (Taubes 2009: 47).

Moreover, this apocalyptic sensibility prophecies that "this world is in [a] state of error" and that "we have not just begun to err, but are constantly doing so, because we have always been in error" (Taubes 2009: 6). The apocalyptic judgement of the world is revelatory insofar as it may diverge radically from every worldly standard of justice. And it is the anxiety-inducing tension generated by the possibility of an ultimate perspective of apocalyptic judgment that diverges radically from our own merely relative and worldly moral perspectives which forms the core of what I shall call "apocalyptic guilt." I will claim that this apocalyptic sense of guilt is sketched out in *Either/Or* and other pseudonymous writings, and I will suggest that it is exemplified in *Carbon Ideologies*—albeit in less explicitly Christian terms.

In his sermon, which is suggestively labelled "Ultimatum," Kierkegaard's pseudonymous pastor points towards an apocalyptic sense of guilt obliquely through an extended meditation on Luke 19:41, in which "the Savior went up to Jerusalem and wept over the great city that did not know what was best for it" (Kierkegaard 1987b: 39). The pastor focuses at first on bringing out just how provocative God's "destruction" (Kierkegaard 1987b: 343) of Jerusalem should seem to his congregation by plaintively asking:

> "Must the righteous, then, suffer with the unrighteous? Is this the zealousness of God—to visit the sins of the fathers on the children to the third and fourth generation, so that he does not punish the fathers but the children?"
>
> *(Kierkegaard 1987b: 324)*

Now, the pastor acknowledges that such philosophically unfashionable questions of theodicy may leave us moderns cold: "the shriek of anxiety from those days sounds very faint to us now" (Kierkegaard 1987b: 343). After all, as A puts it:

> the dialectic that connects the iniquity of kindred or of family to the individual subject in such a way that this one not only suffers under it … but also bears the guilt, participates in it—this guilt is alien to us [moderns], contains nothing constraining for us.
>
> *(Kierkegaard 1987a: 218)*

And yet, against this "cozy conclusion" (Kierkegaard 1987b: 343), the pastor ruminates:

> If it happened once in the world that the human condition was essentially different from what it otherwise is what assurance is there that it cannot be repeated, what assurance that that was not true and what ordinarily occurs is the untrue.
>
> *(Kierkegaard 1987b: 343)*

Can we be sure that we now have nothing to learn from God's zealous destruction of Jerusalem, the city that had lost its way? Should the parable not still cause us anxiety?

It must be acknowledged that these ruminations are somewhat obscure. In order to see how they may point towards an apocalyptic sense of guilt, we should ask: what "cozy conclusion" is the pastor supposed to be unsettling? Thinking in ethical terms, the Judge believes that the pastor is challenging the "cozy conclusion," supposedly exemplified in A's discussion of tragedy, that we cannot be held morally responsible for what we could not have really avoided. Whereas A emphasizes the "soothing" aesthetic ambiguity of tragic action and tragic guilt, Judge William believes that we should "repent" (that is, accept guilt) for even those "painful things" (Kierkegaard 1987b: 216) that we could not control, such as one's unfortunate inheritance. After all, they are still constitutive of who we are:

> Only when I choose myself as guilty do I absolutely choose myself, if I am at all to choose myself absolutely in such a way that it is not identical with creating myself. And even though it was the father's guilt that was passed on to the son by inheritance, he repents of this, too, for only in this way can he choose himself, choose himself absolutely.
>
> *(Kierkegaard 1987b: 216–217)*

In short, Judge William views the pastor as an ally in his dispute with A concerning the morality of hereditary sin. The pastor's sermon is here seen to unsettle A's "soothing" and "cozy conclusion" concerning the ambiguity of tragic guilt.

I noted above that Vollmann frames *Carbon Ideologies* ironically as a strange kind of sarcastic apology to future generations. We can now add that this framing device is well illuminated in terms of the Judge's moralistic take on the idea of hereditary sin just mentioned. In fact, at the beginning of the second volume of *Carbon Ideologies* Vollmann spins a parable that bears a striking resemblance to the events of Luke 19:41. "Once upon a time," he writes,

> There was something dangerous that could not be seen, felt, heard or smelled. ("Because it is invisible …," sighed my Japanese taxi drivers.) Making use of its associated fuel had been convenient, but terribly mistaken. The best plan would have been to get away from this nearly unknowable thing, but such a course of action appeared so utterly inconvenient that we preferred to continue on as before, which might entail killing our children.
>
> *(Vollmann 2019b: 3)*

At length, Vollmann draws our attention to the fact that we know that we are living beyond our planetary means, that a "cost in greenhouse emissions remain[s] unpaid" (Vollmann 2019b: 297). But we stubbornly avoid or deny this knowledge because it is inconvenient. Thus, commenting on our penchant for leisure travel, Vollmann explains:

> You see, in our time we took it for granted that as long as we (and you) could pay, we had the right to be carried wherever we pleased. That our non-monetized

costs would be payable by you, whose various potential distresses never troubled our sleep, made travel all the better.

(Vollmann 2019a: 145)

And, rehearsing one typical way that we evaluate our energy sources, Vollmann summarizes:

Our cost–benefit analysis of nuclear power and competing carbon ideologies went as follows: Get subsidies wherever possible, calculate the ratio of immediate income *versus* immediate costs, such as employee compensation, and expel from the equation such future costs as global warming, air pollution and nuclear waste disposal. Again, we figured that *you* could pay for all that.

(Vollmann 2019a: 239)

Intellectually, we know that our descendants will suffer the consequences of our complacent luxury. But this knowledge is practically impotent. Cruelly inverting the painful situation of hereditary sin to our benefit, Vollmann shows with bracing candor that we are living as though it is *right* that future generations should bear the guilt of our iniquity. The hypocrisy only deepens our shame. Vollmann thus comes to feel that we will be rightly convicted under the harsh—indeed, radio-active—light of the future for the complacency and carelessness in which we live. Is our situation all that different from that of ancient Jerusalem, the city in a corrupted "state of error"?

In short, when Vollmann places our way of life before the tribunal of those in the future, the aesthetic ambiguities of our actions seem to pale to insignificance, and our unambiguous moral guilt intensifies. In this way, we may observe a striking affinity between Vollmann's treatment of carbon ideologies and the Jylland pastor's treatment of hereditary sin—at least as Judge William understands it. However, I now want to suggest that the affinity between Vollmann and the pastor runs deeper than Judge William's moralistic worldview can recognize—beyond moral guilt to apocalyptic guilt. In order to bring this deeper affinity to the surface, we may note that, for Vollmann, it is not just morally painful to attempt to justify carbon ideologies before those imagined "inhabitants of a hotter, more dangerous and biologically diminished planet than the one in which [Vollmann] lived" (Vollmann 2019a: 2), it is also distressingly difficult to so much as imagine how those future ones in the "hot dark future[,] ek[ing] out what Pasolini called *remnants of a life that mirrors its shadow*" (Vollmann 2019b: 656) may judge us.

Vollmann admits, in this context, that "When I think of you in the future reading this, my error is to imagine that you are similar to me" (Vollmann 2019b: 598). And he acknowledges that his imagination may be further distorted by a fervent hope that *Carbon Ideologies* will persuade those in the future—those who have not yet "starv[ed] or chok[ed] to death" (Vollmann 2019b: 555)—to view Vollmann's own generation sympathetically. Thus, he writes:

Being nurtured and pampered by electric power, I owned full leisure to emote, consider your point of view and even feel goodwill toward you. I wish to understand you, and to help you understand me—as if that could somehow distract, console or at least enlighten you about the hot dark world in which you dwell. If you could end up saying, "well, yes, we might have made the same mistakes as you, if we'd been lucky enough to live when you did," I'd feel that *Carbon Ideologies* had accomplished its purpose. How you judge us can mean nothing to us who are dead, but to *you* it might mean something, to accept that we were not all monsters; and forgiveness benefits the forgivers, so why wouldn't I prefer you to call our doings mistakes instead of crimes?

(Vollmann 2019b: 598)

Unfortunately, Vollmann finds that he must immediately pour cold water on any hopes he may hold in this direction, as "weary common sense, and [Vollmann's] experience in war zones and other desperate places, drags down these narcissistic expectations" (Vollmann 2019b: 598).

Realistically, Vollmann surmises that "most likely, you [in the future] are a hard, angry person" (Vollmann 2019b: 598), and he conjectures that

By my standard you must be somewhat uneducated, since we expended most of the magic that would have kept your lights on. How well can you read? For that matter, how well can you see? Do spectacles still exist? Beset by floods, droughts, diseases and insect plagues, unable to jet around the world as I did, but probably aware nonetheless that humanity's habitable islands keep shrinking; engaged in wars about goods, and united merely in tribal hatred (your strongman against theirs), radiation-tinted, fearing for your children in the face of multiplying perils, how can you feel anything better than impatient contempt for my daughter and me? How could you possibly possess the luxury of worrying about *your* future? You reduce demand by murdering your enemies. Like me, you hide from reality (in your case, from sunshine) in a cave. Having skimmed bits of this, you might understand some of our why and how, but does that appease or merely inflame your hatred?

(Vollmann 2019b: 598)

Vollmann's dark musings here elicit a sense of extreme normative disorientation that lies at the core of apocalyptic guilt. He presumes that he will be condemned by the future, but he is unsure by what standards and how harshly he will be judged. It is as though in considering our human future he faces an inscrutable and vengeful God. At least we can say he suspects that whatever moral or spiritual standards those in the future uphold—if any—they will likely depart radically from his own. How can mutual understanding be achieved amidst the turbulence of such disorientation?

To see how Vollmann's musings here resonate with an apocalyptic theme in Kierkegaard's pseudonymous authorship, consider the following. In *Philosophical*

Fragments, the pseudonymous humorist Johannes Climacus argues that a definitive feature of Christianity must be the belief that "essential truth is beyond the reach of human understanding" (Watts 2016: 85). Specifically, Climacus suggests that the Christian concept of "sin" must involve the thought that not only is the sinner in "untruth," lacking proper orientation towards God, but also that they lack "the condition" for recognizing that fact: the sinner, that is to say, is radically disoriented in relation to God.[3] Perhaps another "cozy conclusion" that the pastor seeks to unsettle in his sermon, then, is precisely the thought, to which Judge William notably subscribes, that "God puts the stamp of the eternal on all [the Judge's] feelings and thoughts as well as contemporary social arrangements" (Perkins 1995: 224). Plausibly, this is the force of the pastor's comment that "if it happened once in the world that the human condition was essentially different from what it otherwise is what assurance is there that it cannot be repeated, what assurance that that was not the true and what ordinarily occurs is the untrue." What if we, like the citizens of ancient Jerusalem, are so radically mired in untruth that we lack even the "condition" even to recognize our error? Maybe we do not know what is best for us? Faced with this thought, it would seem that no normative perspective we take up may be radical enough to grasp the wrongness of our lives as a whole, or to encompass how far we have strayed into error. This, I will suggest, is the source of the anxiety involved in apocalyptic guilt.

Notably, the apocalyptic anxiety that this last thought may provoke can be found in abundance throughout *Carbon Ideologies*. Early in the first volume, Vollmann laments:

> In much of *Carbon Ideologies* it becomes my sad task to multiply discouragements and bewilderments: We destroyed ourselves not simply because we burned too much steam, coal or heavy commercial fuel oil in our power plants, but also because we followed sound agricultural practices, and because we made beautiful things of all sorts. We composted garbage, which was surely a good deed, and in the case of rain we wisely carried our nylon raincoats. We published books, which kept *me* employed. So many customary behaviors imperiled us that the most radical solutions might not have been radical enough.
>
> *(Vollmann 2019a: 133)*

And, again, relating a familiar dilemma concerning household appliances in the second volume, Vollmann observes:

> If the householder who owned [an aging refrigerator] cared about "some ecosystem somewhere," she might strive to do good by nursing it as long as possible, delaying the time when climate-changing CFCs and HCFCs from its dismantled guts rose up into the atmosphere, and likewise putting off the necessity to smelt more steel and produce more glass for the sake of a new toy. Oh, [the householder] meant well—but she might be doing harm. And if the

intuitively laudable approach of retaining one's workhorses, living frugally, declining to escalate demand, were *wrong*, who could blame her for throwing up her hands? … [K]eeping an old fridge running was the right thing to do in 1956 and the wrong thing in 1992.

(Vollmann 2019b: 66)

Read in one way, these considerations seem to go beyond the shameful thought that we expect the future to bear the consequences of our carelessness. In addition, we feel the future may condemn us not just for our blatant criminal negligence, but also for our seemingly well-meaning and innocent actions, which turn out to be ultimately misguided and harmful. The anxiety and the frustration here stem from the shadow of doubt that is suddenly cast on the justice of our lives as a whole: it now feels as though anything we do—even those things that are, for all the world, wholesome and inconsequential, and especially those things that we do with ecological concerns in mind—may ultimately prove to have been badly, reprehensibly, mistaken.

It is understandable that, in the face of this thought, Vollmann's friend became dejected and resigned. How can she possibly predict the long-term consequences of her actions in this way? "Who could blame her for throwing up her hands?" She is doing what she can. Unfortunately, Kierkegaard's fictional pastor would suggest that such resignation can never be final:

Was it to your comfort that you said "One does what one can"? Was not the real reason for your unrest that you did not know for sure how much one can do, that it seems to you to be so infinitely much at one moment, and at the next moment so very little? Was not your anxiety so painful because you could not penetrate your consciousness, because the more earnestly, the more fervently you wished to act, the more dreadful became the duplexity in which you found yourself?

(Kierkegaard 1987b: 345–346)

Although it is tempting to try to comfort oneself by thinking of oneself as innocent for such blindness before God (or the future) on the basis that they would surely not hold you responsible for "human frailty" (Kierkegaard 1987b: 97)—the finitude of human understanding and agency—this consoling thought will never quite neutralize the distressing anxiety that one could nonetheless be living in the wrong. Vigilius Haufniensis, the fictional author of Kierkegaard's *The Concept of Anxiety*, captures this thought succinctly when he talks of "anxiety at its most extreme point where it seems as if the individual has become guilty, if not yet as guilt" (Kierkegaard 1980: 98). It is as though through the anxiety over one's finitude—specifically, the possibility that from some absolute perspective one's life may be wholly in the wrong—one finds oneself verging on guilt.

Vollmann's stance in *Carbon Ideologies* exemplifies this anxiety. Straining to imagine how those in the future may judge his life, he is gripped by the thought that no assessment he makes on their behalf may be radical enough to do justice to the wrongness in which he might be living. But this limitation is not consoling for Vollmann. Rather, he goes on laboriously "groping for the sin" (Kierkegaard 1988: 429) as Frater Taciturnus, another of Kierkegaard's pseudonyms, puts it. The result is that Vollmann "cannot begin to repent, because what it is he ought to repent of seems to be undecided as yet; and he cannot find rest in repentance" (Kierkegaard 1988: 451). He is struck by a nagging anxiety that even though he tries to live conscientiously with environmental concerns in mind, acknowledging guilt and taking responsibility where he can, this conscientiousness may ultimately be radically outstripped by the wrongness of his way of living as a whole, leaving him even guiltier than he can imagine. This sense of wrongness lies just on the outer limits of his normative world: he can, in a sense, feel it, but he can't grasp it and incorporate it into his life. Just as Kierkegaard's various pseudonyms explain, Vollmann is torn between a distant but ominously unsettling perspective in which his whole way of life may be cast in an unfamiliar but unmistakably and uncompromisingly incriminating light, and a more familiar and consoling perspective which takes account of his human limitations and in which at least some of his actions must be judged to be harmless—and—even commendable.

Vollmann's Guilt Complex

In this chapter, I have been developing the idea (with the aid of Kierkegaard) that Vollmann's work on the climate crisis sheds light on our sense of guilt for the ruined environment we will leave to our descendants. Although Vollmann is clear that we should be morally ashamed of our way of living, his "subjective case studies" also suggest that this moral guilt is often conditioned by tragic and apocalyptic dimensions, making it a complex, multifaceted phenomenon. In some cases, our complicity in the climate catastrophe is fatefully conditioned by substantial determinants such as lineage, society, and misfortune, which give our guilt a tragic dimension. More generally, Vollmann shows that our guilt may become charged with a radical, apocalyptic anxiety concerning our form of life as a whole, as viewed from the "ultimate" perspective of the last humans. One of the finest features of *Carbon Ideologies* is the way that it exemplifies the difficult thought that the moral aspect of our guilt for the climate catastrophe stands in a fraught, tension-filled, but mutually conditioning relation with tragic and apocalyptic aspects, thereby making it a guilt complex.

Notes

1 See Frye (1957: 209–211) for a similar point made in a different register.
2 See especially Howland (2017).
3 See Kierkegaard (1985: 14–18).

References

Frye, N. (1957). *Anatomy of Criticism: Four Essays.* Princeton, NJ: Princeton University Press.

Howland, J. (2017). "The Explosive Maieutics of Kierkegaard's *Either/Or.*" *The Review of Metaphysics*, 71(1): 107–135. https://www.jstor.org/stable/44807012.

Kosch, M. (2006). *Freedom and Reason in Kant, Schelling and Kierkegaard.* Oxford: Clarendon Press.

Kierkegaard, S. (1980). *The Concept of Anxiety.* Edited and translated by R. Thomte. Princeton, NJ: Princeton University Press.

Kierkegaard, S. (1985). *Philosophical Fragments.* Edited and Translated by H.V. Hong and E. H. Hong. Princeton, NJ: Princeton University Press.

Kierkegaard, S. (1987a). *Either/Or, Part One.* Edited and Translated by H.V. Hong and E.H. Hong. Princeton, NJ: Princeton University Press.

Kierkegaard, S. (1987b). *Either/Or, Part Two.* Edited and Translated by H.V. Hong and E.H. Hong. Princeton, NJ: Princeton University Press.

Kierkegaard, S. (1988). *Stages on Life's Way.* Edited and translated by H.V. Hong and E.H. Hong. Princeton, NJ: Princeton University Press.

Perkins, R.L. (1995). "Either/Or / Or: Giving the Parson His Due." In S. Kierkegaard, *Either/Or*, ed. R.L. Perkins. Macon, GA: Mercer University Press, 207–231.

Taubes, J. (2009). *Occidental Eschatology.* Translated by D. Ramotko. Stanford, CA: Stanford University Press.

Vollmann, W.T. (2019a). *Carbon Ideologies, Volume 1: No Immediate Danger.* New York: Penguin.

Vollmann, W.T. (2019b). *Carbon Ideologies, Volume 2: No Good Alternative.* New York: Penguin.

Watts, D. (2016). "Kierkegaard and the Limits of Thought." *Hegel Bulletin*, 39(1): 82–105. https://doi.org/10.1017/hgl.2016.36.

Watts, D. (2021). "Between Action and Suffering: Kierkegaard on Ambiguous Guilt." *Journal of Humanistic Ideology*, 11(1): 169–197. http://www.socio.humanistica.ro/Public/2021v11n1/IJHI_vol.%20XI,%20no.1,%202021.pdf.

11

APOCALYPTIC TIME AND THE ETHICS OF HUMAN EXTINCTION

Stefan Skrimshire

How are our moral responsibilities to present and future generations affected by the possibility of near-term human extinction? And what does the act of imagining such a scenario tell us about our moral intuitions? This chapter critically explores these questions by bringing together two types of enquiry. The first considers the philosophical question of whether doomsday scenarios incentivise or ought to incentivise types of moral behaviour, and what this tells us about ourselves. 'Doomsday' here indicates the abrupt and definitive end of all human life. It is a thought experiment traditionally served by fictional accounts of asteroid strike, devastating pandemics, or other global catastrophes but which also raises pertinent questions closer to our real-life climate and ecological emergencies. The second enquiry is theological and asks whether there is anything about ancient Jewish and Christian apocalypses that offers an alternative insight into the moral imagination of the end of humanity. 'Apocalypse' is here taken to refer to the 'distinctive conceptual structure' (Collins 2014: 7) of texts that emerged in the intertestamental period and whose biblical canonical examples are the Book of Daniel in the Hebrew Bible and the Book of Revelation in the New Testament. Amongst their distinguishing features are the disclosure, via dream or vision, of the hidden purposes and destiny of the cosmos. In some cases, these revelations include an eschatological cosmic drama: a time of final judgement and the end of creation. Whilst apocalypses do not describe human extinction explicitly, they do narrate the cosmic and historical finitude of the human world.

Whilst both types of enquiry have enjoyed much discussion within their respective disciplines, they are, surprisingly, rarely considered together. Part of the intention of this chapter is thus to allow philosophical and theological interpretations of the apocalyptic imagination to speak to each other. Are there elements of the structure of biblical apocalypses that could shed light on philosophical doomsday arguments? And in what ways might such insights inform moral discussion

DOI: 10.4324/9781003189190-14

about catastrophic climate change and other existential threats to human and other-than-human life on Earth?

What's Bad about Doomsday?

What sort of wrong, or harm, does human extinction represent? Amongst the surprisingly small number of philosophical projects devoted to this question,[1] the most intriguing comes from the moral philosopher Sam Scheffler's (2013) Tanner Lectures: *Death and the After Life*. Scheffler considers what it is that makes the prospect of the end of human life so disturbing. He proposes the following: (1) it is reasonable to expect that the anticipation of an imminent end to all human existence would render much of what we do in the present meaningless or at least greatly diminished in value; (2) if we accept this claim, this reveals important things about human value. More specifically, it reveals an implied relationship between ethics and temporality, because it means that the value that we ascribe to projects in our lives is dependent upon some notion of their continuation, their afterlife. By 'afterlife', Scheffler means the continuation of human lives, in the plural, after I die. He is *not* referring to the religious belief in personal immortality. In fact, Scheffler wants to argue that belief in a collective afterlife on Earth is so important to people that it ought to dissuade them from seeking consolation in the personal afterlife of religious traditions. So his argument might be seen to provide grounds for atheism. But this is a separate issue from the following conclusion which he draws, and which is more interesting to my study: (3) the fact that humans care so much about their collective afterlife (in fact more, he assumes, than they do about their personal afterlife) is good news for those who are concerned about moral motivation to combat such species-threatening phenomena as climate change. It is good news because it means that humans are probably a lot less selfish and individualistic than we perhaps assume them to be. And it is reasonable to expect them to do everything in their power to prevent an event that would precipitate their untimely extinction.

Let's look at how Scheffler arrives at this conclusion. We are asked to imagine a scenario in which we knew that, 30 days after our own natural death, the Earth would be utterly destroyed. With this knowledge, would we be able to conduct ourselves in the same way that we do now? That is, assuming, with reasonable certainty, that the world will continue to exist in a more or less habitable state indefinitely into the future? I will come back to the contemporary significance of an epistemological variant of this, in which there is some *uncertainty* in our knowledge about whether and for how long the human race will continue. For now, it is important to note that this is explicitly *not* Scheffler's concern. His scenario assumes unequivocal epistemic certainty; it is beyond doubt that there will be no world 30 years after our personal life ends. The point of the scenario is that it forces us to consider whether our present projects, values, and commitments are dependent upon temporal continuity. And Scheffler concludes that, for a significant amount of them, this is absolutely right. We would find less value in, or perhaps give up entirely, such activities as researching a cure for cancer, having children, or campaigning for the radical reduction of global carbon emissions. So this is one claim he thinks one can make about the ethics of human extinction. It is not simply the fact that it is very sad for us to contemplate a world in which humans no longer exist. Knowledge about the end of the world would also lead to a kind of nihilism. Our projects would have no value once they are divorced from some guarantee of their continuation by humans like us, existing indefinitely into the future.

Scheffler demonstrates his point by appealing to the scenario sketched in P.D. James' (1992) novel *Children of Men*. That story presents a different scenario to the sudden, cataclysmic perspective of a nuclear attack or an ineradicable virus by focussing more on the status and quality of the last humans. In James' novel, the world's human population becomes suddenly infertile, and societies must adjust to being the last generation. The characters become apathetic and despairing. They struggle to know what is worth fighting for and which projects are worth continuing. People lose the motivations they once had for seeking pleasure. More so in the novel than in Alfonso Cuaron's (2006) film adaptation, this cultural anomie is not only traumatic at the personal level. The absence of a future also leads to total social collapse, a capitulation to fascism, and the nihilistic pursuit of rituals of mass suicide. If we are convinced by James' harrowing scenarios of collapse, Scheffler thinks, then his moral claim is vindicated. The worth of most human projects is only coherent alongside the belief in a continuation of the species. Therefore, we ought to care for the continuation of the human species above all things. Note also that Scheffler believes that this type of valuing is not a consequentialist calculation, since prolonged human existence will produce more suffering as well as more pleasure, and not necessarily/obviously a net increase in the latter. So the surprising implication of this discovery, he thinks, is that morality may be both less consequentialist and less egoist than we might otherwise have assumed. That there will be a community of other humans with whom I bear no immediate affective or reciprocal relationship seems to matter to me prior to, or in spite of, any utilitarian calculations about maximising happiness.

Philosophical criticisms of Scheffler have been many and varied.[2] Though it is not my focus here, one of his arguments' most serious weaknesses is its anthropocentrism. Scheffler assumes that only the continuation of *human* projects gives human projects meaning. Thus, he disregards the fact that for many people their actions have value in so far as they contribute to the continuation of life on Earth whether human or more-than-human. And even were this not generally the case, one might still argue that it ought to be the case. One ought to cultivate a sense of the continued value of life on Earth even in the light of strong evidence that humans will not be around for long to reap its benefits.

Important though such criticism is, it is not my focus here. I want to explore the suggestion that an entirely opposite intuition about human behaviour to that proposed by Scheffler is both possible and demonstrable. Plenty of our activities do *not* rely upon an afterlife to sustain themselves at all. For instance, why not imagine that, after some initial shock and a sense of lost (future) opportunities, even the last generation might be able to find meaning in most of its previous pursuits? Perhaps the last people would find gratitude for their remaining lives in ways that gave their actions different value from before. This was a possibility imagined multiple times in Olaf Stapledon's (1930) unique and bizarre science-fictional 'deep history' novel, *First and Last Men*, in which successive species of humans across billions of years of the history of the Earth contemplate their own inevitable demise, without thereby losing a sense of moral and existential commitment to their present existences. It is also at the heart of Susan Wolf's critique of Scheffler, which is included in the published version of his lectures. Wolf argues that the extraordinary ability for humans to find value in caring for one another and continuing a good society even in scenarios of impending doom is both conceivable and has some historical evidence to back it up. Moreover, even if we find this scenario implausible – perhaps we think that few of the people we know are that stoic and selfless, and we would imagine ourselves as falling quite easily into despair, self-destruction, or total denial – we can still argue that the correct response to imminent destruction *ought* to be along those lines. We ought not to suppose that the threat of imminent extinction would render our present commitments meaningless.

A related critique, and important for the sake of my thesis, questions Scheffler's assumption that it is only an *imminent* doomsday, anticipated within our own lifetime, that causes anxiety and troubles our sense of meaning in the present. Such anxiety is not, he thinks, prompted by the evidence that in the very long run the extinction of human life (at least as we know it) is guaranteed via the heat death of the solar system, scheduled for around 4.5 billion years in the future. And yet, perhaps many of us *do* reflect on this fact of the finitude of human existence from time to time, and as a result have 'put things into perspective'. Perhaps contemplating the inevitably of human extinction has allowed us to let go of some of our anthropocentric delusions about human exceptionality, or the permanence of our human projects. We might agree that these are not the same *kinds* of concerns as those of imminent loss considered by Scheffler. But Wolf's point is that it isn't as clear as Scheffler thinks that one type of worry about death (the proximal type) is 'rational' whereas another (the long-term type) is not. *Some people* do in fact worry about the meaning of their existence because of the long future scenario as well as any short-term scenarios.

Why is this objection important for my argument? Because Wolf's question can be asked in the other direction: since it is clearly the case for many people that the long-term doomsday scenario does *not* seem to cause them to cease caring about present commitments, why then should a short-term doomsday cause them to cease caring? Scheffler (2013) thinks that it is enough to note that people tend *not* to suffer existential crises from the knowledge of our certain extinction in solar heat death, and surmises that the spatio-temporal scale involved in such reflection is simply too big to get our heads around. This seems at first sight like common sense. It comes as no surprise that most people do not have such anxieties as the famous scene in Woody Allen's *Annie Hall* (1977), which Scheffler cites. A young

Alvy Singer is refusing to do his homework because he's just found out that the universe is expanding and will one day 'break apart'. 'What has the universe got to do with it?' yells his mother. 'You're here, in Brooklyn. Brooklyn is not expanding'. The comedy derives from the juxtaposition of mundane cares with what would otherwise be known as 'cosmic pessimism'. But it is also funny because the flipside is equally ridiculed. The mundane cares of his mother and teacher are juxtaposed against a more existential angst that confronts that cosmic pessimism very seriously – namely, an awareness of the finitude of all things, their common destiny in oblivion which ought to humble and chasten their otherwise grandiose moral gestures, and perhaps also the 'indefinite continuation' of the human project. Cosmic pessimism has a case to make, after all. If everything will one day certainly be annihilated, what is the ultimate meaning of our actions in the present? Such considerations were behind Albert Camus' philosophy of the absurd. They are also expressed more recently in Ray Brassier's argument that knowledge of our own extinction means that in some sense we are already extinct (Brassier 2007). And if such perspectives are reasonable, then there is no particular reason that knowingly being the last of the species ought to lead to nihilism.

Such considerations matter a great deal when considering our very real concerns of catastrophic climate change. For if the notion of being the 'last generation' is morally and existentially disastrous, it demands that we *do* clarify more precisely in what sense a 'long future' or a 'short future' can be the basis for our moral deliberation. Reports of scenarios of catastrophic climate change are assumed to be impactful to the extent that catastrophe is credibly imminent within the lifetime of the reader, or the lifetime of their children. We seem to need to know, in other words, what qualifies as too long and too short. Rarely do reports focus on the effects that global warming will have on life on Earth in hundreds of thousands or millions of years, and where they do they are presented as scientific curiosities rather than moral calls to arms.[3] We should want to know why stating that humans won't be around much longer after '100 years' matters and morally motivates in ways that '10,000 years' does not. It would seem to be the case that in order for Scheffler's argument to work he needs a minimally clear sense of what would count as a future that is 'indefinitely long enough' to valorise the commitments of the present. But I suggest that there is a profound ambivalence within this 'healthy and indefinitely long future' which is problematic. Why the aversion to a future that we know will end 'at some point'? The characters in the *Children of Men* scenario do have a future – they have the rest of their lives. But if we decide that this is too short to motivate meaningful projects, we might want to specify where and how that line could be drawn. Would extinction in two generations' time provoke the same loss of meaning? How about three generations' time?

In essence, my criticism of Scheffler centres on the temporal model that his doomsday scenarios requires. Scheffler calls the act of valuing a 'proprietary' attitude in so far as it is a demand for temporal continuity. He says: '[V]aluing is a way of trying to control time [...] to resist the transitoriness of time'. This smacks of an approach to time as control of the future by which sociologists have characterised industrial modernity (Adam 1998) – that is, the imposition (through technology, science, or economy) of industrial, homogeneous time over the rhythms and uncertainties of the natural world. What if we questioned this core assumption upon which the thesis appears to rest – that valuing present activities requires a sense of control over their indefinite continuation? How would that alter our picture? We would need an alternative narrative in which the vision of imminent and unavoidable extinction was not seen as incompatible with a politics of absolute ethical commitment to the present. Let me turn now to the concept of apocalyptic time to consider precisely this alternative.

Apocalyptic Time

What insights into our enquiry can be drawn from the temporality of apocalypse? Essentially, we might think of apocalypses as appeals for an anti-proprietary concept of the future. Whereas the *After Life* thesis presumes that life is meaningful in as much as our own vision of the good life can be guaranteed a good innings of future existence, the apocalyptic imagination inverts that relationship: value is to be found in present existence precisely because its continuation in human terms not only can never be guaranteed, but moreover it has its 'end' always in sight. The apocalyptic imagination can be seen as an attempt to provide depth to one's commitment to the present *in the light of*, rather than *in spite of*, its transitory and finite nature.

At first sight, what I am suggesting might seem contradictory. How can one derive motivation to preserve life on Earth if its annihilation is written in the stars? To understand this, we need to subtly distinguish apocalyptic from everyday forms of determinism. Apocalypses are often claimed to be deterministic (see Popović 2014) in ways that distinguish them from prophetic texts. Some Hebrew prophets (for instance the authors of the Book of Jonah and Deuteronomy – see Skrimshire 2013) sought to swerve the course of human history by predicting the cataclysmic consequences of present actions. By contrast, the apocalypse is essentially a done deal. One cannot change the events that are revealed to the apocalyptic seer. Apocalyptic narrative is often heavily criticised for this very aspect: a tragic, fatalistic narrative of the world's history unfolds, rendering any sense of historical agency nonsensical and people mere passive observers of the chaos.

But this tragic and passive aspect of time could only be at best one of the 'functions' or desired effects of apocalyptic literature. For a more complete picture, we must remember that, according to many theological traditions, apocalypses are at their heart narratives written for, and on behalf of, the dispossessed and oppressed of the Earth. The apocalyptic vision of the undoing of the world is consoling in as much as 'world' is interpreted here as a given (political, social) order. From that particular perspective, the catastrophic vision of the world lifts the veil on the illusory security and certainty of human projects. Such certainty is perhaps the very sense of permanence and security which Scheffler (2013) takes to be a condition of continuing to value our existence. Theologians have spoken about apocalyptic as responding to collective and personal trauma in this sense. In the catastrophic vision, the normal experience of the world, its supposed benevolence and order, is turned upside down and its catastrophic nature is revealed in our midst. More than determinism, apocalypse is the opportunity for the traumatised community to see reality anew in a way that their trauma once masked. Catastrophe is not some unfathomable event in the future whose arrival or imminence would nullify the meaning of existence. Rather, catastrophe is revealed to be both unavoidable (from the perspective of the traumatised, it is already 'here') and radically uncertain (we cannot know in advance how catastrophic futures will unfold). This is an expression of anti-proprietary time because it acknowledges that even an indefinitely long future has its catastrophe before it. Human attempts at securing the future are illusory, both in the sense that they deny the catastrophe that is already in their midst, and in the sense that they are not the true source of meaningful existence (as the *After Life* thesis maintains). There is therefore value to letting go of a proprietary sense of the future, because it leaves our meaning-making capacity vulnerable to the ambivalence about whether or not our continued existence is to be 'long' or 'short'.

This combination of revealed catastrophe in our midst and uncertain, utopian hope for the future is, in essence, the paradox by which contemporary eco-apocalyptic thinking has become known. Politically, apocalypse can seem like a double-edged sword, inspiring fatalism and utopianism depending on the context. Writing from the protest camps of the 2009 United Nations Climate Change Conference in Copenhagen, the political philosopher Michael Hardt neatly summarised this dilemma. The two 'temporal antinomies' – of 21st-century protest movements were the utopian and the catastrophist (Hardt 2010). On the one hand, he says, anti-capitalist mobilisation has borrowed a certain amount of rhetoric from those radical millenarian movements that interpreted apocalypse as a form of historical becoming. Millenarians foretold the unfolding birth of a revolutionary transition culminating in the utopian society or the New Jerusalem – typically the utopian egalitarian claims of Gerrard Winstanley's Diggers in England, or Thomas Müntzer's rebellious peasant in Germany. On the other hand, climate protest has tended to emphasise the point of apocalyptic prophecy as making real the imagination of *final* catastrophe. The latter signifies not the birth of the new world, but only the end of the present world. It is a catastrophist view of time that is assumed to go against the principles of eco-Marxist thought because it presents history deterministically and in post-political terms – with ecological catastrophe and extinction the natural outcome and symptom of *homo sapiens*' rapacious nature. Hardt's distinction looks outdated in the light of those movements' rapprochement in the years since Hardt wrote his 'letter'. Climate justice campaigners from across a now quite wide political spectrum adopt a stance that is arguably both catastrophist in its prognosis and utopian in its political demands. A manifesto such as Naomi Klein's (2015) *This Changes Everything* strikes me as a good example. The aesthetic and rhetorical posture of Extinction Rebellion, whose call to arms in the summer of 2021 was called 'Impossible Rebellion', is also arguably an attempt to bring catastrophism and utopianism together.

But Hardt (2010) also misses the powerful dynamic that is created when those temporal options are juxtaposed in a single, apocalyptic narrative. It is one of the most rhetorically powerful and symbolically complex aspects of ancient apocalypses that constantly juxtapose a sense of the time of the world and the time that lies outside of the world: eternity, or God's time. Both Christian and Jewish apocalypses reveal a thick description of history and historical time, sometimes with coded references to specific periodisations. But at the same time, they explode secular notions of temporal progress and the time of imperial history, making a mockery of human projects by revealing the 'new time' of God's reign. In most cases, this narrative scheme takes place with overt, though coded, references to the political context of the time it was written. Apocalypses are thought to be a literary genre that heightens the tension between the utopian aspirations of its target audience and the oppressive reality of their actual existence by envisaging an imminent revolutionary reversal of those imperial realities. The Book of Revelation for example, was written at the end of the 1st century CE at a time when the followers of Christ were a persecuted minority under Roman imperial rule. Its messages are replete with references to the corruptions of empire and its necessary downfall. There is a sequential fulfilment of eschatological prophecy, which calls forth the entire sweep of human history as receiving its final fulfilment in the revelation that is given to John. One thing follows another: tension and crisis, disaster, judgement, fulfilment, and redemption, which completes God's work. The vision that John of Patmos receives declares itself to be the confirmation and supplement to all previous biblical prophecy. It is thus a reliving of the very origins of human time and its capping with the promise of the end of time itself. At the centre of the narrative, in Revelation 8, there is the time of the 'Kings of the Earth' which manifest the full pomp, exuberance, and debauchery of imperial (i.e. Roman) power. Such displays serve to heighten the contrast between God's time (i.e. eternity) and the temporal regime of empire, which falsely asserts itself to be eternal (hence Rome, the 'eternal city') but is in fact fallible and impermanent. The temporality of empire is also presented as the attempt to assert power and control over life. Like the Book of Revelation, the Book of Daniel was also written during a time – 2nd century BCE – in which a foreign empire (the Seleucids) was oppressing Jews in Judea. The book's apocalyptic revelations refer to epochs that must play out: seventy weeks are allotted in which iniquity must cease; there are 1,290 days before the desolation is brought to an end. These are presented in the context of a series of visions in which Daniel foresees the succession of empires that will rise up against Jerusalem before the time of the end. Here, as with Revelation, time is not mere chronology. Rather, time measurement represents competing discourses of power.

Contrary to Scheffler's version of the usefulness of doomsday scenarios, therefore, I want to affirm a version of apocalyptic time as a juxtaposition of both catastrophe that is lived *now* and opportunity for ethical reorientation that is radically open to (because it is uncertain of) the future. Apocalyptic is a way of interpreting in one manner or another the way in which God ruptures historical time. In the

Book of Revelation, the world itself is transformed and brought low, providing imaginative depth to this sense of rupture. Mountains are flattened, stars blink out of the sky, life is decimated, kingdoms are brought down, and a new city descends from heaven to replace the current order. But these events occur in the vivid and visual imagination of God operating within the temporal order, in the vision of the *now*. Two consequences arise from this orientation. First, the apocalypse intensely retemporalises a theology of history by placing the boundary of time in our midst. In the Christian vision, time will come to an end, and this includes human extinction. Second, however, and here Scheffler's moral logic is overturned, the end as 'time's boundary' occurs not to devalue the commitments of the present but to infuse them with eternal significance. Lieven Boeve (2001) calls this the 'turning of catastrophe into crisis'. Every revealed end of apocalypse – the destruction of the created order, including the natural order – serves in the dualistic schema to present humanity with an existential choice. One must be either on the side of eternal goodness and justice, or on the side of fallible and transient empire.

Apocalyptic Ethics: Living Well at the End

I now want to return to the specific question of what doomsday scenarios reveal about moral intuition. Apocalyptic texts, I have suggested, certainly conjure catastrophe as unavoidable and imaginatively present in our midst. But they also frequently contain a strong ethical injunction within that picture. A common assumption is that the audiences of eschatological literature such as the letters of Paul, Revelation, parts of the Synoptic Gospels, and non-canonical sources such as the Dead Sea Scrolls feared the end of the world in their lifetime. It is thought that they must have therefore abandoned normal ethical/social commitments in favour of extreme asceticism, piety, or righteous anger. These were precisely qualities embodied in eschatological figures such as John the Baptist and Jesus, and in the ascetic demands of the Essene community at Qumran. Such an assumption appears to be compelled by the same logic as Scheffler's (2013) thesis. But the theologian Dale Allison argues that human behaviour in the face of visions of humans' impending doom actually reflected values and beliefs already held and did not in fact introduce new ones. Apocalyptic literature, whilst intensely ethical, does not create a new moral tradition so much as intensify the imperatives of a tradition already held (Allison 2014: 298). Thus for some apocalyptic audiences, nothing at all changed. The imperative was to continue holding on to one's observance of religious laws and customs in the face of catastrophe.

One of the argued functions of biblical apocalypses is that by bringing the end into imaginative proximity with present life – essentially by allowing the faithful to live 'in' the end times – responsibility to one's ethical and religious code is heightened. In a sort of reversal of Scheffler's thesis, Allison notes that pious apocalypse audiences would have interpreted them as exhortations to ready themselves by increasing vigilance. The imminence of the end times is a safeguard against 'complacency and procrastination'. This is precisely the eschatological sense of Jesus'

Parable of the Wedding Guests, which divides those who are ready and awake from those who are caught napping at the return of the Messiah, come to claim his bride (Allison 2014: 296). If, as I have claimed above, the apocalypse is a text directed first and foremost to the dispossessed, then this heightening of one's moral code ought also to mean the ability to hold fast to one's vision/revelation of a corrupt world turned upside down whilst the world around them refuses to pay attention.

So the ethics of living in apocalyptic time represents in theological terms a coming face to face with one's obligations in anticipation of the eschatological end. Apocalyptic ethics states: prepare yourself *as if* the end was coming now, as if the 'end times' were now. Apocalypse is the revelation of a common mortality, including not only personal extinction but collective, human death and the death of the world as we know it. Envisioning mortality in the present makes one more conscious of one's faults and the need to repent. An essential component of this traumatic revelation is, in some texts, what we might call its 'coded temporality'. There is in some cases an explicit prohibition against predicting the date and time of the eschaton. In Paul's letter to the Thessalonians, he admonishes believers not to consider dates or times but to be aware only that the Day of the Lord will come at a time not of their choosing, like a 'thief in the night' (Thessalonians 5:2). In Daniel's final vision, concerning the resurrection of the dead at the end of time, he sees a man ask the one clothed in linen: 'How long shall it be before the end of these wonders?' Daniel cannot understand the coded answer that he is given, and is told:

> Go your way, Daniel, for the words are to remain secret and sealed until the time of the end. Many shall be purified, cleansed, and refined, but the wicked shall continue to act wickedly. None of the wicked shall understand, but those who are wise shall understand.
>
> *(Daniel 12:6–11)*

Time and again in apocalyptic literature, we find this attempt to combine a sense of living in the end times, calling us to radically question social and ethical priorities given that death could be on our doorstep, with an awareness that the final end will come of a time not of our choosing. This speaks directly to the critique of the 'proprietary time' of doomsday scenarios I have been developing. The apocalyptic drama operates within a zone of ambiguity with regard to the length of the future that remains to humans. Humanity lives under the sign of its own destruction – its days are numbered – but ethics does not derive from an assumption about any precise notion of those numbers to be considered.

Doomsday Revisited

Let me now bring this to bear on the imagination of human extinction in the service of climate change activism. An apocalyptic world view is one that wishes to inaugurate apocalyptic time. That is, it wishes to bring the reality of catastrophe in our midst, in order to turn the temporal order on its head. The political philosopher Jean-Pierre Dupuy has adopted something of this argument in a principle he advocates called 'enlightened catastrophism'. Adopting the discursive tactics of Jewish prophetic texts, Dupuy claims that moral discourse about risk should invite a radical epistemic perspective in which the future catastrophe that one fears is assumed to exist already in the present. Only by giving it this ontological character will its ethical injunction upon our lives be apparent.

All of this is of relevance to climate change rhetoric, and particularly the narrative of mass extinction, which immediately introduces a difficult interplay of timescales within moral considerability. In fact, the moral problem of climate emergency presents itself as a perfect test case for the relevance of an ethics of apocalyptic time. At both ends of the temporal spectrum – the inconceivably distant future and the dangerously proximal future – the possibility of extinction presents itself as a moral conundrum. What moral sense are we to make of scenarios in which the future of humanity is put perilously close to extinction? A key limitation of Scheffler's thought experiment is its epistemological assumption. To recap, Scheffler thinks that an encouraging moral intuition is revealed by imagining that I am certain that doomsday will occur 30 years after my natural death, or, to take the infertility scenario, humans will be extinct when the last of us alive now has died. But a much more instructive thought experiment is one in which I know that human extinction will occur at some point in the future, but I am not certain whether it will be in hundreds, thousands, or hundreds of thousands of years' time. Or, imagine a doomsday scenario in which perhaps not all but most of humanity is wiped out, or one in which the quality of life of the remnant of the population is vastly diminished or rendered, in the eyes of some, 'inhuman'. This is in fact the scenario close to some realistic climate change scenarios. Predictions of the extinction of the *entire* human race are extremely speculative and furthermore they evade a much more painful likelihood that in the century to come the survival of those who can afford to protect themselves will be at the expense of those who cannot. This is a scenario that is in an important sense *already* true. Climate change has already so catastrophically changed the lives of the most vulnerable to extreme weather events that they may rightly consider that the end has come for them. In as much as entire cultures are under imminent threat of disappearance, whether directly or indirectly as a result of climate chaos, the end of the world is being experienced at multiple geographical and historical scales.

Perhaps Scheffler does have some sort of an answer to these sorts of considerations. Presumably, he could refer back to his claim that what makes 'our' projects valuable is their being sustained by some minimal guarantee of their continuation. Even a minimal likelihood of their continuation, somewhere on the Earth might motivate my actions. So the survival of some portion of humanity would be morally motivating, even if politically suspect (for my reasons suggested above). But my scenario asks for something more than this. It claims that Scheffler's criteria for affirming what humans really value about present concerns become weak in the face of the *uncertain* future scenarios that I have presented, which arguably are actually close to climate change reality. What if the future of humanity is being certainly curtailed, but within a timeframe that is uncertain and in which our present actions have a hand? How would we manage the translation of that form of catastrophic imagination into the possibility of ethical commitment? Scheffler's argument seems ill equipped to be able to affirm the moral value of my commitment to projects in such a veil of uncertainty.

The alternative model of doomsday which I have sketched above via (Jewish and Christian) apocalyptic literature has the potential to address some of these shortcomings. It combines various elements of a determinist temporal framework, but goes on to supplement this with an intensification of the sense of temporal ambiguity sketched above. Time is certainly 'contracted', or made short, to use Paul's term (ὁ καιρὸς συνεσταλμένος; 1 Corinthians 7:29). The first appearance of the Messiah sets the apocalyptic clock ticking even if he does not give us the time. Apocalypse is thus literally all about how humanity's days are numbered. But they are numbered in a way hidden from humans, according to an entirely other-worldly scheme. The Book of Revelation certainly continues this tradition, but in the earlier Book of Daniel it is even more acute.

Apocalypses thus present a rather unique framework for thinking about human extinction. In parallel fashion, narratives of future disaster may combine *both* the sense that there is some tragedy in knowing that humanity's days are ultimately numbered, *and* that we do not ultimately know how to count those days. And this means – here is the crucial point for my argument – paying no attention to whether the future that lies ahead is a very long one, or a very short one. Human existence, in the Christian scheme, comes out of nothing, and goes to nothing. Thus the end of all things, as far as humans are concerned, is devalued, and the moral value attached to present projects is untouched by references to 'long' or 'short' futures. As Augustine says in *The City of God*:

> The course of life is nothing but a race towards death, a race in which no one may stand still or slow down even for a moment, but all must run with equal speed and never-changing stride. For to the short-lived as to the long-lived, each day passes with un-changing pace [...] On the way to death the man who takes more time travels no more slowly, even though he covers much more ground.
>
> (*Augustine*, City of God *13:10*)

This interpretation of Christian apocalyptic draws on a kind of parallel with the individual's confrontation with their death. Just as directing my attention towards my own death forces me to consider my own life and the meaning that I make for it, so the human race makes the same judgement without any obvious sense in why a 'long' future ought to motivate more meaningful actions than a 'short' one. *This* is the good news, if we can call it that, for the contemporary climate activist. There is every reason to act morally, meaningfully, and passionately in these uncertain and catastrophic times, and what guarantees those reasons are not the certainty that the human project has a long future ahead of it. Motivation for action must be sought elsewhere than a hope for survival.

Conclusion

I believe that the argument offered by Sam Scheffler represents a common naïveté about how and why we should expect narratives of climate catastrophe as the imminent end of the species to galvanise a kind of moral revulsion and survivalist instinct. It is a naïveté that assumes that disasters revealed to our imagination are threatening because they threaten the indefinite longevity of our present moral concerns. Apocalyptic is different to this, and operates according to an ethically more constructive model because it challenges the proprietary notion of time endorsed by Scheffler. Apocalyptic time is the reverse of proprietary time, radically denying human reason the part of moral arbiter over what counts as a long enough future to care about. Its catastrophic imagination shows not the calculated shortness of the future but the radical ambivalence of the time that remains, and the exhortation to live faithfully in whatever remains of that time. Augustine's sentiment above captures this perfectly in theological terms. It is not for us to gauge how long or short the future is to be as the basis for deciding which aspects of moral life are worth living now. And Allison's point about apocalyptic ethics rings true for doomsday ethics as well. Introducing a time limit to human existence will not introduce a new moral tradition. It *may* heighten the ones that we already have. But equally, it may deaden them. What needs re-inventing are the traditions themselves.

Scheffler claims that the thought experiments he provides do not relate to any lived experience that humans have ever gone through. This admission leaves the task of imagining likely reactions to it speculative or, in his case, reaching for science fiction to propose reasonable hypotheses. But the adjustment I have made to his scenario, one in which future extinction is anticipated within an uncertain time frame, is both close to lived experience *and* consistently explored in works of fiction as thought experiments for an ethics in catastrophic circumstances. One would find it less in the blanket infertility scenario of *The Children of Men* and more in the devastated landscape of Cormac McCarthy's (2019) *The Road*. In that novel, the protagonists – father and son – reflect on whether they could ever know whether they were the last humans alive. This is a radical thought experiment to invigorate climate ethics. The important part of the question is to reflect on why we care about the answer. What would it mean to live in a world in which the answer was permanently uncertain? This is the challenge of deriving ethical value in a world whose future is certainly catastrophic, but whose capacity to generate meaningful continuation of human projects is cast radically into question. McCarthy's narrative rightly poses different perspectives on that situation – from the recourse to suicide, to an almost messianic belief in the kindness of some future form of humanity. But the value of his thought experiment is that it leaves open the possibility that even in such circumstances where the future of the 'human project' is cast radically in doubt, meaningful, loving, and hopeful action can still be found (Skrimshire 2011). Scheffler's scenario, and the moral assumptions that he derives from it, does not provide such a basis.

Scheffler believes that if the *Children of Men* scenario is plausible (i.e. that, faced with certain and imminent demise, current moral commitments lose their value), then this is good news for climate change ethics because it suggests that individualism, egoism, and short-term thinking might be less compelling on human behaviour than we thought. I will leave aside the obvious response that today the evidence points strongly towards both individuals and collectives acting selfishly *regardless* of the knowledge about how their actions may contribute to a catastrophic ecological and social collapse. In this chapter, I have been more interested in challenging the thesis on the basis that it misconstrues what it is about doomsday scenarios that is both morally revealing (telling us something about what people do in fact care about) and also morally edifying (providing us with a framework for right action in the face of doomsday scenarios). I have offered an insight from apocalyptic theology that I believe better meets the expectations that Scheffler has for such thought experiments. Apocalyptic theology, in its juxtaposition of deterministic revelation of the 'end of all things' (the revelation that all things will one day pass away) and radical injunction against predicting the manner and timing of such end has the added value to moral thought that it more closely speaks to our current ecological situation – that is, a world in which catastrophic predictions proliferate and doomsday scenarios appear as a temptation to despair, but in which the question of meaningful, moral action in the light of such predictions is far from answered in advance.

Notes

1 A notable exception is Robin Attfield's (1999) chapter 'The Ethics of Extinction' in his *Ethics of the Global Environment*. In fact, many of Scheffler's allegedly 'original' contributions to this field appear to echo Attfield's much earlier study.
2 *Death and the Afterlife* includes critiques of the thesis by different philosophers and Scheffler's responses to them.
3 For a discussion of the moral uses of 'deep future' imagination, see Skrimshire (2019).

References

Adam, B. 1998. *Timescapes of Modernity: The Environment and Invisible Hazards*. London: Routledge.
Allison, D. 2014. 'Apocalyptic Ethics and Behaviour', in Collins, J.J. (ed) *Oxford Handbook to Apocalyptic Literature*. Oxford University Press, 295–311.
Attfield, R. 1999. *Ethics of the Global Environment*. Edinburgh: Edinburgh University Press.
Boeve, L. 2001. 'God Interrupts History: Apocalypticism as an Indispensable Theological Conceptual Strategy'. *Louvain Studies*, 26(3): 195–296. https://doi.org/10.2143/LS.26.3.912.
Brassier, R. 2007. *Nihil Unbound: Enlightenment and Extinction*. Basingstoke, UK: Palgrave.
Collins, J.J. 2014. '*What Is Apocalyptic Literature?*' in Collins, J.J. (ed) *The Oxford Handbook of Apocalyptic Literature*. Oxford: Oxford University Press, 1–18.

Cuaron, A. 2006. *Children of Men*. Film. Universal Pictures. USA.

Hardt, M. 2010. 'Two Faces of Apocalypse: A Letter from Copenhagen', *Polygraph*, 22: 265–274. https://dukespace.lib.duke.edu/dspace/bitstream/handle/10161/9295/Antinomies%20-%20Polygraph.pdf.

James, P.D. 1992. *The Children of Men*. London: Faber and Faber.

Klein, N. 2015. *This Changes Everything: Capitalism vs. the Climate*. New York: Vintage.

McCarthy, C. 2019. *The Road*. London: Picador.

Popović, M. 2014. '*Apocalypse and Determinism*', in Collins, J.J. (ed) *The Oxford Handbook of Apocalyptic Literature*. Oxford: Oxford University Press, 255–270.

Scheffler, S. 2013. *Death and the Afterlife*. Oxford: Oxford University Press.

Skrimshire, S. 2011. 'There Is No God and We Are His Prophets: Deconstructing Redemption in Cormac McCarthy's *The Road*'. *Journal for Cultural Research*, 15(1): 1–14. https://doi.org/10.1080/14797585.2011.525099.

Skrimshire, S. 2013. 'Challenging the Skeptics: False Prophecy and Climate Activism'. *Religion and Society*, 4(1). https://doi.org/10.3167/arrs.2013.040110.

Skrimshire, S. 2019. 'Deep Time and Secular Time: A Critique of the Environmental "Long View"'. *Theory, Culture & Society*, 36(1): 63–81. https://doi.org/10.1177/0263276418777307.

Stapledon, O. 1930. *First and Last Men*. London: Methuen.

12

ESCHATOLOGY AND TELEOLOGY IN THE ENVIRONMENTAL ETHICS OF HANS JONAS

Robert G. Seymour

Hans Jonas (1903–1993) is a major figure in 20th-century thought whose philosophical significance is only beginning to be appreciated in the Anglophone world. His career began with a contribution to the history of religion, a pioneering, phenomenologically informed interpretation of the history of Gnostic eschatological religion in late antiquity. It drew to a close with the far-sighted development of an environmental ethics, which drew on his long reflection on the possibility of a philosophy of nature. Jonas's most famous work, *The Imperative of Responsibility* [*Das Prinzip Verantwortung*; 1979],[1] which has remained a touchstone in German public debate, stands out as one of the most sophisticated contributions to environmental ethics. Jonas's basic argument is that the central questions of ethics have changed as a consequence of the massively increased ability of humanity to affect its environment. The initial task of environmental ethics is to determine the moral status of human agency under conditions of technological modernity. It is, perhaps surprisingly, in this context that Jonas makes use of ideas from the apocalyptic tradition. This is clearest in the categorical distinction he draws in his ethical magnum opus between an adequate doctrine of environmental ethics, which he characterizes as an ethics of responsibility grounded in a renewed teleological account of nature, and what he calls "secular eschatology," a worldview which is based on the exclusion of teleology and which reveals itself to be paradigmatically irresponsible.

In order to understand why Jonas frames the choice in these terms, it is necessary to contextualize his account. To this end, I distinguish two ways in which Jonas's uses of the concept of eschatology reveal his approach to the basic questions of environmental ethics. According to the first use, rather than designating an external event, the concept of eschatology is used by Jonas to bring attention to the futural dimension of technologically modified human agency, thereby emphasizing the novelty of the ethical problem which this form of agency poses. The recourse to religious language is not arbitrary, but is meant to underline a specific deficiency in

DOI: 10.4324/9781003189190-15

previous ethical thought in grasping our moral condition. Jonas's central insight is that technological power requires that climate ethics involves a shift from a system of symmetrical moral obligations to one which must include radically non-symmetrical obligations. If the first use of the term attempts to characterize the ethical situation faced by environmental ethics, the second use concerns its positive content. Here, I argue that Jonas employs the concept of eschatology critically to designate an interpretation of technological agency which has potentially disastrous consequences. In contrast, the basic orientation of an environmental ethics which acknowledges the qualitative change in human agency diagnosed by Jonas should be explicitly "non-eschatological" and instead operate according to a teleological conception of the relationship of human action and nature.

The Eschatological Situation of Climate Ethics

Jonas argues that climate catastrophe opens a "new phase in history" (Jonas 1993: 47) and simultaneously a new chapter in the history of ethics (Böhler and Brune 2004: 72–73). This is not, at least in the first instance, because environmental ethics requires a re-evaluation of traditional ethical *values* as such (Böhler and Hoppe 1994: 204). Rather, as Jonas presents it, what has changed is the scale of ethical action under conditions of climate catastrophe. Ethics is now confronted with an unprecedented "apocalyptic situation" which entails that it must assume "quasi-cosmic dimensions" (Jonas 1993: 97). These grand pronouncements are not merely rhetorical, but are meant to indicate that ethics has become unmoored from its traditional orientation toward the organization of human community, or what Jonas calls its "anthropocentric bias."

This shift in orientation fundamentally concerns the timescale of human action. In a programmatic formulation, Jonas states that environmental ethics is an "ethics of the future" (Jonas 1994: 128). By this, it must be stressed that he does not mean that it is a doctrine to be elaborated in a future age or in view of a merely hypothetical future event, minimizing the urgency of present dangers or the necessity of present action. Rather, it means that ethics must face up to the vastly increased potentialities of human action, taking into account extended causality along timescales far exceeding those presupposed in standard cases of moral reasoning. These far-reaching considerations change the nature of ethical thought, ultimately raising the question of human responsibility for the continuation or destruction of life as such. In this sense, Jonas states: "Its object becomes the greatest it is possible to conceive, one indeed which has hitherto never been conceived except in religious eschatology: the future of mankind" (Jonas 1987: 66).

This precipitous raising of the stakes of ethics has to do with the change to agency, and hence humanity's relation to nature, brought about by specifically modern technology. Jonas's basic point is that technological development, the power of human action to affect its environment, has massively increased in quantitative terms. While this might appear self-evident, Jonas notes that this increase in power has not been accompanied by a corresponding increase in reflection on the nature of that power. The thoroughgoing augmentation of agency by modern technology is a comparatively recent event. Full consciousness of its effects only emerged after the establishment of the Western ethical canon. This is significant, as here a quantitative increase in human power, assuming almost infinite dimensions with respect to the finitude of nature, leads to qualitative change in agency (cf. Jonas 1993: 34). Underlying this change, Jonas argues, is an "eschatological" dynamic which fundamentally distinguishes modern technology from its antecedents.

Despite its occasional destructiveness, premodern technology was unable fundamentally to upset the balance between human ends and nature. The former could not threaten to overwhelm the latter and the distinction between the technological and the natural remained clear-cut. Furthermore, the timescale of human self-assertion against its environment was circumscribed and punctual. None of this holds in the case of modern technology. To understand why, Jonas argues that we have to go beyond the emergence of technology as the fundamental historical driver of change during the Industrial Revolution to examine its theoretical basis. This he takes to be the reconceptualization of nature which accompanied the birth of modern science in the 16th century – an epochal break from a teleologically constituted account of nature toward a fundamentally dynamic worldview which continues to inform modern naturalism.

The account which Jonas gives of the genesis of the theoretical underpinnings of modern dynamics can only be alluded to here. Conventionally, he traces the rupture back to the philosophies of René Descartes and Francis Bacon; the fundamental theoretical parameters are established by the former, while the "ideological" self-understanding of the dynamics of scientific discovery is derived from the latter.[2] In schematic terms, one can say that for Jonas this signals a transition from what one might call an "inclusive" to an "oppositional" understanding of the relation of nature and human reason (cf. Hösle 1994: 48). This change can be grasped in terms of its significance for the relation of nature to the orientation of human action. The earlier view is primarily contemplative, which is to say that it subordinates practical questions of the manipulation of nature to a broader metaphysical account which explains the place of the human within the whole of nature. This metaphysical account is *theoria* in the classical sense, the highest occupation of human reason. Practical sciences are by definition concerned with the domain of the mutable, and thus derive from knowledge based on experience, not speculative laws of reason; as such, they are understood as *arts* (*technai*) (cf. Jonas 1984b: 67f.). The relation between these two branches of knowledge is incidental; theory does not interfere with the separate, parallel development of technics. What it can do, however, is orientate the use of the latter by informing the way technics are applied – that is, by providing knowledge of good final ends. In the simplest terms, the overarching metaphysical picture which informs *theoria* is thus one in which human action is regarded as fitting into a teleologically structured whole.

By contrast, according to the worldview underlying modern science, nature is understood in purely quantitative terms; causal explanation is limited to efficient causes; final causes are excluded. For Jonas, the fundamental and most coherent expression of the worldview is that of Cartesian dualism, which locates human subjectivity and nature in ontologically separate domains. It necessarily follows from this exclusion of subjectivity from nature that teleological forms of explanation are ruled out axiomatically. On this account,

> final causes have relation to the nature of man rather than to the nature of the universe – implying that no inference must be drawn from the former to the latter, which again implies a basic difference of being between the two. This is a fundamental assumption, not so much of modern science itself as of modern metaphysics in the interest of science.
>
> *(Jonas 2016: 71)*

Jonas shares with other philosophers of technology the notion that the denaturalizing of subjectivity on the one hand and the emergence of a purely quantitative account of nature on the other are linked not only to theoretical innovation, but also to the domination of nature (cf. Spaemann 1983: 53). Specifically, Jonas claims that the constructivist structure of modern science ensures that theory is inherently connected to practical application, and by implication to technology. As he puts, it, "[m]y thesis is that…[modern] science is technological by its nature" (Jonas 1984b: 76). In this context, it must suffice to stress the point that with the loss of a teleologically structured account of nature the accompanying theory of judgement becomes obsolete. Practice now no longer refers to the orientation of action, save in the service of the perpetual increase of power via the manipulation of nature. Paradoxically, this entails that, by emancipating human action from dependence on nature, modern technology also introduces a new set of constraints on it.

We can get a sense of what Jonas means by these constraints via a comparison with premodern conceptions of instrumental ethical reasoning.[3] The original concept of a *techne* is by definition ethically neutral, simply designating a specific ability or function. The moral status of this ability is dependent on whether it is applied to good or bad ends. This, in principle, admits of clear adjudication: in addition to knowing what function a given *techne* performs, all that is needed is knowledge of what the good and bad ends are, which are provided by *theoria*. The neutral function can in this way unproblematically be subsumed to the judgment of the agent. By contrast, in evaluating modern technology this calculation is complicated not only by the fact that its effects transcend a readily identifiable agent, but also that these effects are potentially ambivalent. In other words, it is impossible to rule out that what appear to be precisely legitimate, "good" applications of technology may in the long run have highly negative effects due to their massively extended causal reach. Thus, if we only focus on their proximate effects, ethical calculations which appear to be positive may in fact be self-undermining. A massive increase in power enabled by theory is unaccompanied by any immanent standards for its use. As Jonas puts it: "Never was so much power coupled with so little guidance for its use. Yet there is a compulsion, once the power is there, to use it anyway" (Jonas 2010: 177).

In consequence, our understanding of what it means to act by means of technology requires fundamental revision. The traditional concept of *techne* involves a distinction between potentiality and actuality. Merely possessing a technical ability doesn't imply its actual use: whether an ability is actualized or not is a question of judgement. Jonas argues that modern technology nullifies this distinction: the logic of technological development is cumulative and self-perpetuating. This is to say that if a new technology is successful, it is automatically integrated into a complex system of mutually dependent elements, and then becomes the basis for further development. In this way, a successful technology is automatically actualized and its use becomes in a sense irreversible, which has obvious implications for our ability to set limits or control technological innovation. Thus, in contrast to traditional *techne*, whose application is temporally circumscribed, Jonas describes modern technology as a permanent activity.

The fact of a massively extended causal scale coupled with an apparently inexorable tendency toward increase in power raises the question of the ethical interpretation of the kind of agency made possible by modern technology. It is in this context that Jonas's repeated references to the "eschatological dynamic" and "apocalyptic potential" of modern technology should be understood (Jonas 1984a: 22; Jonas 1987: 48; Jonas 2010: 87). Jonas notes that initially the logic of technological development was understood in entirely optimistic terms due to its unprecedented success in emancipating humanity from dependence on nature. This account of the logic of technology as linear progress crystallizes in what he calls the "Baconian spirit," an unconditional faith in the advantages that can be reaped from the manipulation of nature for human purposes. This one-sided optimism, taken from the nascent process of scientific discovery but still tangible in contemporary technological "solutionism," is a correlate of the abandonment of the classical theory of judgment. If technological development is a good in itself, then there is no basis on which one can impose limits on it. The recourse to religious characterization in order to qualify the future orientation of technology is meant to underline the inadequacy of its optimistic self-interpretation of this process (for a complementary account, see Hösle 2013: 144). Apocalypse is revealed as the obverse of secular meliorism; the great accumulation of power enabled by modern technology tends not toward a commensurate great good but toward the possibility of bringing about that which is "absolutely inadmissible" [*das absolut Unzulässige*]: the destruction of the condition of possibility of future human action as such.

It might be objected that Jonas's drastic evocation of the worst-case future scenario is not only hyperbolic but also exculpatory in suggesting that human action has not already undermined the biosphere to catastrophic effect. In fact, neither is the case; rather, we might think of Jonas not as providing an empirical prediction of the process of climate breakdown but as identifying the ultimate potential of technologically enhanced agency, one in which the end of human action is potentially destructive of all ends as such (Jonas 1987: 54). From this perspective, the justification for qualifying modern technology in eschatological terms is apparent. Furthermore, only by becoming conscious of the ultimate nature of the choice hypothetically enabled by modern technology – essentially between willing being or non-being – can the question of the goal-directedness of human action be posed with the requisite urgency. It is at this point that the central question of environmental ethics poses itself; for all the horror that the possibility of annihilation evokes, it is not straightforward to specify the normative basis of this reaction. Be it unanimously adopted as a goal of action or the merely passive result of moral solipsism, the outcome of self-destruction may strike us as absolutely morally inadmissible – yet it is not equally clear that its avoidance can actually be grounded as an unconditional obligation.

In order to understand why this is, we must examine the consequences for traditional ethical thought that arise from the eschatological potential of technological agency. The fundamental postulate which underlies Jonas's ethics is the imperative that "there be a mankind" in the future (Jonas 1984a: 43). Put another way, the fundamental moral problem of environmental ethics concerns the responsibility for the future coming-into-being of agents who do not yet exist. The root of the problem comes into view if we think of the most basic elements of our moral vocabulary – concepts such as sympathy, justice, love; it is clear that in the first instance they arise out of and concern reciprocal relations in shared ethical life. This focus, Jonas claims, is also ingrained at the theoretical level; for if one abstracts from the various conceptions of the good, previous ethical doctrines invariably presuppose an unproblematic reference to a shared present. In other words, irrespective of the level of abstraction at which ethical postulates are formulated they contain an implicit limit to their temporal applicability: "the ethical universe is composed of contemporaries, and its horizon to the future is confined to the foreseeable span of their lives" (Jonas 1984a: 5).

This is of course not to say that "traditional" ethics is entirely unconcerned with long-term temporality, examples of which can be found in accounts of statecraft, or with non-reciprocal forms of relation. Rather, Jonas's claim is that previous ethics takes reciprocal relations between agents of a fixed nature, whose stable relationship to their environment is unproblematic, to be the standard form of relation in traditional ethical thought. This presupposition of traditional ethics is of course not arbitrary but a corollary of the limited causal power of human action in pre-technological society. Although Jonas takes this to be a feature of all previous ethics, the paradigmatic status of reciprocal or symmetrical ethical relations is particularly obvious in modern theories of natural right. Here, the derivation of duties is the result of a mutual recognition in which the right which I ascribe to another agent is the mirror image of the right ascribed by this agent to me. However, the relation required by our imperative – to an inexistent being – is radically non-symmetrical, involving neither equivalence of power nor of recognition, and hence does not seem to fit this model. An agent can only make a rights claim once that agent exists, that is once it enters a moral community based on mutual recognition. But according to Jonas, "a *right* of the unborn to being born (more precisely: of the ungenerated to generation) is simply not arguably from any principle whatsoever" (Jonas 1984a: 40).

It is this point which motivates the central claim of Jonas's work, namely, that the non-reciprocal relation subtending ecological ethics cannot be captured – at least originally – in terms of rights but should instead be thought of in terms of an obligation or duty arising from the ontological constitution of nature. Given the manifestly controversial nature of this claim, it is worth examining in more detail.

The crux of the issue can be brought into clearer view if we consider the unconditionality of the imperative. In the first place, Jonas takes inspiration from Immanuel Kant; specifying the modality of the imperative as categorical, rather than hypothetical. The reference to Kant is not arbitrary but reflects the fact that Kant's principle of generalizability provides a universal criterium for judging the correctness of moral norms and action. The principle as is well-known is merely formal, simply determining that a particular maxim can only be raised to the level of a universal law of action if it can be affirmed by all agents involved. It should be clear, however, that the procedure of universalization underlying this principle is strictly symmetrical. That is to say, it is limited to the sphere of mutual reciprocal relationships between rational agents – "fellow citizens, so to speak, in the cosmos of reason" (Böhler and Hoppe 1994: 40) – but not obviously extendable to as yet non-existent agents. The principle of generalizability is abstract in the sense that it can be performed *at any time*, but the resulting generalization will always have an implicit relation to the present context of action. Jonas thinks this is enough to show that the relevant imperative is unconditional in a different and deeper sense to that postulated by Kant, for it does not merely concern the *formal idea* of an action, but rather an idea which requires the existence of its content (cf. Jonas 1984a: 44), that is, the existence of future humanity.

The conclusion Jonas wishes to draw from this has been resisted by a group of ethicists in the Kantian tradition. These thinkers share Jonas's conviction that environmental crisis must precipitate a new form of moral consciousness.[4] However, they wish to avoid Jonas's option for a metaphysical grounding of environmental ethics and are suspicious of what they see as his abandonment of the priority of rational autonomous agency most stringently formulated by Kant. Common to these responses to Jonas is the idea that the Kantian principle of generalizability can be salvaged as the foundation of a future-oriented planetary ethics if it is given an appropriately intersubjective reformulation. Most prominently, Karl-Otto Apel has claimed that Jonas's ethics *must be* made to conform to this model to ensure the obligation that future humanity live according to requisite standards of dignity dictated by rational autonomy (Apel 1990: 196). According to Apel, and other advocates of the school of discourse ethics, what actually guarantees Kant's principle of generalization are certain *a priori* rules governing the necessarily always intersubjective structure of communication. The norms of communication implicitly commit us to the ideal of an unlimited community of communication which is governed by the criterion of coherence and by the principle of maximizing consensus. This gives content to Kant's merely abstract and formal imperative. Accordingly, the generalization provided by the ideal of an unlimited community of communication must also take into account the hypothetical interests of future humanity when adjudicating the morality of action. This claim is unproblematic insofar as it makes sense to consider the hypothetical interests of future generations when scrutinizing action which would impinge on these hypothetical interests. However, Apel also takes this as providing the normative basis for the imperative that there be humanity in the future. The reasoning for this appears to be that the continuation of the community of communication is necessary in the interests of maximizing both coherence and consensus.

It has been pointed out that Apel is not warranted in drawing this conclusion (Dews 1995: 163–165). Apart from the bizarreness of the claim that the *a priori* conditions of belonging to an ideal community of communication necessitate the duty to generate actual future participants in it, the reciprocal principle of generalization adopted by discourse ethics cannot plausibly be thought of as grounding Jonas's imperative for the following reason. This principle stipulates that a norm is valid when its universalization can be affirmed as being in conformity with the interests of all members of the community. Yet, the concept of interest can only intelligibly be applied to beings which already exist. It is hard to see what sense can be made of the idea that one could, even hypothetically, "consult" the interests of non-existent human beings on the relative merits of a maxim which would lead to their never coming into being (cf. Dews 1995: 163–165). Hence Jonas's basic point stands: "The claim to existence begins only with existence" (*Der Anspruch auf Sein beginnt erst mit dem Sein*) (Jonas 1984a: 39) – and this leads us into the domain of metaphysics.

The Non-Eschatological Content of Climate Ethics

We can now establish that it is the eschatological situation of environmental ethics that dictates Jonas's argument that axiology must be integrated into ontology. The reason for this is that ethics, by assuming its new cosmic role, is posed a question of unavoidably metaphysical significance, one unanswerable in the terms of traditional ethics. Recalling Leibniz, Jonas suggests that our technological power forces us to take a position on the question "why should there be something rather than nothing?" (Jonas 1984a: 47–48). As mentioned above, the idea that the potential destruction of life could be actualized by human action is intuitively rejected as something which absolutely must not come to pass. Jonas's starting point is that this rejection has its basis in an ontological sympathy with nature "in itself." Any defence of this hypothesis depends on a willingness to entertain a metaphysical account of the relation of human agency and nature, something excluded *a priori* on the modern scientific worldview but whose consideration is prompted by the possibilities ultimately enabled by it. Jonas's answer to the metaphysical question posed to ethics determines his critical rejection of what he calls "secular eschatology" as a potential response to climate disaster. As such, his reasoning needs to be examined on its own terms.

It goes without saying that the form of metaphysics proposed by Jonas cannot attain to apodictic certainty. Nonetheless, he believes that faced with ultimate questions reason is inevitably drawn to explain its position with respect to the whole of being, albeit now – subsequent to the great criticisms of metaphysics – with a due sense of its fallibility and the provisional status of its explanations (Jonas 1994: 137). It in fact follows from a teleological explanation of nature, as Jonas conceives it, that it cannot present itself as a theory in the terms of reductive physicalism, as it is precisely a form of *sympathetic* knowledge of nature (cf. Spaemann 1983: 44). This is to say, it draws on human experience of subjectivity in order to understand how it fits into being as a whole. If the premise of a teleology is that subjectivity emerges from nature, then this is clearly not something that can be theorized in objectivizing terms without falsifying the nature of human co-dependency with nature (cf. Dews 1995: 160). As mentioned above, Jonas argues that the conception of nature as "value-free" derives from an axiomatic exclusion of any "anthropomorphism" from nature, which itself does not have any *prima facie* evidential basis but received its justification as it were *ex post* due to the predictive success of mechanistic theory. Against this, Jonas repeatedly points out that reductionist, "value-free" accounts of this type are incompatible with our own deeply ingrained sense of ourselves as agents.

This incompatibility is thrown into sharp relief by the fact that the basic characteristics of subjectivity appear to be exhibited to some degree by all living beings. This fact imposes itself in a phenomenological analysis of the difference between life and non-life. Although the value-free account holds that there is no essential qualitative difference between the two states, we cannot help but be struck by an essential distinction between the anorganic and the organic. Whereas physical matter is inert and indifferent to its surroundings, organic life by contrast actively maintains itself. This much is evident in the metabolic process, which Jonas takes to be the defining characteristic of organic life. The organism has an environment on which it is materially dependent but from which it is formally distinct; organic life consists in self-differentiation of an individual form with respect to its environment and the sum of its material parts, something already evident in the first stirrings of irritability. These two characteristics suggest that a differentiation of the inner from the outer is a form of transcendence, insofar as the form of the organism is irreducible to its material constitution at any one moment in time. However, the genesis of a transcendent structure is ontologically fragile; given its rudimentary structure, the organism is engaged in a struggle, one in which it is constantly threatened by annihilation, a relapse into non-being.

Jonas puts the fundamental contrast in this way: organic life has an "absolute interest...in its continued existence" (Jonas 1994: 26). The inert uniformity of matter, on the other hand, is simply *indifferent* to its existence; it is not governed by any immanent form of organization, and its relation to other entities is only achieved on the basis of combination with other elements, leaving its basic nature unaffected. This is expressed in the measurement of matter by quantification, that is, in terms of a purely external criterion of identity. Contrastingly, organic life's interest in its own being is expressed teleologically; its form of existence is *purposiveness*, orientated toward the end of maintaining its form. Furthermore, this purposiveness takes a particular form: the organism is an end in itself; its orientation toward its end derives from and expresses its natural constitution. In other words, the organism has an internal purposiveness distinct from the externally imposed purposiveness of an artefact or a technological function. This is significant, as it allows us to posit an ontological basis for the existence of ends, something axiomatically excluded from any value-free account of nature: according to the latter, the ability to have ends goes together with the ability to bestow value, both of which are the exclusive preserve of human subjectivity.

This analysis is not intended simply as a regional ontology of the organic, but to reveal a fact of metaphysical significance which goes beyond the mere drive toward self-preservation. What Jonas thinks he can point to is being's affirmation of its own value – an affirmation which has an essentially normative dimension. Indeed, he states that at a fundamental level being and normativity cannot be separated: "'Value' – this is contained in its concept –immanently makes a claim to actualization [*Wirklichkeit*]. It states, that is better that it *is* [sei] than that it is not" (Jonas 1994: 132). Here, Jonas appeals to fundamental intuition: the emergence of purposiveness opens up a metaphysical contrast which imposes the idea of the absolute ontological priority of living being: "the mere fact that being is not indifferent toward itself makes its difference from nonbeing the basic value of all values, the first "yes" in general" (Jonas 1984a: 81). The intuition of the absolute difference of purposiveness from indifference is thus the intuition of something good in itself, for it strikes us as overwhelmingly plausible that the ability to have ends is *as such* axiologically superior to the lack of such an ability.

It is on this precise point that teleology and ethics converge; whereas in the case of specific given ends normative properties are not coterminous with descriptive properties, the mere fact of the existence of beings which are ends in themselves can be seen as itself an end of nature, that is as affirmed by being as objectively good.[5] As Jonas puts it:

> Life is an end in itself [*Selbstzweck*] that is an end which actively wills and pursues itself. Purposiveness as such, which through its zealous affirmation of itself is infinitely superior to that which is indifferently purposeless, *can for its part very well be considered to be an end*, as the surreptitiously desired goal of the otherwise so empty undertaking of being.
>
> *(Jonas 1994: 221, emphasis added)*

A detailed account of Jonas's ontological grounding of axiology would have more to say about how he elaborates his philosophy of nature. In this context, it must suffice to state that Jonas argues that it is when the purposive form of life reaches its most sophisticated form – in human freedom – that the proto-normative self-affirmation of the organism acquires obligating power. Here, purposive action is accompanied by self-conscious awareness in the formulating of ends, as well as the recognition of the claims made by the self-affirmation of other purposive beings. We are asked to consider that the pangs of our conscience which arise from the premonition of our destruction of nature, in the wake of our emancipation from it, derive from the echoes of the primordial "Yes" of being to its own existence with which our being too is ultimately in sympathy.[6] For practical agents who can freely and self-consciously modify their ends, the sympathetic recognition of something good in itself must translate into responsibility for fostering the continued existence of purposive being. Before turning to this central concept, it must be noted that if the future relation characteristic of environmental ethics is to be governed by responsibility, this concept can only be derived from the inherent value of being. A value-free nature, indifferent to human ends, presents no normative claim which could bind our wills and is thus not a possible object of responsibility.

The non-symmetrical relationship which Jonas has argued must define ecological ethics is one of responsibility, which in turn derives from the intrinsic value of being. Jonas's elaboration of this idea may well strike contemporary readers as idiosyncratic; much of the latter part of *The Imperative of Responsibility* is taken up with a critical evaluation of the ideological underpinnings of emancipatory conceptions of history: "utopianism" or what he calls "secular eschatology." These theories function as a foil for his conception of responsibility, and part of the reason for Jonas's interest in them is as models for collective action, a central issue for any environmental ethics. In our context, however, their relevance derives from their orientation toward the future. From Jonas's perspective, while it is the case that traditional ethics leaves us in the breach when it comes to orientating planetary action, there is one serious, theoretically grounded, alternative account of future-orientated action which likewise has its origins in reflection on the possibilities afforded to human action by technological advancement and which might seem to correspond to the challenge of climate ethics. This is provided by an outgrowth of modern progressivism: utopianism, the most significant development of which Jonas identifies with the Marxist tradition. As he puts it, secular eschatology offers a conception of "duty to the future" whose attraction is powerful but stands in direct contrast to the "noneschatological conception of duties which we believe to emanate from the emerging global plight." As such, "[d]etermining their relationship is not a question of abstract correctness but of concrete urgency" (Jonas 1984a: 143).

It should be noted that Jonas's discussion is multi-faceted, extending from ideology to world-historical impact and is far from straightforwardly hostile. It should be seen against the background of his identification of a total failure of liberal capitalism even to pose the relevant questions. At one level, *The Imperative of Responsibility* [*Das Prinzip Verantwortung*] is conceived as a response to Ernst Bloch's magnum opus *The Principle of Hope* [*Das Prinzip Hoffnung*], whose philosophy of utopia operates at a similarly level in terms of its concern with metaphysics. Here, it is only possible to focus schematically on why Jonas qualifies this tendency as "utopian," a label explicitly rejected by Karl Marx, and even in more trenchant terms as "secular eschatology."

The central issue arising from Jonas's analysis is one of ends: industrial civilization simultaneously abandons a substantive conception of end while being governed by a logic which dictates the actualization of short-term increases in its power irrespective of the long-term consequences, including the undermining of the future ability to have ends as such. Jonas's ethics of responsibility is designed to respond to this situation. It should be stressed that for Jonas responsible future action is not to be guided by the illusionary ideal of reverting to a pre-technological civilization. Rather, the ethical challenge of climate catastrophe is dialectical: what is required is a higher degree of power not the quietist relinquishing of it. It needs to be shown, how, after the will has been emancipated from metaphysically sanctioned constraints and its power unfettered, it is possible to *limit* its power, how to break with the compulsive dynamic to actualize a seemingly infinite increase in potentialities.

Jonas's sketch of the dialectical development of technology betrays certain parallels to the Marxist account of the development of productive forces. In the first stage, human power is limited – static, rather than dynamic – but incapable of constituting a serious long-term threat to itself; in the second stage, this power is increased by orders of magnitude but in this process becomes alienated, acquiring a compulsive dimension which becomes radically self-undermining. What is needed is what Jonas calls a "third power," a power over second-order power (Jonas 1984a: 141–142). However, in each case the conception of this power differs according to whether the underlying process of technological development is understood in optimistic terms as a fulfilment of the potential of human agency, or in negative terms as a means of avoiding the worst.

It is Jonas's contention that there are two converging intellectual preconditions of utopianism. First, whereas traditional ethics neglected the issue of technology, the main currents of 19th-century utopian thought were nourished by the outlook afforded by rapid technological advancement. The theoretical evaluation of technological change was thus fundamentally optimistic, aligning the Baconian ideal of subordination of nature with the achievement of social justice. On Jonas's interpretation, the social content of utopianism is essentially dependent on the process of technological rationalization. As he puts it:

> the technological impulse is built into the essential nature of Marxism, and to resist it will be all the more difficult as it is bound up there with a stance of extreme anthropocentricism, to which all of nature (including human) is but a means for the self-making of still unfinished man.
>
> *(Jonas 1984a: 156)*[7]

Second, this future-oriented vision can be understood as eschatological in a secular sense due to its coupling of an optimistic account of technological change with a metaphysical understanding of the agent effecting this change. Here, the intellectual precondition of utopianism is diagnosed as a displacement of the classical notion of the Idea of the Good from the "vertical," or transcendental, metaphysical plane, to the "horizontal" plane of immanent, historical development. The decisive shift, on Jonas's account, occurs with Kant's notion of the Highest Good, which is to be achieved in a future state of the world – albeit one which is infinitely deferred, due the ultimate metaphysical unreality of time in Kant's philosophy. Jonas's further story is a familiar one: with Hegel's criticism of Kant, reason's immanence to history becomes constitutive, thereby setting the stage for the Marxist inversion of Hegel's account, in which the "cunning of reason" becomes self-conscious in the action of the absolute class. Hence the ever-greater development of technological power for the purpose of the "rationalized" refashioning of nature to suit human purposes is utopian not merely in the sense of allowing progressive emancipation but also in the sense that it is itself raised to the level of a metaphysical principle. The confluence of these two ideas lends utopianism its intellectual force, but entirely blinds it to the compulsive dynamic of technology analyzed above.

The qualification "eschatological" thus has its justification in the specific nature of the future orientation of utopianism. Accordingly, the Highest Good as a future end is the source of all normativity. Its future attainment is the *ultima ratio* of all present human action, which can only claim provisional status – subject to judgement according to its achievement or otherwise of final emancipation. For Jonas, underlying this account of history is a metaphysical account of reality expressed most succinctly in Bloch's proposition "S is not yet P" – the subject is not yet its predicate, humanity *as such* awaits its future actualization (Jonas 1984a: 199). In its most consistent form, therefore, utopian orientation toward the future must be governed by a binary understanding of final ends: fulfilment of utopian ideals, or total failure. The form of practical action which this understanding of history dictates consequently is one guided by the necessity of optimism, the "principle of hope" in Bloch's terms.

Jonas suggests that this intellectual configuration can be considered to be history's first truly "active" form of eschatology. For example, the future orientation of premodern messianic eschatology is passive, in the sense that it depends on a divine dispensation which is radically independent of the believer; it does not presuppose an agent with the ability to bring about the future so conceived. By contrast, modern utopianism unites a doctrine of action whose object is a future state with an agent whose destiny is to realize this state: "Here, for the first time, *responsibility for the historical future in collusion with history's own dynamic* is put with rational intelligibility on the ethical map" (Jonas 1984a: 127).

From the foregoing discussion, there are two things that can be said in conclusion concerning Jonas's determination of an ethics of responsibility as "anti-eschatological." The first relates to the situation of ethics under conditions of a catastrophic climate emergency. Although distinct from traditional ethics, utopianism similarly derives its prescription from what Jonas calls an "abstract" ideal. His assessment of technological power, however, forces us to relativize ethical priorities. At a meta-ethical level, this can be expressed by asserting that the customary configuration of "ought" and "can" must be reversed: responsible ethical reasoning cannot start from abstract consideration of ideals of human nature and action, according to which human ability must conform. Rather what takes precedence is that which "mankind in fact already does because he has the power and with it the incentive thereto; and the duty springs from the deed already underway: it is made his duty by the stretching causal fate of his actions" (Jonas 1984a: 128). If it is the case that humanity is responsible for the continued existence of nature and humanity's place within it, and that this stands in contradiction with the further development of technological power, a limitation of the latter becomes the singular overriding duty incumbent upon ethical and political subjects. This means that projects of radical human improvement and emancipation must, at least temporarily, recede behind our new ethical horizon. If the cosmic scale of environmental ethics means that its conception unavoidably has religious overtones, Jonas counterposes to an eschatological vision of the perfection of man the more modest idea of the "guardian of creation" who is the "custodian of all ends-in-themselves" [*Treuhänder aller Selbstzwecke*] (Böhler and Brune 2004: 83).

The second point is not subject to such temporal qualifications but concerns metaphysics and the nature of freedom. The utopian's hopeful orientation toward the future is eschatological, in the sense that it strips the present of its metaphysical dignity. That this is inadmissible follows from the normative content of Jonas's teleological understanding of being; the Ought is already contained within the Is. The basic future orientation of environmental ethics is thus already present in the teleological structure of life; the fundamental duty is one of conserving life's self-affirmation, not its supersession. Yet this is precisely the form which utopian hope takes; in Jonas's reformulation, utopian hope is essentially a wager on a final consummation in which being as a whole is at stake.

One might suspect that this bet can only be entertained as the stakes are implicitly lowered due to the implicit eschatological denial of reality to present being. Such metaphysical safeguards are not offered by Jonas. Climate disaster confronts human freedom with a realization of the extent of its power as well as of the inherent worth of being. Our choices are ineluctable and, Jonas argues, the only mediation between free human action and the intrinsic value of being itself is provided by the consciousness of responsibility – this is the starting point for a doctrine of ethical action in the era of climate catastrophe, and its first dictate is the absolute immorality of the wager.

Notes

1 Sections quoted from *The Imperative of Responsibility* are cited according to the English translation (Jonas 1984a); all other translations are my own.
2 Jonas lays out his account of the revolution in rationality in detail in Jonas (1984b).
3 In the following, I draw on Jonas's essays "Warum die moderne Technik ein Gegenstand für die Philosophie ist" and "Warum die moderne Technik ein Gegenstand für die Ethik ist" in Jonas (1987: 15–41, 42–52).
4 Cf. especially the contributions of Apel and Dietrich Böhler in Böhler and Brune (2004) and Böhler and Hoppe (1994). For a more detailed critical discussion, see Dews (1995: 151–168).
5 As Vittorio Hösle and others have pointed out, there is a difference between teleology as it refers to an end in itself and an end of nature (Hösle 1998: 73). Jonas's argument rests on demonstrating that in this exceptional case these two forms of teleology converge.

6 As Jonas put it once in conversation: "There is just something self-evident about the idea that it cannot be the end of spirit [*Geist*] to make future spirit impossible" (Böhler and Hoppe 1994: 208). It is also in this sense that Jonas argues that the mere fact that we have an intuition of our responsibility for nature proves that we do in fact have it; the intuition tracks our ontological attunement with nature.

7 There is no doubt that Jonas's presentation of the relationship of Marxism to technology is crudely schematic and fails to do justice to Marx's many explanations of how productive forces can be transformed into destructive forces. Nonetheless, this point is incidental to his general characterization of "secular eschatology."

References

Apel, Karl-Otto, *Verantwortung und Diskurs: Das Problem des Übergangs zur postkonventionellen Moral*, Suhrkamp, Frankfurt am Main, 1990.

Böhler, Dietrich; Brune, Jens Peter (eds.), *Orientierung und Verantwortung: Begegnungen und Auseinandersetzungen mit Hans Jonas*, Königshausen & Neumann, Würzburg, 2004.

Böhler, Dietrich; Hoppe, Ingrid, *Ethik für die Zukunft: Im Diskurs mit Hans Jonas*, C.H. Beck, Munich, 1994.

Dews, Peter, *The Limits of Disenchantment: Essays on Contemporary European Philosophy*, Verso, London, 1995.

Hösle, Vittorio, *Philosophie der ökologischen Krise. Moskauer Vorträge*, C.H. Beck, Munich, 1994.

Hösle, Vittorio, *Objective Idealism, Ethics, and Politics*, St. Augustine's Press, South Bend, IN, 1998.

Hösle, Vittorio, "Être et subjectivité: les implications métaphysiques de la crise écologique," *Laval théologique et philosophique*, 69(1), 135–158, 2013. https://www.erudit.org/fr/revues/ltp/2013-v69-n1-ltp0818/1018359ar.pdf.

Jonas, Hans, *The Imperative of Responsibility: In Search of an Ethics for the Technological Age*, University of Chicago Press, Chicago, 1984a.

Jonas, Hans, "The Practical Use of Theory," *Social Research*, 51(1/2), 65–90, 1984b. https://www.jstor.org/stable/40970931.

Jonas, Hans, *Technik, Medizin und Ethik: Praxis des Prinzips Verantwortung*, Suhrkamp, Frankfurt am Main, 1987.

Jonas, Hans, *Dem bösen Ende näher: Gespräche über das Verhältnis des Menschen zur Natur*, Suhrkamp, Frankfurt am Main, 1993.

Jonas, Hans, *Philosophische Untersuchungen und metaphysische Vermutungen*, Suhrkamp, Frankfurt am Main, 1994.

Jonas, Hans, *Philosophical Essays: From Ancient Creed to Technological Man*, Atropos Press, New York, 2010.

Jonas, Hans, *Organism and Freedom: An Essay in Philosophical Biology* (eds. Beckers, Jens Ole, Preußger, Florian) (Online-Appendix zu Bd. I,1 KGA), 2016. http://hans-jonas-edition.de/?p=142.

Spaemann, Robert, *Philosophische Essays*, Phillip Reclam, Stuttgart, 1983.

PART 4

Beyond the Environmental Apocalypse

13

THE IMPROPER APOCALYPSE

Vitalism with and against a Psychoanalytic Approach to the End of the World

Timothy Secret

In these fractious times, one claim that reasonable people ought to be able to agree on is that the belief that anthropogenic climate change will lead to the apocalypse is an illusion.

The statement above appears unlikely to make me many friends amongst the readers of this volume, though perhaps I can turn my fortunes around with the swift addition that I am using the word 'illusion' in a strictly Freudian sense. When explaining this psychoanalytic concept in *The Future of an Illusion*, Sigmund Freud memorably uses the example of a middle-class girl who believes that one day she will marry a prince (Freud SE.XXI: 31).[1] He immediately clarifies that it is possible that the girl will be proven correct: such marriages have indeed taken place, and it is this possibility that distinguishes an illusion from a delusion. However, irrespective of the eventual result or its current likelihood, her belief would still be an illusion insofar as 'a wish-fulfilment is a prominent factor in its motivation' (Freud SE.XXI: 31). While there is little mystery in a girl's wish to break with her monotonous life of labour and care for one of luxury and romance, nor in the religious illusion that Freud's example set out to clarify – that of belief in an unconditionally loving, paternal deity who protects us and brings justice – it seems significantly more puzzling that anyone would actively wish for apocalyptic closure in association with the climate catastrophe that is already taking place.

That said, if we were to try to explain such a desire using the resources of psychoanalytic theory, we would rapidly discover an embarrassment of riches. One might hazard that the absence of a solid and settled psychoanalytic theory, either of the human actions driving contemporary climate change or of the long tradition of apocalyptic prophecy, is not due to the difficulty of accounting for such material in terms of unconscious desire but rather due to the ease of constructing a host of such theories. Some of these might prove compatible and overdetermine each other in specific cases, while others might remain contradictory yet nevertheless

DOI: 10.4324/9781003189190-17

each be true of the actions or visions of certain agents or groups. It is thus easy to imagine a volume akin to this one, written entirely within the bounds of psycho-analytic theory, in which dozens of thinkers would offer entirely novel explana-tions of the same phenomena. The clear danger for any investigator in such circumstances is that the choice of approach becomes as much a display of personal idiosyncrasies as a representative response from the discipline.

To illustrate this claim, let us ask: is the universal annihilation of the present world, irrespective of the birth of another world promised to some, to be theorised as a nostalgically willed-for return to the subjectless and objectless simplicity of auto-erotic or even uterine existence, or as the product of a future-oriented nar-cissistic ego that can only accept the prospect of death if everyone and everything follows it into the dark? Is the environmental devastation that piles up around us an inevitable price rationally paid for *Kultur* as protection against a Hobbesian state of nature, a return of the repressed driven by obsessive actions that psychoanalytic awareness might alleviate, a product of collective psychosis that can barely be understood and never cured, or the result of a rampant death drive in which a primary masochism has turned outwards to attain its end? Furthermore, if there is some human tendency to desire the end of this world and to produce images of linear or cyclical apocalypses – ones found far beyond the Abrahamic faiths, from Ragnarök to the Stoic *ekpyrōsis* – then what peculiar psychic effects might emerge when a creature prone to such fantasies encounters them concretely realised within their lifetime?[2] Alternatively, what would we say of the psychology of those cul-tures that seem to have no such apocalyptic discourses?[3]

Instead of simply pursuing one strand in this ever-growing list of potential pro-jects, in this chapter I will attempt something more disruptive. Rather than allowing psychoanalysis to remain a pure theory that can be applied to decode the underlying meaning of our actions or fantasies, I will argue that the psychoanalytic death drive in its Freudian formulation is itself inhabited by an unruly apocalyptic illusion when thought at an ecological scale. This illusion, I claim, renders it 'metaphysical' in the pejorative sense with which Freud himself would have wiel-ded that term. Even if such contaminations are inevitable and it proves impossible to reformulate a stable, functional notion of the death drive without this meta-physical crutch, there is a clear danger when a theory designed to expose illusions shows itself to be founded on one. Furthermore, even if it is neither a matter of jettisoning this concept nor repairing it, there remains the chance that a decon-structive engagement with this apocalyptic inhabitation might provide the opportunity for the emergence of a more modest discourse. Assuming that the repetition compulsion of the death drive is at the heart of the actions driving climate change (a claim I will support but not establish here, insofar as that would require evaluating and rejecting the other options outlined above), perhaps this more modest discourse might allow us to address climate change more success-fully by being attentively suspicious towards figures of apocalypse, and indeed of what I will call 'proper death' more generally, even if we cannot rid the theory of them.

Put bluntly and briefly, if there is something we wish for in the religious fantasy of the apocalypse, then tying the contemporary climate catastrophe and our response to it to images of that apocalypse being realised or averted might, for quite obvious reasons, prove counterproductive. In this basic claim, we are positioning ourselves against a dominant theoretical discourse that holds there to be some motivational value in drawing the apocalyptic parallel, as for example in Castro and Danowski's claim that 'The Anthropocene is the Apocalypse, in both the etymological and eschatological senses' (Castro and Danowski 2017: 22). We instead would like to drive this fantasy out of the drive itself, aiming to be left with climate catastrophe as a non-apocalyptic, senseless and merely horrible form of devastation – one in which we have no libidinal investment. The hope is that this will provide a stronger fulcrum around which to enact a shift in our repetitive acts, that is, a shift in the drive's pressure, aim, object or source.[4] Without offering a positive model of precisely what that change ought to be theoretically or how it might practically disrupt the global patterns of action that are driving climate change, we might through this at least have made a little progress in dispersing a veil of metaphysical ideology to allow clear-sighted creative action. Indeed, we would suggest that the very urge to identify the apocalypse with climate catastrophe, which we pitch as obscuring and debilitating rather than revealing and empowering, is a consequence of the metaphysical grounding of the death drive in the notion of 'proper death'. Against this, we would proclaim an 'improper apocalypse', which is, more properly speaking, not an apocalypse at all.

In the background of our method is a broader engagement with the notion of the 'proper' in connection with death, guilt and debt, an ancient association that was being repeated in its definitive form in the work of Martin Heidegger in the same post-war period that Freud launched his speculation on the death drive.[5] While Heidegger's account will form an implicit background to these reflections, we will focus here on some common threads and counter-threads of these discourses that were being woven into the work of nineteenth-century vitalist thought, particularly by the anatomist Xavier Bichat. As a positive model, I claim that Bichat at the turn of that century was offering a non-metaphysical discourse on the end of life (individual and by extension collective) that breaks with the classical conception of death in association with debt and that might allow us to respond to catastrophe more effectively. However, his theory is nevertheless incapable of addressing the specific nature of climate change. Freud and Heidegger, both in different ways, made genuine and irreversible progress but at the same time plunged us back into this classical conception of death. To begin with, however, let me must briefly introduce one problematic dimension of the metaphysical tradition that will concern us below.

The Proper Apocalypse

If we were to provide a snapshot of the so-called 'marketplace of ideas' since Friedrich Nietzsche's seminal reflections on nihilism, we might say that it has been dominated by a familiar pitch in which the Western metaphysical tradition is the long-term culprit for the perilous state of the world, yet also the soil from which a narrow, fragile cure might grow. It is this cure that each great thinker or movement claims to be the unique custodian and dispenser of. Thus, in a classic version of this move that we find in Heidegger, one that went on to play a foundational role in many ecological discourses, the metaphysical tradition going back to the very beginning of Western thought is the root of our impoverished modern technological *Gestell* ('enframing'), in which every object can only appear as a resource for human exploitation, 'that revealing through which the real everywhere, more of less distinctly, becomes standing-reserve' (Heidegger 1977: 24). However, while this situation in which objects can no longer reveal themselves as themselves might call for us to break with 'metaphysics' or even 'philosophy', it is certainly not a call for us to stop 'thinking' in some new and different sense – indeed, it is perhaps what calls genuine thinking into action.

One common critical move within this marketplace is, of course, to prove that some rival thinker or school remains at least partly entrenched in the metaphysical tradition that has led to the particular problem (whether it be nihilism, genocide, or environmental catastrophe), rather than to accept that the work of the school or individual constitutes the radical departure they might claim to represent. Again, Heidegger provides a classic example of this move, asserting that Nietzsche was not an anti-philosopher standing outside of the tradition that had culminated in nihilism, but was instead the last great metaphysician in the Platonic line (Heidegger 1991: §1–10).

For such critical moves to hold substantive interest rather than being mere acts of academic one-upmanship, it is important to show that this metaphysical inheritance has tangible deleterious effects. To show that this is the case with Freud, I will need to give a quick account of one dimension of what it means to be problematically 'metaphysical'. I should stress that this is certainly not the only way in which a theory can remain 'metaphysical'; indeed, it is not even the dominant way that this label is wielded critically.[6] Nor am I declaring that there are no other ways of using the label 'metaphysical' outside of this form of critical move. For the sake of clarity and space, I will need to reduce this facet of the tradition to a relatively simplified schema that will allow us to recognise it at work not only in metaphysics but also in contemporary climate activism and in apocalyptic thought.

The oldest preserved fragment of Western thought, the so-called 'Anaximander fragment', states the following:

> Whence things have their origin, there they must also pass away according to necessity; for they must pay penalty and be judged for their injustice, according to the ordinance of time.
>
> *(Quoted in Heidegger 1984: 13)*

Although there has been a great deal of sophisticated philosophical reflection on this claim and its nuances in the Ancient Greek language, if we take a simple approach to its interpretation then the message is quite clear. Indeed, not only is the message quite clear, but the majority of climate-related thought seems to take place to the beat of this ancient wisdom. After all, it is rarely disputed that a climate catastrophe would be anything other than us finally paying a penalty and being judged for our injustice according to the ordinance of time. Our planetary destruction aimed against other animals and ecosystems, humankind living reck-lessly beyond the means of a finite globe, and our technological vision of the world as a mere resource for exploitation – these would all be posed as outstanding debts long accrued and still growing exponentially. As such, it is not only their intang-ibility that means one cannot legitimately dispute with nature or necessity when these debts are finally called in, however unfair it may be that those debts will be largely paid by future generations and the natural world rather than by the gen-erations who have been the main beneficiaries of this destruction. It is not without cause that we would commonly label someone ruefully reflecting on the justice and appropriateness of our coming suffering in light of our ongoing behaviour towards the natural world 'philosophical'.

Anaximander's reflection, which inaugurates Western metaphysics and still structures our most basic ethical and political reasoning (from the conviction that dictatorial tyrants will eventually be brought to justice or that they are at least miserable in their tyranny, to the common wisdom that 'there is no such thing as a free lunch'), sits at the heart of the Christian faith. Here, we have a God who knows not only our actions but our intentions and hearts, who loves us yet because of this love punishes us appropriately to bring about justice, and who reveals all of this in magnified form in the apocalyptic moment when his nature reveals itself in its fullness. Humankind, as a whole, will then finally pay penalty and be judged for our injustice. All seven bowls of God's wrath have been destined for us since creation and are perfectly apportioned to be deserved by the sinners they rain down on.

Although many theological currents would see the salvific elements of the apocalypse as unmerited by us, as a gift of grace freely given by God to his unde-serving creation, the key point here is that the negative aspects of the apocalypse are entirely appropriate. For all its lavish detail and apparent contingency – four horsemen, seven trumpets, seven bowls, one beast, seven heads, ten horns – not a thing in this great vision is too much or too little. The apocalypse is the great balancing of accounts that closes the very business of accountancy that has been going on in the Book of Life. Perhaps life itself is revealed here as nothing other than this accruing of debt in anticipation of the great repayment that ends history. Of course, each of our lives has already been guided by such balancing of accounts and justice was never reserved for the end times, yet in this process there have always been problematic remainders – unpunished crimes and unmerited suffering. It is only with the apocalypse that all remainders are paid and every loose end tied up.

Adopting a Heideggerian vocabulary, we might say that the apocalypse is 'proper' to humankind, its 'proper' end – that this is humankind's 'ownmost' in the same way that death is the individual's ownmost and proper end (Heidegger 1962:

279–311). Like death, the apocalypse is ever present throughout humankind's life, as with each sin the apocalypse swells. Against the now famous Parisian graffiti, another end of the world is *not* possible. Alternatively, if another end of the world is possible, it would need to be fought against constantly to bring about the ordained end – the only end appropriate insofar as it settles all debts without remainder. Insofar as our psychic suffering according to psychoanalysis consists of such unpaid remainders and unlaid ghosts, those who 'cannot rest until the mystery has been solved and the spell broken' (Freud SE.X: 122), the prospect of such a final settling of accounts is equivalent to the prospect of freedom from such psychic torment, even if it is justly paid for by great physical tribulation.

For the apocalyptic mind, there are many vulgar ends of the world but only one authentic end. It is within this context that we need to worry when climate catastrophe is elevated to apocalyptic status and judged as appropriate to us and our many injustices. Just as seven bowls of wrath rather than six or eight was the precisely appropriate number for our sins, similarly the climate catastrophe is pitched as the perfectly proportioned response to the chain of actions that might be dated back to the first industrial use of fossil fuels or beyond.

To give flesh to just one way in which the above association might have tangible impacts: if we hold that climate catastrophe is our proper and appropriate end, that we are paying a penalty and being judged for an injustice rooted in our technological attitude towards the natural world as a standing-reserve that exists only for the imposition of our productive will, then this stance provokes a default opposition to any form of geoengineering. From such a perspective, the only proper response to climate catastrophe that might ward off catastrophic suffering is a fundamental remaking of the human–nature relationship in terms of humility, while any response based on a great technological gambit and assertion of human dominance over the planet can only be a further act of hybris for which we will eventually pay an even greater penalty in accordance with metaphysical law. Of course, it may well be correct that all proposed geoengineering solutions are to be rejected, yet there is nevertheless an obscuring, ideological dimension to this rejection. The illusion that climate catastrophe is our proper apocalyptic end itself provokes a host of further illusions, as actions are appraised and beliefs held not on the basis of rational assessment but as wish-fulfilments.

The Century of General Anatomy

It has become a scholarly cliché when introducing psychoanalysis to mention that the twentieth century itself was inaugurated by the publication, supposedly in 1900 even if it was really in late 1899, of Freud's *On the Interpretation of Dreams*. From there, one can refer to that century with its modernist anxieties, sexual revolutions, generational rebellions, historical traumas and irrational violence as 'the century of psychoanalysis', a label that once proclaimed its contemporaneity even if today it seems to damn psychoanalysis to being old news – although without the label, this cliché, including the inaccurate dating of the publication of *On the Interpretation of*

Dreams, was perhaps first practiced by Freud himself in his chapter for the *Encyclopaedia Britannica*'s wonderfully titled 1924 volume, *These Eventful Years: The Twentieth Century in the Making as Told by Many of Its Makers, Being the Dramatic Story of All that Has Happened throughout the World during the Most Momentous Period in All History.*[7]

Today, one might try to start the search for a text conveniently published around the turn of the millennium that similarly captures the twenty-first-century's core dramas, though it is perhaps still too early to recognise what they will have been. Nevertheless, if one were placing bets based on the wildfires and melting glaciers increasingly dominating our news feeds, it seems realistic to suspect that it will be something oriented more towards the Earth than the psyche. However, what about the nineteenth century?

Conveniently near the cusp of that century, we have apparently the first use of the phrase 'industrial revolution'. This occurred in a letter dated 6 July 1799, by the French envoy Louis-Guillaume Otto, announcing that France was entering the race to industrialise (cf. Crouzet 1996: 45). Perhaps that captures what the 19th century will have been – not the start of the carbon-burning industry itself, certainly, but the start of a global race to imitate that model. By 1809, the Ruhr Valley in Westphalia was proudly dubbed 'Miniature England' for its lines of factories that mirrored Manchester, and by the end of the century much of the world could boast horizons indistinguishable from England's dark satanic mills. Nevertheless, we must surely admit that Otto's letter hardly constitutes a theory.

As such, perhaps the work that opens the 19th century is Johann Fichte's 1799 *The Vocation of Man*. Could there be a better slogan for the nineteenth century's ambition than its claim that 'Nature leads men through want to industry; through the evils of general disorder to a just constitution; through the miseries of continual wars to endless peace on earth'? (Fichte 1965: 162). Here, we have an atheist theodicy, if not a diabolodicy. Want, lack and need, along with the industry they necessitate, are no longer curses placed on humankind for the original disobedience of reaching beyond our allotted place; instead, they are the very tools nature uses to drive us to gloriously reach beyond that place and become akin to God in our productive capacity.

However, for all of Fichte's appropriateness, I would propose a slightly less obvious candidate for the distinctive publication that opened the 19th century. This comes from the French anatomist Xavier Bichat, who, having spent much of the previous decade dissecting the bodies of victims of the French Revolution's guillotine, opened his 1800 *Physiological Researches on Life and Death* with the famous definition: 'Life consists in the sum of the functions by which death is resisted' (Bichat 1827: 11).

With this apparently simple claim, Michel Foucault saw heralded 'The Age of Bichat' and the birth of modern medicine (Foucault 2003: 122). Elizabeth Haigh writes that, while superficially Bichat's claim might sound like a truism, at another level this opening expresses the basic conviction of later French vitalism – that 'a living organism must be understood as a focus of reaction against the forces of decay and decomposition which continuously assail it and finally prevail at death' (Haigh 1975: 72). The key point here is that these forces of decay and decomposition are situated outside of the living organism, at least insofar as it is living, and as such the forces it is exposed to are contingent, a question of the environment a particular body is exposed to. Furthermore, without those external forces of death, there would be no reaction and hence no life.

It would not be difficult to miss the novelty in this and only to hear a repetition of Baruch Spinoza's account of the conatus, according to which everything, 'insofar as it is in itself, endeavours to persist in its own being', that it 'is opposed to all that could take away its existence', and that this endeavour 'is nothing else but the actual essence of the thing in question' (Spinoza 1997: IIIp6, IIIp6d and IIIp7). However, in Bichat's vitalism there is not only no abstract *anima* or *élan vital* that animates all living matter, there is also no singular essence that might play a causal role in the manifold biological actions going on within an organism. As he put it in his *General Anatomy* two years later – he utterly rejects 'a single principle, purely speculative, ideal, and imaginary, whether designated by the name of *soul, vital principle*, or *archeus*' (Bichat 1822: viii). In Georges Canguilhem's words, such a vitalism 'is first of all the rejection of all metaphysical theories of the essence of life' (cited in Wolfe et al. 2020: 221).

The diverse reflex actions of the 21 types of tissue that Bichat patiently identified, or more accurately their *re*-actions against the forces of decay and decomposition outside of them, do not have a common source or ground. They are unified and identifiable as 'living' forces only through their common result, which can only weakly be construed as anything akin to an Aristotelean telos. The biological forces of resistance acting in living matter work conservatively to maintain and allow the repetition of those very forces, which react against the ever-present and ever-pressing threat of a collapse back into the state of inorganic matter in which only non-living forces remain. The living locus, an inside, is constantly reestablished and protected by those very living forces that struggle to keep the forces of death outside. The metaphors here are drawn from combat: attack and defence, power and resistance, the living organism as a citadel constantly under siege and eventually overrun. In terms of this battle, life is not a positive power or a fixed substance but can only be measured in the form of a differential: 'the difference which exists between the effort of exterior power, and that of internal resistance' (Bichat 1827: 12). Life emerges as a calculation, the momentary result of an equation of forces.

In summary, we might say then that the passing of urine, the beating of the heart, ovulation, the formation of blood, and all the thousands of other biological activities carried out by very different tissues in very different parts of our bodies and to very different rhythms constitute a single organism only insofar as they contribute to a common function – the delay of death.

As Foucault memorably wrote in *The Birth of the Clinic*, Bichat's transformation of vitalism was the illumination that gave birth to modern biology, in which the 'living night is dissipated in the brightness of death' (Foucault 2003: 146).

However, this illumination by death comes from a construction of death and its relation to life that is unlike any encountered before in the history of philosophy. In Foucault's words:

> Bichat relativized the concept of death, bringing it down from that absolute in which it appeared as an indivisible, decisive, irrecoverable event: he volatilized it, distributed it throughout life in the form of separate, partial, progressive deaths, deaths that are so slow in occurring that they extend even beyond death itself.
>
> *(Foucault 2003: 144–145)*

This relativisation, volatilisation and pluralisation manifests in several different ways. Certainly, one of these paradoxical deaths that 'extend even beyond death itself' (a problematic formulation insofar as the consequence of this theory is the rejection that there is any privileged *death itself*) is in the manner that after the guillotine brings an end to *animal* life many *organic* life processes continue at their own various rhythms in different parts of the headless body (for example, chemical processes in the stomach will continue to digest food for a little while) – a pluralisation and distribution of both life and death. However, another is the way in which, throughout its whole existence, the assemblage of tissues was only ever a living assemblage insofar as it was reacting to – being shaped and sculpted by – the forces of death outside it. As Gilles Deleuze pointed out, Bichat is breaking here with the classical conception of death by simultaneously presenting death as coextensive with life and as made up of 'a multiplicity of partial and particular deaths'. Deleuze even names as 'Bichat's Zone' this domain of 'partial deaths, where things continually emerge and fade' (Deleuze 1988: 95, 121).

Death is no longer the end of a life but its constant provocation; death is the environment within which life temporarily arises when a systematic resistance to death establishes an inside. The moment of death is, by definition, the outside overrunning the citadel that had been resisting it. As Bichat describes: 'In living bodies [...] whatever surrounds them, tends to their destruction. They are influenced incessantly by inorganic bodies [...] they could not long subsist, were they not possessed in themselves of a permanent principle of reaction' (Bichat 1827: 11).

In this break with the classical conception of death, we find that every death is contingent, external and improper to the organism. Although life itself is a reaction to the forces of death and is only found in relation to its prompting, there is no singular death proper to an entity, even if eventually a death is inevitable as the internal forces of life that maintain a particular inside exhaust themselves, diminish and are swept away. Indeed, there is no such thing as a natural death within this modern picture; while doctors may euphemistically refer to death from old age, in reality there is always something in particular that the diminished forces of life can no longer resist.

Returning to the initial conceit that opened this section, aside from the birth of modern biology for all its importance, why might Bichat's thought characterise the nineteenth century in general? We can say this because his picture of the establishment and constant maintenance of an inside within and against an environment of destructive forces was rapidly taken up beyond the realms of biology, particularly in the understanding of political society. This occurred in various ways across the 19th century – first in Henri de Saint-Simon and following him in August Comte, where these ideas played an important role in the foundation of sociology. However, they found their fiercest advocate in Arthur Schopenhauer, who lauded Bichat's *Physiological Researches on Life and Death* as 'one of the most profoundly conceived works in the whole of French literature', going on to state that 'his reflections and mine mutually support each other, since his are the physiological commentary on mine, and mine are the philosophical commentary on his; and we shall best be understood by being read together side by side' (Schopenhauer 1966: 261).[8] It is in this context that it is unfortunate that many texts on the philosophy and metapsychology of psychoanalysis address Schopenhauer's account of the will while making no reference to Bichat, even though one could make a strong case that Freud is approaching these ideas from reading physiological texts rather than philosophical ones.[9]

Although Freud does not refer to Bichat's texts directly, just as few scientists would refer to works within their rapidly advancing fields that are a hundred years out of date, he is the inheritor of this great nineteenth-century biological tradition that has its direct roots in his works. Although his most famous and direct biological speculations take place in 1920 with the sixth chapter of *Beyond the Pleasure Principle*, in which Freud engages with a key follower of Bichat's biology, August Weismann, we can fairly say that the entire metapsychology from the early formulations of 1895's *Project for Scientific Psychology* through to key papers such as 1915's 'Instincts and Their Vicissitudes' is rooted in a model where the psyche, as further inside within the inside of the biological organism, is created and differentiated in response to external forces from body and world.[10] Freud's systematic account of the drives begins from 'the angle of *physiology*', the reflex arc and the model of a hostile stimulus being applied to a living tissue '*from* the outside' that causes it to withdraw to protect itself, and it is from this Bichatian starting point that he considers the drives as a stimulus coming from the inside that necessitate more complex action insofar as they 'make far higher demands on the nervous system and cause it to undertake involved and interconnected activities by which the external world is so changed as to afford satisfaction to the internal source of stimulation' (Freud SE.XIV: 118, 120).

One key idea that Freud takes from this tradition is the claim that the infant, prior to achieving the organisation and systematisation of its various drives to constitute an inside equivalent to the complete biological entity, is a field where various component or partial drives act independently: 'the subject's component instincts, each on its own account, seek for the satisfaction of their desires in his own body' (Freud SE.XII: 321). Within this original auto-erotic disorder, the independent component drives (which are only 'components' insofar as they will later be systematised, and not out of any lack in themselves), which have their source in different erotogenic zones, are equivalent to multiple lives constituting different insides under the provocation of multiple forces of death. The systematic organisation and subjugation of these as a child develops into a singular being that has mastered its drives sufficiently to be capable of rationally ensuring its survival and reproduction as a complete entity is itself only something that exists only for the most part and in times of health. What we witness in Bichat's headless body that continues its digestive functions, which is that the unification of those processes for the purposes of a single living creature is only a part of the truth, is equally seen in how the localised drives display their independent functioning in moments of weakness.

Freud and the Many Ends of the World

Although the claim might sound bombastic, if, as Bichat claims, 'Life consists in the sum of the functions by which death is resisted', then perhaps we can say that our collective, political life 'consists in the sum of the functions by which the end of the world is resisted.'[11] For all its apparent extravagance, this is perhaps only a way of capturing the core of what Saint-Simon and Comte took from Bichat in their thinking about society. Indeed, one might propose it as a manner of putting forward another great nineteenth-century idea: the Marxist theory of economic and social reproduction. That which unifies the thousand different actions that take place within a political society and that which distinguishes them as distinctively socio-economic actions – from working in the fields to raising children, from joining the military to running for electoral office – it that they act to conserve the collective human society itself. Indeed, even the most dramatic revolutionary violence remains conservatively opposed to the total collapse of the collective world that would come about if all such activity simply ceased.[12]

The purpose of such a formulation, in light of the above reflections on the non-classical account of death in Bichat, is to propose a break with the traditional model of the apocalypse as the end of the world that we have seen to be modelled on 'proper death' and the closure of accounts. Instead, we would reconceive it as an end of the world that is relativised, volatised, and distributed throughout collective life – that is to say, not really as an apocalypse at all, but rather as a pluralised end occurring at multiple rhythms and in different places. Furthermore, this would be an end of the world that is the very prompt of collective life and that is invigorating rather than debilitating. The metaphysical dream of a proper and appropriate closure – the full and final payment of all debts, the establishment of justice and an existence after the end of the world that involves no further accountancy – has no place in this model. As such, there would be no romanticised apocalypse that might contaminate a disaster such as climate change with a sense of its appropriateness or justice.

At the same time, within this nineteenth-century vision of a social world created by constant conservative activity against the forces of darkness relegated firmly to an outside, anthropogenic climate change would enter the scene as a very peculiar twist. When we think through climate change, we realise that those acts of economic and social reproduction that we had been engaging in to preserve the world against the external forces of death and decomposition are themselves directly leading us towards the end of the world. Where for Bichat we saw that in living bodies whatever surrounds them, the inorganic, pushes them towards death and they cannot subsist without a permanent principle of reaction, at a societal level with the revelation of climate change the principle of reaction itself emerges as a tendency towards destruction.

This demoniacal twist is unthinkable within Bichat's model itself. However, it has a precise name in a particular development of that biological model. Freud, in 1920, named it the 'death drive'. In Todd McGowan's summary: 'According to the theory implied by the death drive, any movement toward the good – any progress – will tend to produce a reaction that will undermine it. This occurs both on the level of the individual and on the level of society' (McGowan 2013: 14).

It is this capacity to think the climate catastrophe as an unavoidable self-undermining of actions performed to stave off catastrophe that explains why I stated above that Freud's theory constitutes a genuine and irreversible step forward, even if what it reveals is an aporia. Not only does Freud offer us an account of this demoniacal action in specific cases, the ground-breaking move that perhaps constitutes a break with Foucault's 'The Age of Bichat' is that this is placed as the basis of all drives: there is not simply a duality of life drives and death drives, of Eros and Thanatos, as certain readings of Freud's text assert. Rather, the life drives are themselves seen as only a diverted form of the death drives. As McGowan puts it, Freud's 'self-proclaimed dualistic conception of the drives – first the sex drive and the self-preservative drive, then the life drive and the death drive – is actually a dialectical conception in which a single drive produces an antagonistic struggle' (McGowan 2013: 11).

How, then, if Freud constitutes an advance on this point, does he at this very same moment also represent a lapse? After he has proposed the death drive and stated that the final goal of all organic striving is death, he goes on to claim that:

> The hypothesis of self-preservative instincts, such as we attribute to all living beings, stands in marked opposition to the idea that instinctual life as a whole serves to bring about death. Seen in this light, the theoretical importance of the instincts of self-preservation, of self-assertion and of mastery greatly diminishes. They are component instincts whose function it is to assure that the organism shall follow its own path to death, and to ward off any possible ways of returning to inorganic existence other than those which are immanent in the organism itself.
>
> *(Freud SE.XVIII: 39)*

It is in this notion that an organism is driven to 'follow its own path to death' against any other 'possible ways of returning to inorganic existence' that we encounter in Freud the notion of the 'proper death' that is distinctive of the conventional account of death from Anaximander to Heidegger's *Being and Time*. The organism is thus driven to pursue a path that would have no debt or remainder insofar as its end is proper to itself, its ownmost, an idea that pushed up to the level of collective society justifies the desire that humankind have a particular end that is appropriate to it. It is this that can easily take the form of climate catastrophe, insofar as today the guilt we feel over our fundamental collective injustice would be based on our relation to the natural world as exploitative beings, just as 50 years earlier nuclear war had appeared as our proper apocalypse in the face of our injustices to other humans as bellicose beings.

Of course, there is no way to simply remove this claim from Freud's theory of the death drive, as without this notion of a singular, 'proper' end it would be unintelligible that the death drive does not lead the human organism to simply throw itself from the cliffs or stop eating, just as politically we would seem to be able to achieve the aims of the apocalypse through general inaction on the model of Herman Melville's (2004) Bartleby. It is the notion that we are working to avoid improper deaths and to bring about our ownmost, proper death that allows us to understand how the death drive is moulded into serving the interests of life, and yet this very distinction of the proper and improper ought to be excluded insofar as it constitutes a step backwards into classical metaphysics.

At this point, there seem to be two options open to us. One would be to discover if there is a reformulation of the death drive, perhaps that of Jacques Lacan who moves against the implicit biologism of Freud's model, that might avoid some of the issues outlined above. The second is to accept the contamination of metaphysics and a certain suturing as a constitutive element in any closed theoretical system, which does not amount to a rejection of that system or its uselessness but rather a call to vigilance and the lingering bad conscience that Jacques Derrida welcomes in the political arena. We can perhaps do no better than to continue to pursue our apocalyptic visions and wander in the dreams of others, but to do so in an attitude of distrust and with our ears pricked for the subtle turnings that constitute the self-sabotage of our attempts to avert catastrophe.

Notes

1 For ease of reference, I will refer to Freud's works using the volume number rather than year of publication, as found in *The Standard Edition of the Complete Psychological Works of Sigmund Freud* (SE), edited by James Strachey.

2 As a reflection on the Oedipus Complex and the events of Sophocles' *Oedipus Rex* illustrates, the actual realisation of one's deepest desires in concrete reality can be highly traumatic.

3 As the anthropological reflections of Eduardo Viveiros de Castro and Déborah Danowski have shown in *The Ends of the World*, rather than simply offering visions of the 'end of the world' we are pushed by the investigation of global myths towards considering structure, including not only the 'world before us', the 'world without us', and the 'world after us', but also such uncanny options as 'us before the world' (Castro and Danowski 2017).

4 Freud's classic presentation of drive analysed into these four dimensions can be found in 'Instincts and Their Vicissitudes' (Freud SE.XIV).

5 Freud tentatively and with much hesitation launched his death drive in *Beyond the Plea-sure Principle*, which was published in 1920 (Freud SE.XIX). Heidegger's most famous engagement with death and the 'proper' takes place at the start of the second division of *Being and Time*, which was published in 1927, though he was engaging with these themes in his teaching earlier in the decade (Heidegger 1962).

6 The dominant form of this accusation relates to the distinction between Being and Becoming and the positing of an unchanging, stabile world behind this world. For the classic criticism of this, see 'How the "True World" Finally Became a Fable' (Nietzsche 2005: 171).

7 The text for this volume, 'A Short Account of Psycho-Analysis', is found in Freud (SE. XIX: 191–209).

8 For an excellent account of these developments, see Esposito (2012).

9 For an otherwise excellent work that engages deeply with the tradition of thinking about the drives in philosophy leading up to Freud, particularly in Schopenhauer, yet without referring to Bichat, see Bernet (2020).

10 These texts can be found in Freud (SE.XIX), Freud (SE.I), and Freud (SE.XIV), respectively. Outside of quotations and references to the title of 'Instincts and Their Vicissitudes', I will follow the standard practice in commentary of translating *Triebe* as 'drives' rather than 'instincts', against Strachey's translation.

11 To refer to the end of 'the world' rather than merely 'our world' is a rhetorical step I cannot justify here.

12 This may be well why Melville's (2004) Bartleby is such a challenging figure to society, a figure who eventually translates his refusal to engage in social reproduction to the level of biological reproduction in refusing to eat.

References

Bernet, R., 2020. *Force, Drive, Desire: A Philosophy of Psychoanalysis*. Evanston, IL: North-western University Press.

Bichat, X., 1822. *General Anatomy, Applied to Physiology and Medicine, Volume 1*. Boston: Richardson and Lord.

Bichat, X., 1827. *Physiological Researches on Life and Death*. Boston: Richardson and Lord.

Castro, E.B.V. de and Danowski, D., 2017. *The Ends of the World*. Cambridge: Polity.

Crouzet, F., 1996. 'France'. In Teich, M. and Porter, R. (eds.) *The Industrial Revolution in National Context: Europe and the USA*. Cambridge: Cambridge University Press, pp. 36–63.

Deleuze, G., 1988. *Foucault*. Minneapolis: University of Minnesota Press.

Esposito R., 2012. *Third Person: Politics of Life and Philosophy of the Impersonal*. Cambridge: Polity.

Fichte, J.G., 1965. *The Vocation of Man*. La Salle, IL: Open Court.

Foucault, M., 2003. *The Birth of the Clinic: An Archaeology of Medical Perception*. Abingdon, UK: Routledge.

Freud, S., 2001. SE.I. *Project for a Scientific Psychology. The Standard Edition of the Complete Psychological Works of Sigmund Freud: Volume I (1886–1889)*. London: Vintage, pp. 283–343.

Freud, S., 2001. SE.X. *Analysis of a Phobia in a Five-Year-Old Boy. The Standard Edition of the Complete Psychological Works of Sigmund Freud: Volume X (1909)*. London: Vintage, pp. 3–152.

Freud, S., 2001. SE.XII. *The Disposition to Obsessional Neurosis. The Standard Edition of the Complete Psychological Works of Sigmund Freud: Volume XII (1911–1913)*. London: Vintage, pp. 311–326.

Freud, S., 2001. SE.XIV. *Instincts and their Vicissitudes. The Standard Edition of the Complete Psychological Works of Sigmund Freud: Volume XIV (1914–1916)*. London: Vintage, pp. 109–140.

Freud, S., 2001. SE.XVIII. *Beyond the Pleasure Principles. The Standard Edition of the Complete Psychological Works of Sigmund Freud: Volume XVIII (1923–1925)*. London: Vintage, pp. 3–66.

Freud, S., 2001. SE.XIX. *A Short Account of Psycho-Analysis. The Standard Edition of the Complete Psychological Works of Sigmund Freud: Volume XIX (1923–1925)*. London: Vintage, pp. 191–212.

Freud, S., 2001. SE.XXI. *The Future of an Illusion. The Standard Edition of the Complete Psychological Works of Sigmund Freud: Volume XXI (1927–1931)*. London: Vintage, pp. 3–58.

Haigh, E., 1975. 'The Roots of the Vitalism of Xavier Bichat', *Bulletin of the History of Medicine*, 49(1): 72–86. https://www.jstor.org/stable/44450204.

Heidegger, M., 1962. *Being and Time*. Oxford: Blackwell.

Heidegger, M., 1977. *The Question Concerning Technology, and Other Essays*. New York: Garland.

Heidegger, M., 1984. *Early Greek Thinking: The Dawn of Western Philosophy*. New York: HarperCollins Publishers.

Heidegger, M., 1991. *Nietzsche: Volumes One and Two*. San Francisco: HarperSanFrancisco.

McGowan, T., 2013. *Enjoying What We Don't Have: The Political Project of Psychoanalysis*. Lincoln: University of Nebraska Press.

Melville, H., 2004. *Bartleby the Scrivener*. New York: Melville House Publishing.

Nietzsche, F., 2005. *The Anti-Christ, Ecce Homo, Twilight of the Idols, and Other Writings*. Cambridge: Cambridge University Press.

Schopenhauer, A., 1966. *The World as Will and Representation, Volume 2*. New York: Dover Publications.

Spinoza, B., 1997. *Ethics*. Murfreesboro: Middle Tennessee State University. https://capone.mtsu.edu/rbombard/RB/Spinoza/ethica-front.html.

Wolfe, C., Penocelli, M., and Wong, A., 2020. 'What Kind of Vitalism in *The Normal and the Pathological?*' In Méthot, P.-O. (ed.) *Vital Norms, Canguilhem's The Normal and the Pathological in the Twenty-First Century*. Paris: Hermann, 219–250.

14

WIPING AWAY THE TEARS OF ESAU

Adorno's Reconciliation with Nature

Agata Bielik-Robson

In the recent debates on ecology, Theodor Adorno's name resurfaces very rarely, despite the fact that his late thought is concerned almost uniquely with the idea of *reconciliation with nature*.[1] In this chapter, I will attempt to reconstruct Adorno's variation on the Hegelian theme of reconcilement (*Versöhnung*) as very different from the idea of return to nature, and because of that as a unique – perhaps even the most convincing – solution to the problem of antagonism between humankind and natural life. In contrast to the post-humanist position, here associated with Martin Heidegger's famous 'turn,' I will call Adorno's unfinished project 'neohumanism' and explain it through the famous quote from Emmanuel Lévinas, according to which "a little humanity distances us from nature, a great deal of humanity brings us back." The *Zohar*, the early kabbalistic *Book of Splendor*, describes the gesture of wiping away the tears of Esau – the biblical emblem of natural life trampled and overcome by a higher form of existence following God and his law, represented by Jacob, Esau's victorious twin brother – as the necessary precondition of redemption, envisaged as a reunion of the conflicted principles of nature and civilization, sensual and spiritual life.[2] Below, it will emerge as the best metaphor (perhaps even an inspiration) for Adorno's philosophical strategy. I will attempt to show that his reconcilement with nature does not aim at atonement/at-one-ment, which would annul the anthropological difference, but at the ethical act of giving justice to nature understood as the Lévinasian other.

Yet, it must be kept in mind that Adorno, despite his frequent use of Jewish *theologoumena* (Adorno 2004: 299), is *not* Lévinas. Or rather, he *is* as much Lévinas as he *is* Heidegger, whose "Question Concerning Technology" Lévinas directly attacks in the little text from which the above quote derives. In what follows, I will thus claim that Adorno's model of reconciliation emerges also as a dialectical sublation of the antithesis between Lévinas and Heidegger, where the former stands for the agonistic attitude towards nature as humankind's eternal other while the

DOI: 10.4324/9781003189190-18

latter represents the view of humans' harmonious belonging to the order of *physis* (*Zugehörigkeit*), where nature and humans are parts of the one 'history of Being' (*Seinsgeschichte*).

Lévinas on Nature: A Call for a New Humanism?

The short essay from *Difficult Freedom*, written by Lévinas to a wide public in the 1960s under the title "Heidegger, Gagarin and Us," deals mostly with the human relationship with nature. In this little highly polemical and polyvocal text, Lévinas positions himself as a thinker going against the flow of the *Zeitgeist*, which is now falling into the Heideggerian *Stimmung* of the cheap critique of modern technology and the prophesy of the imminent apocalypse of nature. In his parody of the apocalyptic tone of the post-Heideggerian 'deep ecology,' he thus ironically states that

> In the future, to exist will mean to exploit nature; but in the vortex of this self-devouring enterprise there will be no fixed point. The solitary stroller in the country, who is certain of belonging [this is obviously Heidegger himself], will in fact be no more than the client of a hotel tourist chain, unknowingly manipulated by calculations, statistics and planning. No one will exist for himself [*pour soi*].
>
> *(Lévinas 1990a: 230)*

And although Lévinas admits, now in his own voice, that "there is some truth in this declamation. Technical things are dangerous. They not only threaten a person's identity, they risk blowing up the planet" (Lévinas 1990a: 230), he perceives the Heideggerian position as merely 'reactionary,' that is, as coming from the 'enemies of industrial society' who

> forget or detest the great hopes of our age. For faith in man's liberation has never been stronger in human souls. This faith [...] identifies only with shaking up sleepy civilizations, eroding the heavy dullness of the past, fading local colour with the fissures that crack all these cumbersome and obtuse things that burden human particularisms.
>
> *(Lévinas 1990a: 230)*

For Lévinas, this revolutionary disenchantment, waking people from their dogmatic slumber and the 'nocturnal sluggishness' of their archaic tribalism, is already inscribed in the ancient text of the Jewish revelation, which he, following Hermann Cohen, understands as the first enlightenment undertaken by the 'religion of reason.' "*Torah* jolts the Real," he declares in his *Talmudic Lectures*, meaning precisely the beginning of the process of disenchantment/enlightenment rebelling against the 'magical forces' of nature (Lévinas 1990b: 39). The reactionary turn against technology, therefore, which Lévinas associates with the name of Heidegger, is not just a case of simple anti-modernism: it is a reaction against the much older paradigm which Lévinas calls 'Hebrew humanism.'³

In Lévinas's account, the name 'Heidegger' stands for the anti-humanist turn in late modern philosophy, which advocates the radical de-centring of the privileged position of man and the abolishment of the so-called 'anthropological difference'. Lévinas diagnoses the rise of Heidegger's popularity in terms of the 'flood of the pagan recesses,' which remained dormant for ages but now became once again awakened in "our Western souls" (Lévinas 1990a: 231). Contrary to the modern development of science which disenchanted natural reality by reducing it in the process of *Entzauberung* to the objectual 'en-framing' of resources, the Heideggerian 'pagan awakening' aims at the re-enchantment of the world (precisely the way it was postulated by the German Romantics, most of all Schiller, the author of the concept of *Entzauberung*):

> This, then, is the eternal seductiveness of paganism, beyond the infantilism of idolatry, which long ago was surpassed. The Sacred filtering into the world – Judaism is perhaps no more than the negation of all that. *To destroy the sacred groves – we understand now the purity of this apparent vandalism.* The mystery of things is the source of all cruelty towards men.
>
> *(Lévinas 1990a: 232, emphasis added)*

The Sacred filtering into the world: this, for Lévinas, constitutes the gist of the danger of repaganization, against which only Judaism can be an efficient weapon, because it was the original monotheism, still preserving the memory of the violent separation from the magical 'immanent sacrum' and the 'lure of immanence' (Lévinas 1990a: 232) – even if that necessarily ruthless struggle involved the 'apparent vandalism' of the sacred groves. Lévinas makes no bones about it: *it was worth it.* The sacred groves, now the favourite victim-objects of idealising ecological nostalgia, were once the places of human sacrifices – the dark centres of the pagan cult, in which humans offered themselves to the cruel goddess of Nature, vainly hoping for her favours.

Having rejected the repaganising nostalgic-reactionary strategy, Lévinas announces: "A little humanity distances us from nature, a great deal of humanity brings us back" (Lévinas 1990a: 232). Does he mean it here ironically? Or is he identifying with this statement? Is he still criticising the Heideggerian call to return to the oneness of Being/*physis* – or is he already proposing his own vision of the human inhabitation of the world, part of which would be a new rapport with nature? Hard to say. The rest of his intervention is devoted to the name 'Gagarin,' which stands for the technological success of the instrumental reason that subjugated the Earth, uprooted man radically from his attachment to the Place, made him nomadic and light, and now readies itself for the conquest of the whole cosmos. 'Gagarin,' therefore, represents the triumph of enlightenment which, for Lévinas, stretches from the Judaic disenchantment of the pagan magic of *genii locorum* – the spirits of sacred groves and places – to the ultimate *lekh lekha* of the late modern human being who managed to 'get himself out' of the Mother of all Places, the Earth, thus repeating the Abrahamic gesture of obedience to the divine imperative of "get thee out of here," bidding him to leave the place of his origin. Poising himself in deliberate opposition to any romantic resacralization of nature, Lévinas defends the iconoclastic purity of Judaism against Christianity, which "integrated the small and touching household gods into the worship of saints, and local cults" and thus regressed monotheism to the pagan 'enrootedness':

> Judaism has not sublimated idols, on the contrary, it has demanded that they be destroyed. *Like technology, it has demystified the universe. It has freed Nature from a spell* [...] But it has discovered man in the nudity of his face.
>
> *(Lévinas 1990a: 233–234, emphasis added)*

The only *sacrum*, therefore, which Judaism would allow is an enlightened humanism, cherishing other human beings not in their glory but in their precarious and vulnerable 'nakedness.' While it engages in the violent struggle of cosmic proportions against Nature's magical spell, it abandons all violence in the encounter of *alterity* – the face of another human being, lonely and naked in its antagonism against the natural cosmos. Judaism's ethical sense would thus be based on the solidarity of human placeless 'others' alienated from participation in the natural world: other to nature's harmony, other to its 'sacred groves,' even 'other to being' itself (*autrement qu'être*).

But, if we followed Lévinas's undeveloped suggestion to rethink our rapport with nature in terms of 'more humanity,' he could have emerged as a more promising dialectical thinker who rejects both antithetical extremes – synonymized by 'Heidegger' and 'Gagarin' – and gestures towards a solution which draws on the sources of Judaism as a non-violent and more reconciliatory variant of the anthropological difference: the human beings, precisely because she is human and thus different from other natural beings, can – *should* – treat their *otherness* with respect and restraint. For, it is not just the human face of the other that grounds our ethical sense of alterity; there is also a powerful experience of *natural otherness* which should become a subject of moral deliberation as well. The tears on the face of this natural other, biblically represented by Esau – the forgotten, red-haired and uncouth twin from whom Jacob, the founder of Judaism, stole his blessing – should also be recognized and wiped away. But Esau is still weeping, because his legacy – the sensual life of a purely natural being – had not been blessed and thus included in the history of Abrahamic religions: Esau remains banned in the wilderness of Edom, the natural realm *extra muros*, outside Holy Jerusalem and its utopian model of human civilization. This is precisely the moment when Adorno the Comforter enters the scene – in order to wipe away Esau's tears.

After Violence: Adorno's Reconciliation

The very term, *Versöhnung* – 'reconciliation,' implies that we are not dealing here with a simple *return*, as in the case of 'Heidegger' chosen by Lévinas to represent a pagan nostalgia for 'sacred groves.' If 'deep ecology' can indeed be accused of some fantasmatic naïveté, which projects natural life as perfectly functional biological machinery and then opposes this absolute efficiency to the piles of waste and ruin caused by human *hubris*, Adorno is completely free of any naturalist *Heimweh* that would present nature as an 'ecosystemic' harmonious home of life, superior to the human, technological mode of existence which dangerously spirals into the unknown. *The return is impossible* – not only because there is nothing to return to (as the 'Anthropocene' thinkers rightly point out, there are no 'pure' natural eco-systems left on Earth that would remain unaffected by human intervention), but also because it is simply not desirable. The term 'reconciliation' suggests that there is a *difference* between human and natural ways of being and that it is not to be erased but merely negotiated. In other words: Adorno emerges as a strong advocate of the *anthropological difference*, but with an equally powerful dialectical twist, which allows us to see the process of man's becoming different from nature and, *once* the difference is securely established, of attempting to form a new relationship with nature as the non-human other. Separation was good and necessary, and here he does not differ from Lévinas, who also saw the *detachment/displacement* from the maternal body of nature as a welcome sign of human emancipation. Following the Herderian school of philosophical anthropology (an *a priori datum* of all German thinkers of his time, Freud included), according to which the human being is a *Mängelwesen*, a 'deficient creature' born prematurely and unable to survive in hos-tile natural environment, Adorno believes that in the case of humankind nature is indeed a 'cruel stepmother.'[4]

While Adorno's 'negative dialectics' is often regarded as a Kantianized Hege-lianism, so is his concept of reconciliation with nature. After Immanuel Kant – and *only* after Kant – Georg Hegel can come, which means that the human being first has to make herself distinct from the natural Kingdom of Necessity and *only* then, no longer influenced by the natural criterion of the mere survival of the fittest, re-establish a connection with the natural world. This re-linking, however, can occur solely on the basis of her own, distinctly human set of values—most notable of which is *justice*. Adorno thus keeps close to the main tenets of Jewish ethics, which were best articulated by Steven Schwarzschild in his theological version of the Kantian dualism of *Sein und Sollen*: "To have an ethics means to have an alternative to reality" (Schwarzschild 1990: 73). In Adorno's rendering, this rule reads as: *to have justice means to have an alternative to nature*. Adorno repeats the Kantian critique of nature as the Kingdom of Necessity: the oppressive form of life oriented towards sheer survival and the maintenance of 'bare life' in itself, where everything becomes a means to an end and nothing can be an *end in itself*, as it should be in the moral Kingdom of Freedom or the *Realm of Ends*. For Adorno, therefore, the Kantian antinaturalist foundation of ethics as the alternative to the natural *Realm of Means* must precede the Hegelian moment of reconciliation, which makes sense solely as a later *act of justice*, an *ethical gesture non-deducible from the natural way of life*. On Adorno's dialectical account, *nature as such knows no justice*; in order to gain an intuition of justice, the human being must leave the world of nature (Exodus), establish her own ethical Kingdom of Freedom (Desert), and only then, already inspired by the new principle of justice, which aims at the universal *Realm of Ends*, turn towards nature again and learn to treat it *justly* – that is, not as an object of instrumental calculation but as an end in itself demanding ethical attention and respect for its *other* way of being (Promised Land). Paradoxically – but only see-mingly so – the human subject must first exit nature, which in itself is a violent act, in order to stop violating it. As long as it is still 'in nature,' the human subject, driven by fear and the natural exigency of survival, exacerbated by its 'deficiency,' fights against other natural beings, which it perceives as danger and competition. But as long as it engages in the Exodus out of nature and leaves the domain of the 'bare necessities of life,' it can re-embrace the other, non-human objects forming natural landscapes and ecosystems as *ends in themselves* – yet without any nostalgic urge to go back to the natural form of life.

The phrase 'Exodus out of nature,' does not appear here by accident. The bib-lical narrative of *yetsiat mitsraim*, 'the getting out of Egypt,' forms a deep theological canvas for Adorno's dialectical story of enlightenment: first, the exit out of 'natural bondage'; then, the violent struggle with nature on her own grounds, which simply turns the tables of natural domination by means of technology and instru-mental reason; and finally, the realization of the promise hidden in the very symbol of the Exodus, that is, the *true* exit from nature, where the subject does not content itself with just reversing the poles of domination (which is still a 'natural thing to do'), but ventures beyond *any* game of mastery and violence into the properly universal and ethical Realm of Ends. The enlightenment, which perceives itself only as disenchantment (the Lévinasian violent destruction of the 'sacred groves') states its goal by saying that we, humans, must leave the realm of nature, which is dominated by the rule of survival and which, as the 'cruel stepmother,' does not privilege the human species in this respect – precisely for the sake of survival. But for Adorno, this forms the major contradiction of enlightenment as such, which, by sticking to the criterion of survival, only proves that it never left the natural realm: "The negative fact that the mind, failing in identification, has also failed in reconcilement, that its supremacy has miscarried, becomes the motor of its disen-chantment" (Adorno 2004: 186). To get out of nature *truly* would thus mean to leave the principle of survival altogether – and not just to seek a possibility of sur-vival outside the natural 'place' that, having shaped humans as 'outcasts of nature,' denied them survival in the first place. Enlightenment, therefore, should not be a *revenge on nature*: it should not proceed along the apocalyptic scenario of 'leaving the Place' (Lévinas 1990a: 233) and turning it into a heap of ashes. It should be a time of justice which will have given justice to nature as the perennial other of humanity. But as long as enlightenment only cares about self-preservation, then, structurally speaking, it does not differ from 'pagan' mythology, the main concern of which was to secure human survival on natural grounds: disenchantment as the main weapon of instrumental reason merely manages to reverse the relations of power, where it is now rational subjectivity which gains domination over nature. Confused with nothing more than disenchantment, *Entzauberung*, enlightenment reproduces the very essence of myth from which it wanted to free itself in the first place: *the cycle of power*. As the definition of myth in *Dialectic of Enlightenment* states: "For in its figures mythology had the essence of the *status quo*: cycle, fate, and domination of the world reflected as the truth and deprived of hope" (Adorno and Horkheimer 2002: 27).

Enlightenment understood as a merely superficial and partial manoeuvre of gaining distance from nature, does not, therefore, prevent humankind from imitating *the worst* aspect of nature itself, namely the principle of domination and its cyclical fatalistic master–slave dynamics of 'ups and downs': rise and fall, glory and shadow, growth and decay, power and ruin, *genesis kai phthora*: "What men want to learn from nature is how to use it in order wholly to dominate it and other men. That is the only aim" (Adorno and Horkheimer 2002: 4). But if that is the only purpose – just the reversal of the poles of domination within the 'eternal cycle' – then enlightenment, as a strategy of getting out from the mythological world, must be doomed: "Just as the myths already realize enlightenment [by confronting nature on its own grounds], so enlightenment with every step becomes more deeply engulfed in mythology" (Adorno and Horkheimer 2002: 12). Only when reminded of the now-forgotten messianic promise may enlightenment relinquish the survivalist games of power and establish itself firmly in its human ethical *otherness*, and only *then* can it come to reconcile with what it left behind: it shall give nature *justice* by recognizing its non-human *other* way of being, yet also without giving in to the nostalgic temptation of renaturalization, of undoing separation, and, in the Heideggerian manner, becoming one with nature again. It is due to this residual promise that Adorno stakes his hopes for a new Exodus: "Enlightenment itself, having mastered itself and assumed its own power, could break through the limits of enlightenment" (Adorno and Horkheimer 2002: 172).

But, does it mean that the subject shall now atone for the violence done to the natural world, yet without rejecting it as a mere error? Or, on the contrary, should she reject it as an error from the start, a severe distortion of the process of enlightenment as 'getting out' of archaic myths that could have gone completely otherwise? The 'Hegel' in Adorno – the dialectician attuned to the historical difference between *now* and *then* – opts for the former by confirming, together with Lévinas, that separation is a *necessarily* violent act and that it can find peace with what it got separated from only later, in time. But the 'Kant' in Adorno – the severe moralist, having no time (literally) for dialectical theodicy – opts for the latter by stubbornly claiming that violence as such must be negated in totality, now and then, and that the recognition of nature as the non-human other could have occurred without the brutal cut. There is thus a certain undecidability in Adorno's notion of the reconciliation with nature. While its 'Hegelian' part sees the separatory violence as the necessary stage of enlightenment, indispensable for the emergence of the ethical value without which there would be no concept of justice and, subsequently, no imperative of reconciliation, its 'Kantian' part calls for a radically new beginning (an equivalent of the Heideggerian *neuer Anfang*) in which the Exodus would have happened again, but this time in full innocence, without inflicting violence on the Egypt of nature: no tears of Esau, no need to wipe them away.

By using the names of Hegel and Kant in inverted commas – as virtual counter-positions of the inner conflict in Adorno's thought – I do not mean to suggest that any part of it can actually be reduced to the historically real Hegel and Kant. Adorno does not simply recover the Kantian moment of rapture between *Sein und Sollen* – the natural necessity, on the one hand, and human ethical freedom, on the other – within the Hegelian context of *Versöhnung*, thus delaying the reconciliatory finale in the indefinite process of negative dialectics. Both, the real Kant and the real Hegel are severely criticized by Adorno as the two major representatives of German Idealism in which the violence of the Spirit over the sensual element is tacitly assumed – and condoned. It is precisely this never-questioned violent ingredient that pushes Kant to phrase the dualism of nature versus ethics in terms of "the separation of the sensual and intellectual realms": "by the *chorismos*, sensuality is designated as a victim of the intellect" (Adorno 2004: 389). Hegel's Idealist thinking, which unites the whole of human history and reduces sensual contingency to one scenario of the triumphant self-recovery of the Spirit, is also full of implicit violence:

> In the midst of history, Hegel sides with its immutable element, with the ever-same identity of the process whose totality is said to bring salvation. Quite unmetaphorically, he can be charged with mythologizing history. The words 'spirit' and 'reconcilement' are used to disguise the suffocating myth.
>
> *(Adorno 2004: 357)*

Adorno's negative dialectics, therefore, is grounded in a complex set of sublations which simultaneously reject the idealist garment of Kant's and Hegel's arguments and preserve their hidden truth which only now can come to the fore: the true value of *chorismos*, which separated human beings from nature, yet not as hindering but as prompting the right kind of reconciliation:

> Reconcilement would release the nonidentical, would rid it of coercion, including spiritualized coercion; *it would open the road to the multiplicity of different things and strip dialectics of its power over them*. Reconcilement would be the thought of the many as no longer inimical, a thought that is anathema to subjective reason [...] none of the reconcilements claimed by absolute [...] has stood up, whether in logic or in politics and history.
>
> *(Adorno 2004: 6, emphasis added)*

Adorno's model of reconcilement wants to avoid the Idealist trap of rational violence, where it is only the human spirit/reason which one-sidedly dictates the conditions of truce. While commenting on *Dialectic of Enlightenment*, Jürgen Habermas stresses the *symmetry* of the Adornian New Deal between the human and her other, nature, as resisting conceptualization and finding a refuge in 'sheer impulse,' the last natural residue in the human psyche, otherwise dominated by the rule of reason:

> they [Horkheimer and Adorno] meet instrumental reason with a 'mindfulness' or 'remembrance' [*Eingedenken*] that seeks out the stirrings of a rebellious nature rising up against its instrumentalization. They even have a name for this resistance: *mimesis*. The name evokes associations that are intended: *empathy and imitation*. It calls to mind a relationship between persons in which the accommodating, identifying externalization of one partner in relation to the model of the other does not require the sacrifice of that partner's own identity, but preserves dependency and autonomy at once [...] But this mimetic capacity evades any conceptual framework fashioned for subject-object relationships alone; so *mimesis* appears as *sheer impulse, the exact antithesis of reason* [...] By way of his *Negative Dialectics*, Adorno tries to circumscribe what cannot be presented discursively; and with his *Aesthetic Theory*, he seals *the surrender of cognitive competency to art* [...] Adorno summons this to be the single witness against a praxis that in the course of time has buried everything once meant by reason [*Vernunft*] under its debris.
>
> (Habermas 1987: 68, emphasis added)

The mimetic-empathic faculty, best developed in modern art, becomes thus a new vehicle of reconciliation in which the natural other does not gain its own voice or an alternative conceptual language, but remains silent, non-communicative, and thus alien – yet, precisely because of that, rescued and preserved in its difference:

> The reconciled condition would not be the philosophical imperialism of annexing the alien. *Instead, its happiness would lie in the fact that the alien, in the proximity it is granted, remains what is distant and different, beyond the heterogeneous and beyond that which is one's own.*
>
> (Adorno 2004: 191, emphasis added)

This happy empathy with the alien requires also an attitude of patience-passivity, which goes beyond the idealist notion of emancipation and freedom. The true liberation of the human subject can thus occur only if the other of nature is no longer felt as a negative *constraint* which impinges on the subjective sense of freedom, but rather as a positive *restraint* which acts as a moral scruple: "The subject would be liberated only as an I reconciled with the not-I, and thus it would be also above freedom insofar as freedom is leagued with its counterpart, repression" (Adorno 2004: 283). To reconcile oneself with nature, therefore, would mean to be able to coexist with the natural other as posing in front of us a moral restraint and not an obstacle to absolute freedom, the manner in which it was perceived by modern *ratio*, forcing the antagonistic attitude in which happy "uncontrolled mimesis is proscribed" (Adorno and Horkheimer 2002: 148).

Habermas, however, presents Adorno's project with deep reservations, pointing to its non-resolved ambiguity: is the 'mimetic faculty' a strong alternative to separation by violence, or is it rather a historical stage to be achieved only in the act of the *nachträglich* atonement, the afterwardness of coming *after-violence*, which would replace the previously hostile moment of separation, yet without negating separation itself? I agree with Habermas: the Lévinasian account of the vehement *chorismos* is closer not only to the historical truth of humankind emancipating itself from the sway of the 'sacred groves,' but also to the psychoanalytical teaching on separation as an inescapably violent act, fully confirmed by Freud (1955) in *Moses and Monotheism*. Violence cannot be cancelled in some magical manoeuvre of undoing (*Ungeschehenmachen*) that is invalidated and foreclosed by the 'mimetic faculty': it has to be confronted as the ineluctable moment of the history of humankind and only *then* counter-acted, so it does not lead to the ultimate revenge on the natural world or the Apocalypse of Nature. Just as the tears of Esau were, sadly, necessary, so is the subsequent act of justice, which will have wiped them away.

Adorno's Messianic Aesthetics: Beauty without Power

I would thus like to read Adorno with the Habermasian correction in mind as a thinker more dialectical than simply antithetical, and to see his idea of reconciliation of nature as not negating but rather as negotiating the gap between man and nature. In 'Elements of Anti-Semitism,' Adorno attributes this milder kind of reconcilement to Judaism, which he sees as free from the violent rule of pure spirituality, which is more characteristic of Christianity and its philosophical avatar in German Idealism:

> Reconciliation is Judaism's highest concept, and expectation its whole meaning. The paranoid reaction stems from the incapacity for expectation.
>
> *(Adorno and Horkheimer 2002: 163)*

Expectation – as opposed to the impatient exercise of the will – belongs too to the new passive-patient attitude towards the other of nature, which stakes on the dialectical work of time: only 'Time, the Healer' can eventually appease the wound of violent separation, which cannot be closed by any voluntaristic *fiat*. To expect and anticipate reconciliation, therefore, means to give up on the decisionistic power acting either way – either in favour or against separation from nature – and open oneself to the capacities of 'disempowerment,' *Entmachtung*.[5]

As Habermas rightly notices, in order to see the other side of enlightenment – not just the violent, 'paranoid' and vengeful destruction of the 'sacred groves,' administered by Western reason, but, more predominantly, the ethical promise of universal justice addressing itself to humans and non-humans alike – Adorno shifts towards the domain of aesthetic theory and its main concept: 'beauty deprived of power.' This Hegelian phrase, deriving from the preface to *Phenomenology of Spirit*, appears originally in the negative context while being contrasted with the 'power of Understanding' which shatters the *machtlos* images and introduces its own disenchanting dominion of concepts. Against Hegel, Adorno takes up his concept in an affirmative manner: the only hope for better enlightenment, which would announce a true Exodus out of the relations of power, lies in the powerless beauty which has *already* given up on violence (not 'always already,' as Jacques Derrida would have put it, but only now, *after* all the violence done by reason):

> Men had to do fearful things to themselves before the self, the identical, pur-posive, and virile nature of man, was formed, and something of that recurs in every childhood [...] The road [of civilization] was that of obedience and labor, over which fulfillment shines forth perpetually – but only as illusive appearance, a beauty deprived of power.
>
> *(Adorno and Horkheimer 2002: 33)*[6]

It is precisely beauty's *Entmachtung*, making it so frail under the scrutiny of the mundane, suspicious, cynical and power-driven modern eye, that now becomes the source of the ethical alternative, pointing to an altogether different world sharply cut off from 'everything that exists,' *der Bannkreis des Daseins*. The *Entmachtung* becomes thus *beauty's strength* from the perspective of the messianic reversal. The only possible form in which the utopian-redemptive fulfillment can manifest in the world as it *is*: "when we are hoping for rescue, a voice tells us that hope is in vain, yet it is powerless hope alone that allows us to draw a single breath" (Adorno 2005: 121).

The hope for reconcilement with nature, so far subject to 'fearful things' on the human's part, shines particularly strongly through Adorno's notes on literature, which were written in the last years before his death in 1969. In the essay devoted to Johann Wolfgang von Goethe's late drama, *Iphigenia in Tauris*, Adorno for-mulates the final definition of what he sees as the epitome of the dialectical Judaic wisdom: "Reconciliation is not the simple antithesis of myth; rather, it includes justice toward myth" (Adorno 1992: 169). Aging Goethe, having lived through the early Romantic Storm and Pressure (*Sturm und Drang*) movement, and now critical of its naïve Rousseauist premises, seeks a "civilized naturalness" (Adorno 1992: 161) – and this synthesis also becomes a goal for late Adorno, who, in 1968, watches the return of the repressed nature in his young students with fear, anxious that it will too result in the turn against culture and what he, after the Weimar master, calls 'humaneness' (*Humanität*).

'Humaneness' constitutes the main theme of the play, which can be regarded as one of the first critiques of colonialism widely understood: while it tells the story of the colonization of 'barbaric Scythians' by 'noble Greeks,' it uses it as a canvas for pondering the drama of the enlightened subject which must engage in the vand-alism of the 'sacred groves' in order to stop human sacrifices (the play begins in the 'old, shady, consecrated grove' devoted to the goddess Diana), and precisely because of that necessity to counter the archaic violence with a new 'civilized' violence, falls short of its own ideal of peace-loving *Humanität*: "Civilization, the stage of the mature subject, outstrips mythic immaturity, thereby becoming guilty toward it and entangled in the mythic web of guilt" (Adorno 1992: 158). The central point of the plot is the monologue of Orestes, Iphigenia's brother, who represents the violent aspect of enlightenment – a 'harsh antithesis to myth' – and articulates it with the kind of sincerity which we already know from Lévinas (and which also found its paradigmatic expression in Kurtz's last words in Joseph Conrad's (2011) *Heart of Darkness*: "exterminate all the brutes!"):

> The deep dialectic of the drama, however, should be sought in the fact that through his harsh antithesis to myth Orestes threatens to fall prey to myth [...] By condemning myth as something he is distant from, if not something he has fled from, Orestes identifies himself with the principle of domination through which, in and through enlightenment, the mythic doom is prolonged. *Enlightenment that flees from itself, that does not preserve in self-reflection the natural context from which it separates itself through freedom, turns into guilt toward nature and becomes a piece of mythic entanglement in nature.*
>
> *(Adorno 1992: 168, emphasis added)*

Convinced that "woman's words can be as powerful as a man's sword," Goethe leans towards Iphigenia's feminine gentleness; she is compassionate towards the barbaric king Thoas, who, taken by her persuasion, stops human sacrifices in the 'sacred grove.' Adorno endorses Iphigenia's strategy of non-violent negotiation and juxtaposes her with another feminine heroine of 'the most enigmatic work Goethe produced': Melusina. Iphigenia and Melusina represent the same principle of civilized renunciation of violence. While the former poses a stark contrast to her brother Orestes, ready to exterminate all the brutes, the latter, wearied of the violent outbursts of jealousy in her lover, shrinks to the size of a doll and begins to live in a chest where there is no longer room for any violent passion. Within this miniaturized fitted space, there is only room for what *befits* life and *fits* life's size, or what can be 'tactfully' *contained* in life's vessel – nothing *more*, no excess or surplus threatening to break the form or become unchained from the right measure/*tactus*/rhythm and self-restraint (as Adorno rightly points out, Goethe was obsessed with the idea of tact). Melusina's miniature existence allows only for gentle and self-restrained forms of civility, leaving no room for anything *incontinent*, also in the sense of the Romantic spirit of the Sublime, whose excessive cosmic proportions transcend any form or vessel; her chest symbolizes self-containment: it gives up on violent acts towards its natural environment. What for Friedrich Nietzsche, therefore, would be an alarming sign of decadence – a 'domestication' of natural instincts put in the stifling 'iron cage' of civilization – is for Goethe (and Adorno) is not a negative limitation: rather, it is a strategy of seeking rest, fulfillment, and peace. The *Sturm und Drang* of enraged drives wearies itself off, something which, in fact, is only natural. Both, Iphigenia's exhaustion (*Ermattung*) and Melusina's self-restraint include the natural process of weariness in their understanding of mature civilization: "Iphigenie's metaphor of exhaustion is learned from nature. It refers to a gesture that yields instead of insisting on its rights, but without self-denial" (Adorno 1992: 170). While Adorno earlier claimed that "what men want to learn from nature is how to use it in order wholly to dominate it and other men" (Adorno and Hor-kheimer 2002: 4), now he sees a possibility of another lesson being drawn from the natural world: weariness conceived not as a failure but as a gentle way of letting-go, a notion closely chiming with the Heideggerian *Gelassenheit*. Thus, just as Melusina resolves the problem of violence by leaving literally no room for it, Iphigenia believes in the peace-making power of exhaustion, which only seemingly is a weakness. While it yields to the other side, it does so without self-denial, thus maintaining separation and difference:

> The little chest in the Melusina story, one of the most enigmatic works Goethe produced, is the *counterauthority to myth*; it does not attack myth but rather undercuts it through nonviolence. In these terms it would be hope, one of Goethe's Orphic *ur*-words and one of the watchwords of *Iphigenie*: the hope that the element of violence contained in progress, the point where enlightenment mimics myth, would *fade away*; that it would *diminish*, or, in the words of the line from Iphigenie, 'become exhausted' (*ermattet*). Hope is humaneness' having escaped the curse, the pacification of nature as opposed to the sullen domination of nature that perpetuates fate.
>
> *(Adorno 1992: 169)*

For Adorno, this is the only meaning of hope: to escape the curse of violence, Cain's stain, and achieve a dialectical balance between the violent rule *of* nature over us (the Herderian stage of the helpless 'deficient being'), on the one hand, and the violent rule *over* nature by us (the Promethean stage of separatory violence), on the other. In that sense, "hope is not memory held fast but the return of what has been forgotten" (Adorno 1991: 120): the repressed empathic-mimetic image of the messianic peace as 'beauty deprived of power,' which also includes the natural world, itself exhausted from the struggle with human species in the age of the Anthropocene, in the gesture of 'yielding, but without self-denial' on the part of the equally wearied humankind, no longer so eager to vandalize the 'sacred groves,' but nonetheless assertive in its civili-zational difference. In that sense, Adorno's reconciliation by exhaustion (*Ermattung*) would be not so far from the position of Herbert Marcuse (2015) in *Eros and Civili-zation*, despite their fervent disagreement around the student revolt in the 1960s: Goethe's Melusina and Iphigenia could appear next to Orpheus and Narcissus, the two Marcusian emblems of the post-Promethean age and its new milder form of civility, which, having learned the lesson of time, has become *wearied of all violence*. The 'expectation' of reconcilement has thus finally matured into its *kairos: now* is the high time to wipe away the tears of Esau.[7]

Conclusion

While Lévinas is still the man of the violent *agon* – represented both by Pro-metheus, who stole fire from gods, and by Jacob, the twin who disobeyed the natural order and challenged God to steal from him a new blessing – Adorno, taking the side of the wronged brother, weeping Esau, moves forward to the final stage of *weary reconciliation* which perceives those sublime struggles of biblical/apocalyptic proportions – us, humans, against the natural universe which fights back in vengeance – as *out of time* and, after Melusina's miniaturising strategy, *out of place*. He truly wants to wipe away the tears of Esau: the 'hairy' uncouth twin, representing the order of nature, left behind in the wild deserts of Edom. But, *pace* Lévinas, this is not at all a betrayal of the monotheistic tradition of the first enlightenment. On the contrary, this is the very 'final scene' of the drama of Exodus, which no longer "attacks myth but rather undercuts it through nonviolence" (Adorno 1992: 169) and in this highly dialectical way achieves a mature formula of reconciliation, which Adorno associates with the intellectual legacy of Judaism: not return *to*, and not negation *of*, but reconciliation *with*, where the natural *other* stands next by, simultaneously in the distance of irreducible difference and in the proximity as our *neighbour*. "A little humanity distances us from nature, a great deal of humanity brings us back" (Lévinas 1990a: 232) or, rather, *closer* to it: Adorno's neohumanism is based precisely on this dialectics of *closeness* as the reconcilement of the 'suffocating myth' of unity (Heidegger), on the one hand, and violent disenchantment of all mythic 'sacred groves' (Lévinas), on the other. The 'great deal' of the Goethean-Adornian *Humanität* is no longer engaged in the sublime *agon* with nature, but represents 'civilized naturalness' that has become *wearied of all violence* and is now ready to yield – but without self-denial.

Notes

1 The most prominent exception here is Deborah Cook's *Adorno on Nature*, which juxtaposes Adorno with the thinkers explicitly concerned with ecological issues: "Arne Naess's deep ecology, Murray Bookchin's social ecology and Carolyn Merchant's ecofeminism" (Cook 2011: 122). However, my take on Adorno's thinking about nature will be different, focusing mostly on its implication in the 'the Jewish *theologoumenon*' of redemption as the no-longer-antagonistic arrangement of all beings.

2 *Zohar* II, Shemot 12b: "[The] Messiah will not come until Esau's tears have stopped flowing."

3 A conjecture, by the way, fully confirmed by Heidegger's own claims in *Black Notebooks*, in which the Jews figure as the "agents of the age of *Gestell*," propagating calculating instrumental reason acting against the sublime "truth of Being": "Within the time-space of the Christian West, that is, metaphysics, Jewry is the principle of destruction" (Heidegger 2015: 20).

4 Johann Gottfried Herder, the founder of philosophical anthropology, was the first thinker to define the human being as a creature tragically maladapted to the order of nature. Unlike all other beings which have their place in the natural system, humans are hopeless misfits, *Mängelwesen*, "the most orphaned children of nature" (Herder 1966: 108). Their only way out of this predicament of the originary lack is to compensate for the natural deficiency: produce culture which will provide her with substitute skills ('prosthetic limbs') and thus appease original anxiety. Herder's solution to the problem of human difference is thus highly dialectical: human being is simultaneously continuous with the natural world as its extremely 'experimental' or even 'erroneous' form of life and discontinuous because in order to survive she must resort to the innovative products of her own 'imagination,' which create the symbolic sphere of culture.

5 This accent on disempowerment puts Adorno in close proximity to late Heidegger, who also criticizes the voluntaristic paradigm of modernity and advocates 'waiting' (*Warten*) as a way out of the domination of will: "By not letting things be in their restful repose, but

rather – infatuated by his progress – stepping over and away from them, the human becomes the pacesetter of the devastation, which has for a long time now become the tumultuous confusion of the world [...] In waiting, we release things precisely into where we – as those who wait – let ourselves into, namely *into that in which we belong*" (Heidegger 2010: 149, emphasis added). Close, but not the same: while Heidegger's waiting is supposed to make us all – humans and nature – one again, Adorno insists on the difference and the sense of mutual otherness, which are not to be sublated but maintained in the emphatic *mimesis*.

6 Translation slightly altered: while John Cumming translates *die entmachtete Schönheit* as 'devitalized beauty' I want to emphasize beauty's dis/connection with power or its after-violent appearance.

7 This 'weariness' is also a theme of Franz Kafka's little parable on Prometheus, well-known to Adorno, to which Hans Blumenberg devoted the last chapter of his *Work on Myth*, "To Bring Myth to an End": when all the protagonists of the millennia-long struggle between man and nature – Prometheus, on the one hand, and the eagles devouring his liver, on the other – become tired of this endless repetition, in which trespass and revenge alternate within the eternal 'mythical cycle,' the wound itself, symbolizing the crux of this hopeless *agon, schliesst sich müde*: "The gods grew weary, the eagles grew weary, the wound closed wearily" (Blumenberg 1985: 635).

References

Adorno, Theodor W., Horkheimer, Max, 2002, *Dialectic of Enlightenment: Philosophical Fragments*, trans. Edmund Jephcott, Stanford, CA: Stanford University Press.

Adorno, Theodor W., 1991, *Notes to Literature, vol. 1*, trans. Shierry Weber Nicholsen, New York: Columbia University Press.

Adorno, Theodor W., 1992, *Notes to Literature, vol. 2*, trans. Shierry Weber Nicholsen, New York: Columbia University Press.

Adorno, Theodor W., 2004, *Negative Dialectics*, trans. E.B. Ashton, London: Routledge.

Adorno, Theodor W., 2005, *Minima Moralia: Reflections on a Damaged Life*, trans. Edmund Jephcott, London: Verso.

Blumenberg, Hans, 1985, *Work on Myth*, trans. Robert M. Wallace, Cambridge, MA: MIT Press.

Conrad, Joseph, 2011, *Heart of Darkness*. London: Palgrave Macmillan.

Cook, Deborah, 2011, *Adorno on Nature*, London: Routledge.

Freud, Sigmund, 1955, *Moses and Monotheism*, trans. Kathrine Jones. New York: Vintage.

Habermas, Jürgen, 1987, *The Philosophical Discourse of Modernity: Twelve Lectures*, trans. Frederick Lawrence, Cambridge, MA: MIT Press.

Heidegger, Martin, 2010, *Country Path Conversations*, trans. Bret W. Davis, Bloomington: Indiana University Press.

Heidegger, Martin, 2015, *Anmerkungen I–V (Schwarze Hefte 1942–1948. Martin Heidegger Gesamtausgabe*, Band 97), ed. Peter Trawny, Frankfurt am Main: Vittorio Klostermann.

Herder, Johann Gottfried, 1966, *Essay on the Origin of Language*, trans. Alexander Gode, Chicago: University of Chicago Press.

Lévinas, Emmanuel, 1990a, *Difficult Freedom: Essays on Judaism*, trans. Sean Hand, Baltimore: Johns Hopkins University Press.

Lévinas, Emmanuel, 1990b, *Nine Talmudic Readings*, trans. Annette Aronowicz, Bloomington: Indiana University Press.

Marcuse, Herbert, 2015, *Eros and Civilization*. Boston: Beacon Press.

Schwarzschild, Steven, 1990, *The Pursuit of the Ideal: Jewish Writings of Steven Schwarzschild*, ed. Menachem Kellner, Albany: SUNY Press.

15

LOOKING BEYOND THE APOCALYPSE

Environmental Crisis, Colonial Environmentalism and Eastern India's Tribal Communities

Vinita Damodaran

Humanity today is staring at an uncertain future. We can however draw lessons from the past, and being a historian today is a particularly advantageous position to be in, especially if you are an environmental historian interested in understanding turning points in history. Urban development 5,000 years ago heralded by the old empires was one major turning point, creating a threshold for the creation of new resource demands and markets; the emergence of the Industrial Revolution from 1800 with its resource implications was another, as was the period after 1945, which saw the rush to exploit fossil fuels on an unparalleled scale. As we can see from our current perspective, this was a period that has dramatically affected our global food production systems, the quality of the air and water we breathe and drink, our exposure to infectious diseases, and the suitability of the places we live in. Recently, the term "planetary health" has been coined to understand the changes to natural life support systems that are impacting our health and are projected to drive the majority of the global burden of disease to the most vulnerable and to future generations (Planetary Health Alliance n.d.).

Looking back as a historian, I could be pessimistic about whether we could halt the juggernaut that we have put in motion since 1800 or evaluate positively the impact of top-down initiatives by colonial empires in the nineteenth century, one example of which was the draconian legislation put in place in India under the British Empire which put one-sixth of the landscape into forest reserves because it was perceived that deforestation threatened the long-term security of the colonial state (Grove 1995). As a historian of India's tribal communities, who sit on some of the best bauxite reserves in the world, which are now being mined for the military-industrial complex of the West, I have seen the effects of relentless capitalist ethics in the region "chasing raw materials and other means of production from … non-capitalist strata and countries" (Luxemburg 1913). "Capital, impelled" in the words of Rosa Luxemburg "to appropriate productive forces for purposes of

DOI: 10.4324/9781003189190-19

exploitation, ransacks the whole world; it procures its means of production from all corners of the Earth seizing them, if necessary, by force, from all levels of civilization and from all forms of society" (Luxemburg, 1913).

But I have also been heartened by the resolve of India's aboriginal communities to resist the despoliation of their lands and territories by the state and corporate capital over a period of centuries.

Where we will be in 2050 will depend on whether we have truly grabbed the significance of this moment as a global society. Everyone today from the top scientists to policy makers are trying to understand the magnitude of what this means to us as a society and to understand risk and resilience in new and different ways. The solutions I hasten to add cannot be top-down or technocratic. They have to come from communities, localities, and be democratic bottom-up solutions, such as those being sought in places like eastern India where threatened groups ask 'development for whom?' as they challenge multinational mining companies polluting their water and their air armed with nothing but their anger and their understanding of different ways of living and being in this world.

Scientists are being seen as the historians of the future in the context of what has been termed "eco-apocalyptic" arguments (Sorlin 2014, Ortega Breton and Hammond 2016). Contemporary eco-apocalyptic arguments have been criticized for being anti-political, lacking in historical imagination and being more about individualized and small-scale actions which are mainly therapeutic in character. There is a danger that pushing the apocalypse could usher in non-democratic social transformation through conservation from above. In this sense, environmentalism can be understood as post-political, reinforcing "processes of depoliticization and the socio-political status quo" (Ortega Breton and Hammond 2016: 106). By rewriting the history of environmentalism from a historical perspective and from below and by presenting a deep engagement with locality and community, one can challenge these understandings and tap into new political and emancipatory futures.

The Anthropocene in recent literature has been seen as a tipping point, a period of civilizational collapse mapped by climate change, bio-diversity loss and phosphorous and nitrogen loss (Crutzen and Stoermer 2000). The dating of it is still unsure and the mapping of anthropogenic change has led some Earth system scientists to designate the last few centuries as the Anthropocene. The term itself is an informal geological and chronological one that marks the evidence and extent of human activities that have had a significant global impact on the Earth's ecosystems (Zalasiewicz 2010).

The Anthropocene has no precise start date, but based on atmospheric evidence may be considered to start with the Industrial Revolution (late eighteenth century) or even earlier (Lewis and Maslin 2015). By ceding the terrain of history to scientists in the debate on the Anthropocene and foregrounding planetary narratives on the fate of humanity, locality, class, gender and race all risk being glossed over. However, by bringing a strong historical and bottom-up perspective and by studying long-term environmental change, I hope here contextualize current eco-apocalyptic arguments within a longer time frame. Furthermore, whilst outlining turning points in global environmental history I hope to emphasize how a focus on the locality in the Anthropocene helps us move away from planetary debates and develop a more pointillist approach to understanding anthropogenic change.

Early Environmentalism and Scientific Responses under Colonialism

Western colonialism marked an important threshold or turning point in global environmental history. For example, in the West Indian plantations after 1492 a major turning point was reached in terms of rates of soil erosion, which were enhanced by the use of slaves. Slavery also brought Yellow Fever and other diseases and the development of an economic package of exploitation and consumption. The historian John F. Richards has outlined these step changes in his classic text *The Unending Frontier*, where he examines the relentless quest for resources with regard to fur hunting, whaling, fishing and pastoralism between 1500 and 1800 (Richards 2006). In brief, early colonialism brought about step changes in the rates of environmental and landscape development and environmental information collection, storage and transmission, and yet a further turning point in rates of change with early island plantation agriculture, hunting, fishing, and colonial pastoralism. In about 1800, the acquisition of very large continental colonies, especially India under the East India Company, brought about another turning point, requiring government departments to manage and monitor environments, forests, climates, biota and fishing, especially after the famine panics from 1877 to 1879. The recognition of the importance of the resources of India to England in terms of the supply of raw materials was recognized, for example, by English naturalist and botanist Joseph Banks by 1787 (Brockway 1979).

From the middle of the eighteenth century onwards, the utilitarian strain in natural history became more pronounced as scientists and collectors interested themselves in the practical benefits which might be reaped from their endeavours. In many respects, the world of nature had not endowed equally the peoples of the Earth. Botanists in particular believed that their technical skills would enable them to rectify the imbalance by transferring certain natural productions from one region to another and acclimatising them in the new environment. Periodic famines in India could be averted by planting sago or another crop. English cotton mills might be supplied with raw cotton from the empire following the transplantation of flax seeds. Such schemes had the mercantilist purpose of destroying the monopolies or predominance of rival nations and substituting Britain's own. Botany and great power rivalries became intertwined as nations endeavoured to guard their precious treasures.

The impacts of economic globalization was noted by scientists in the colonies especially, as Grove (1995) argues, on oceanic islands, but the responses were slow in coming. Before the 1760s, the effects of colonial economic globalization were addressed on a piecemeal basis in order to protect local food, fuel, timber supplies, and what were already recognized as rare island species. However, in the mid-1760s responses to deforestation in particular suddenly changed. This was due to the rapid spread of a theory, first enunciated in France by Pierre Poivre, that linked deforestation to rainfall and regional climate change. Poivre, who had observed the effects of extensive deforestation in the East Indies and on Mauritius, explored the specific climatic impact of tropical forest clearance (Grove 1995).

The wider implications of his remarkable theory were quickly taken up by the British, who had gained control of several heavily forested islands in the east Caribbean under the terms of the Peace of Paris in 1763. These were Tobago, Grenada, St. Vincent, and Dominica, which were all highly valuable islands for plantation crops. A fear now grew, however, that unrestricted deforestation might lead to economically damaging rainfall reduction. As a result, an ordinance was passed in 1764 designating the mountainous part of Tobago a protected forest that was "reserved in wood for rains" (Grove 1995: 271). This protected forest still exists within its original boundaries. The legislation that created it marked a critical watershed in the history of environmental concern, since it applied a universal scientific theory about Earth atmosphere processes (since shown to be substantially correct) to a local environment. It was thus the forerunner to all subsequent national and international attempts to control rainfall and climate change. The 1764 Tobago ordinance specifically recognized the need to restrict profits to sustain an environment in the long term. Moreover, the mechanisms used to set up forest reserves under the ordinance justified the alienation (in the face of much local litigation) of large tracts of private plantation land to colonial state control and implied a permanent role for the state, rather than the individual, in conserving forests and the atmosphere. In 1765, identical ordinances were applied to Barbados and Dominica (Grove et al. 1998).

Environmental transformations in the nineteenth century devolved from India in terms of controlling world forests (see Tucker and Richards 1983) but also set the scene for guano and other fertilizer production, fossil fuel production and oil production, all in turn boosting agricultural production and enabling further transformations at the centre and the periphery and further population growth. In the ensuing century, forest reserve legislation responding to fears of deforestation-induced climate change slowly began to spread around the world, especially throughout the French, British, and Dutch Empires. It was in the French colony of Mauritius, however, that the most far-sighted and comprehensive environmental measures were adopted, as they were passed soon after the arrival of Poivre as administrator of the island in 1767. Unlike the Lords Commissioners for Trade and Plantations, which had instituted the forest-reserve measures in the Caribbean, Poivre was as ideological as he was economic or scientific in his environmentalism. He was specifically hostile to mercantile capitalism, absentee "rentiers," and what he called "bullion accumulation" (Grove 1995). Instead, he favored the deliberate encouragement of the latest agricultural science to increase food production, boost the well-being of the rural economy, and preserve the landscape aesthetics of what he termed the "Eden" of Mauritius.

The life of Poivre is useful in drawing attention to the profound connections between early environmentalism and the development of social reform and popular, even revolutionary, movements. This connection grew steadily stronger during the next two centuries, allowing the proliferation of both state and nongovernmental environmental discourses. Poivre also radically enlarged the philosophical and practical ambit of the notion of the state and its possibilities. Bernardin de Saint-Pierre, Colonial Engineer of Mauritius and a disciple of Jean-Jacques Rousseau corroborated Poivre's stance in his elaborate environmental discourse on Mauritius. Saint-Pierre's writings stand out as the first fully developed and evidenced critiques of the European impact on the world environment. Saint-Pierre made perhaps one of the most profound remarks concerning what he saw as the innate destructiveness of Europeans: "To contemplate the progress of a rising colony is a spectacle worthy of [a] philosopher, for it is there that the culture of man forms a striking contrast with that of nature" (cited in Grove 1995: 251). In the eyes of Saint-Pierre, man had become destructive and had allowed commerce to run its destructive path because he had himself become denatured. As man had come to manipulate nature cruelly, he increasingly acted "the part of the tyrant of Sicily, who fitted the unhappy traveller to his bed of iron; he violently stretched, to the length of the bed, the limbs of those who were shorter and cut the limbs of those who were longer." It is thus, Saint Pierre added, "that we apply all the operations of nature to our pitiful methods, in order to reduce the whole to our common standard" (cited in Grove 1995: 251).

The environmentalist initiatives of Poivre and Saint-Pierre on Mauritius were exceptional. They legislated and theorized about deforestation, climate control, pollution control, fishery conservation and tree planting. Their initiatives were apparently imitated in the Caribbean, where the Kings Hill Forest Act was passed in 1791 on St. Vincent, again setting up a "rain reserve" in an upland part of the island. Similar measures were passed on the Atlantic island of St. Helena in 1794. Both islands had been affected by the El-Niño-induced drought of 1791, a drought recognized as global in its impact by East India Company scientists as early as 1816. Significantly, the inspiration for the forest-protection legislation on St. Vincent was a Scotsman, Alexander Anderson, the first of a long series of Scottish scientists who formulated much of the earlier environmentalism of the British Empire during the period 1780–1900 and influenced that of the US Empire in the Americas. This Scottish influence dominated at least until the heyday of Rooseveltian progressivist conservation in the United States, and the pioneer of that influence was Anderson, an Edinburgh-trained surgeon who became the curator of the St. Vincent botanic garden in the eighteenth century.

Further fears about the global consequences of deforestation and climate change resulted from research on vegetation–atmosphere relations by Alexander von Humboldt, a German geographer, and Joseph Boussingault, a French chemist. Both men advocated large-scale state intervention in forest protection. Once again, these ideas were taken up in Mauritius, this time by botanist Louis Bouton, whose advocacy led to strengthened forest protection on the island and hence to the survival of the remarkable endemic birds of the island, apart from the Dodo (Grove and Damodaran 2006). By the late 1830s and the 1840s, the reiteration of climatic environmentalism by Humboldt and Boussingault was being acted upon by environmentally minded scientists and officials working not just on the islands but on the large land masses of India, Southeast Asia, Southern Africa, and Australia, where the demands of European colonial empires were now bringing about deforestation at an unprecedented speed (Williams 2003).

The emergence of environmentalism in India was especially important simply because of the huge area of land involved. Since the 1780s, there had been sporadic attempts by the East India Company to annex private or community forests for state use to ensure a sustainable supply for ship-building purposes, both for the Royal Navy and for the Company Marine. Most of these schemes had failed, however, and the remnants were abandoned by Governor Thomas Munro, an advocate of indirect rule and the restoration of Indigenous land ownership. After 1823, therefore, deforestation proceeded at a prodigious rate, such that between 1823 and 1850 up to 50 per cent of India's forested area may have disappeared, a rate only exceeded by deforestation in India after 1947 (Damodaran 1997). A succession of famines in the 1830s, as well as the writings of Humboldt and Boussingault, forced a change in this laissez-faire approach. Moreover, there was a growing interest in the forest reserves already set up by Indigenous rulers in Sind and the Bombay Presidency, which appeared to provide a suitable model for a colonial forest-reserve system. In 1843, a Sind Forest Department was established, which was soon followed by a Bombay Forest Department in 1847 and a Madras Forest Department in 1855 (Grove 1995).

Three Scottish surgeons in the company service—Alexander Gibson, Hugh Cleghorn, and Edward Balfourconducted a vigorous propaganda campaign to further extend the early forest-protection efforts to cover the whole of the Indian subcontinent. Their arguments were summed up in a paper for a meeting of the British Association for the Advancement of Science and published in 1851; it was entitled "Report of the Committee Appointed by the British Association to Consider the Probable Effects in an Economical and Physical Point of View of the Destruction of Tropical Forests" (Grove and Damodaran 2006). To the present-day reader, this report seems remarkably modern in its concerns. Deforestation, its authors warned, threatened damaging reductions in rainfall and increases in regional temperature. Potentially important drugs might be lost as little-known trees and plants were cut down, while fuelwood shortages would become serious. Famines would become more frequent. The loss of perennial streams would encourage diseases thriving in the stagnant watercourses left after deforestation.

This report came to the receptive ears of the Governor-General of India, Lord Dalhousie, who used it as the scientific justification to set up an India-wide Forest Department in 1864 (Barton 2002). Dalhousie, a Scotsman, was, like so many of his Scottish compatriots, imbued with the desirability of the tree-planting fashion that had swept his native land. The foundation of a Forest Department, motivated by his concern both for maintaining a sustainable timber supply and for curbing drought, was his crowning achievement and one of the most durable achievements of British rule in India. By 1880, the Forest Department controlled one-fifth of the land area of India. The exclusionary forest-reserve system, which often shut out hunters and farmers from their traditional resources, has caused chronic social conflict since its foundation. Without it, however, no significant forest cover would have survived in South Asia, which, like most of the tropics, was subject in the nineteenth century to an entirely new kind of globalizing economic penetration which traditional common property arrangements could not have survived. In China and Thailand, where no significant reserve systems developed, the forests have now largely disappeared. By contrast, in India the forested area was stabilized for nearly a century after about 1870.

Much of the responses outlined above can be analyzed as a form of green imperialism with little input from local communities; they also underline the fact that ideas of environmental crisis are not new and can be dated to the late seventeenth century. However, the responses to these early modern environmental crises were piecemeal and failed to challenge the capitalist roots of our current environmental crisis. Current climate change debates and big science in a similar fashion also risk being top-down and steering clear of radical solutions such as changes to capitalism and to the ever-increasing demand for raw materials and resources. Lessons from India's tribal heartland need to be incorporated into how we conceptualize anthropogenic change, and it is to this issue that I now turn.

Forests Destruction, Conservation, and Communities in Eastern India

With progressive liberalization and trade links with world economies, India is being transformed at an unprecedented rate. Landscapes and livelihoods are being significantly impacted on by this pace of change. One such area is the predominantly tribal area of Eastern India (Orissa, Chhattisgarh and Jharkhand), which is undergoing extensive mining development including by companies such as Vedanta and Tata that are listed on the London Stock Exchange. The most mineral-rich areas of Central India are also the areas of greatest forest diversity and tribal population (Damodaran and Padel 2018). As tribal communities are displaced, their lands and resources taken over for mining and metal factories, their lives are changed at every level. From a livelihood based largely on self-sufficient subsistence agriculture, supplemented by hunting and gathering in the forest, they are forced to become industrial laborers living in resettlement colonies in swiftly industrialising areas, where poverty takes a radically different form from anything they had ever known. There have been few studies to date of the effects of mining on livelihoods and environment in South Asia. By examining the relationship between social structure, environment and cultural history and the impact of mining in these localities, this research poses important analytical as well as empirical questions concerning the effects of industrial developments and globalization in the *longue durée* on "displaced livelihoods" in the context of the Anthropocene.

Travelling through the mining town of Noamundi, in Jharkhand, in 2013, I was struck by the presence of red oxide dust everywhere: on my clothes, on houses, on people, on the once brightly painted advertisement for Tatas noting that it was a company which valued its corporate social responsibility (Priyadarshini 2008). Noamundi has a long history as one of the centres of the mining industry set up by Tata in Eastern India in the early part of the twentieth century. Above the local town with its withered trees and red oxide dust was the officer's colony with its bungalows and its tennis courts and its magnificent views of the reserved Sal forest of Saranda, whose proportions set in the colonial period were rapidly eroding under pressure of development following the cleansing of the Maoists in the region (Damodaran 2017). This is a forest which is extremely important in bio-diversity and cultural terms both for local communities and for the Indian imaginary. Not only is it part of the cultural world of *adivasi* communities but it has fuelled the imagination of Bengali intellectuals such as Bibhutibhushan Bandyopadhyaya, who wrote *Aranyak* (1976) around it (Bandyopadhyaya 2002). In 2005, Saranda Forest, which is part of the core elephant reserve, was in a bid for world heritage status that was organized by the Ministry of Forests. The World Wildlife Fund was also very interested in its status. It later became a Maoist stronghold in the long fight between disaffected intellectuals, local communities and the state over rights to resources and the growing poverty and inequality in the region. Following the purging of Maoists in the region, the Saranda development plan sought to hasten the carving up of the reserve into mining leases. Saranda is part of the story of Eastern India's rapid transformation in the post-colonial period and the resistance to this by both Indigenous communities and political groups such as the Maoists.

This study on the *longue durée* of landscape change caused by economic globalization documents the empirical aspects of deforestation, state forest control, dam-building, mining and land tenure/ownership changes and conservation and their impact on tribal communities. It looks at globalization as it has affected and continues to affect tribal communities in East India in three phases, 1800–1947, 1945–1991 and 1991 to date. The documentation on the environmental history of Eastern India for the earlier phases of globalization has been obtained from archival sources as a basis for understanding the much more rapid changes which have taken place since 1945 and especially since foreign direct investment began in India in 1991 (Damodaran 2017). It is important to note that a detailed analysis of the environmental impact of unprecedented capital inflows into my research region in the contemporary period has global relevance.

More recently since 1991, as the last pretence of tribal protection has been given up, Eastern India has become subject to new kinds of internal and external colonization, which have been far more traumatic than its pre-1947 colonization.

The Locality in the Anthropocene

India and China are becoming the "Asian Drivers" of the globalizing world economy. The consequences for people and environments in both countries are profound as a scramble for resources takes place to feed the demands of industrialization in both; while environmental governance has failed to keep pace with the speed of change (Kaplinsky 2005).

The states of Orissa and Jharkhand possess some the world's best deposits of the bauxite used in aluminium production, a process which requires the construction of dams to provide electricity. Over 20 mountain ranges in this region are now planned for exploitation by global mining companies. Many of these ranges have complex sacred meanings attached to them by Indigenous communities or *adivasis* and are locations of IUCN-defined biodiversity 'hotspots'.[1] The inland areas of Jharkhand and Western Orissa can be considered a colony of the coast. Thus, the Hirakud Dam has its submergence zone in the west, but the putative benefits from flood control and irrigation go to the coastal plains. Since 1945 up to 5 million people (mostly *adivasis*, which make up 25 percent of the region's population) have been forcibly removed and "resettled" to enable dam-building and mining/industrial development, a forced migration that has rarely been equalled globally in the twentieth century. These movements were facilitated by the failure of the Orissa authorities to redistribute land or rights over 60 percent of the state, which had originally been incorporated from the Princely states in 1947. Research on the history of the ecological distribution conflicts in these areas, as analogues to much broader conflicts, is essential in gaining insight into the forces of globalization in rural India as a whole. As the forces of globalization have accelerated since 1800, vigorous contestations for space and resources have taken place between *adivasis*, peasants, the state and mining and other commercial companies.

Since 1945 and much more since 1990, these contests have involved an increasing level of state and corporate violence against *adivasis* and other peasants, and have been coupled with a rise in violent and non-violent resistance as well as armed "Naxalite" insurgency throughout the central interior of India (Lok Sabha Report 2005). The Kalinganagar massacre on January 1, 2006, in the Jajpur district in neighboring Orissa, which also has a large tribal population, appears as a turning point in the breakdown of governance. I question whether the central government is motivated at all to effectively control the activities of multinational companies and their agents, who are intent on alienating Indigenous land.

Currently, over 100 memoranda of understanding (MoUs) have been signed in the region between mining multinationals and state governments to promote large-scale open-cast mining of bauxite and iron ore as well as other less important minerals, and to build processing plants and port export facilities (Kumar 2009). Most of these MoUs ignore Schedule 5 of the Indian Constitution prohibiting the alienation of tribal lands and a 1996 act strengthening its provisions, and most of them also ignore the legal protection of reserved forests under the Forest Acts. Despite these laws, several companies commenced mining operations in both states, including two British mining companies which proposed open-cast mines on sacred mountain sites at Niyamgiri (the Hill of Law) and Gandhamardhan Mountain. The latter is the most sacred site in Orissa, and is closely associated with the gods Ram and Hanuman. Thus, global mining interests now confront core values in the mainstream Hindu and tribal religions (Tatpati et al. 2018). Despite the success of the Gond communities in challenging the mining of Niyamgiri, the confidence of the multi-nationals in such plans indicates the seriousness of the plight in which *adivasi* communities and environmental/human rights campaigners find themselves. Under new government plans with the Modi regime, these pressures on the environment can only get more severe. In documenting the history of this confrontation in environmental, landscape and human terms, my research plays a critical role.

Most studies of the region's landscape concentrate on the arable parts of the landscape, while the non-arable forests, mountains and marshes were neglected. Despite the colonial discourse of tribal protection and the Nehruvian legacy of constitutional guarantees for scheduled tribes, the decades after independence saw large-scale tribal land alienation. There is no study which maps the history of land alienation and environmental destruction in the 1950s and 1960s in Eastern India. The forests of the uplands of Chotanagpur and Northern Orissa were contested between local communities, the Ministry of Forests, landlords, mining companies, and politicians during the nineteenth and twentieth centuries.

British interests in the forests of India had a long history. An escalating demand of Indian teak was evident in the increase of tonnage of British merchant ships from 1,278,000 tons in 1778 to 4,937,000 tons in 1860. In view of the growing deficiency of oak in England, a timber monopoly was established in Malabar in 1806. From 1793 to 1815, the period of the Napoleonic Wars, England was saved by Malabar teak. The process for the acquisition of new woods was intensified by the creation of the railways, which led to an extensive search for these resources throughout the sub-continent followed by an attempt to conserve them in a haphazard fashion. The first 30 years of forest conservancy in Bengal, for example, were mainly guided by a desire for a constant timber supply and profits from the forests (Damodaran 1997).

British attempts to dominate the forest, mineral and water sources in the interest of production and profit were to have long-term ecological consequences. They reshaped the landscapes of Eastern India from the 1800s, especially in the region of Chotanagpur with important consequences for the local inhabitants. It can be argued that the landscape was redefined in these terms through the masculine discourse of scientific forestry. In the process, it threatened the small and relatively autonomous economies of the tribal communities with catastrophic effects on the subsistence ethics of all groups, and particularly those of women. The forests were in the early part of the nineteenth century primarily regarded as a resource. The policy of the colonial rulers was to extend the cultivation at the expense of forest tracts and to exterminate all wild and dangerous game. In Ranchi, the district gazetteer recorded the unchecked destruction of forests in the district in the latter half of the nineteenth century. The major cause of the destruction of jungles in most districts was the sale or lease of the forests to contractors for the supply of railway sleepers. Large areas of the forest were destroyed to supply the timber necessary for railways (Damodaran 1997).

The forests of Singhbhum district for example were subject to heavy fellings, and it was reported in 1898 that the 'selection fellings for the supply of broad gauge sleepers from trees over 6½ feet in girth amounted to over 20,921 trees at the average of 10.4 sleepers per tree' (Annual Progress Report of the Forest Administration in Bengal 1898). These fellings were reported to have greatly impoverished the forests. Singhbhum was the district with the largest proportion of forest in the 1880s and with over 80 percent of its population still 'tribal'. This was now threatened. An attempt was launched in the 1880s to acquire all private forests with a view to exploiting timber in even the remote areas. The private forests of Dumka sub-division were thus acquired as were those of Parasnath and Gobindpur, which were seen as particularly valuable because they lay between the railway line and the grand trunk road (Schlich, W, Report 1885).

To combat the extensive deforestation, the provincial government and the Ministry of Forests embarked on a wholesale programme of forest reservation. However, the implementation of this forest policy in India was extraordinarily slow and for most of the eighteenth century the overall policy of the government was to extend agriculture and destroy forests. It was only by the early nineteenth century that the increased demands on forests were viewed gravely and attempts were made to conserve them. Even then, the seriousness of the situation was not recognized, and though a forest conservancy system was established in some states it was not till much later that it rose above the level of revenue administration. By the late nineteenth century, the discipline of scientific forestry had led to a systematic programme of forest management derived from German and French continental systems, where the principles of minimum diversity, sustained yield and the balance sheet held sway. As Ravi Rajan notes: "by the end of the nineteenth century this utilitarian sentiment became a development ideology in its own right" (Rajan 1998: 351).

The discourse of scientific forestry was no more accepting of Indigenous ideas of conservation than the earlier ethics of exploitation had been. It can be seen as a mechanistic science where nature, the human body and animals could be described, repaired and controlled, as could the parts of a machine, by a separate human mind acting according to rational laws. In Carolyn Merchant's words, the scientific world view, within which debates on scientific forestry were embedded, viewed the "world as dead and inert, manipulable from outside and exploitable for profits ... living animate nature died ... increasingly capital and the market assumed the organic attributes of growth ... nature, women and wage labourers as human resources for the modern world system" (Merchant 1980: 44). The results of these changes had a far-reaching impact on the lives of the local people. In many places, the landlord and the state battled with each other to secure large areas of jungle land, extinguishing the traditional common rights of the people. In Ranchi district, several of the landlords looked upon the jungles as a providential asset to be exploited for the payment of debt. They were prevented from fully exploiting this asset only by the difficulties of communication, so that the more remote jungles survived. However, most of the latter were taken over by the state for forest reserves under the rigorous policy of forest conservation. By the 1890s, the total area of reserved forest in Chotanagpur was 5,839 square miles. Of these, over 5,431 square miles were closed for grazing purposes. In 1894, all state lands within the five districts of Chotanagpur division were declared to be protected forests, further controlling hitherto unclassed forests. Where patches of jungle survived, the grip of both the state and the landlords could be felt, as the spread of the railway system further aided the process of destruction. The opening up of the Purulia–Ranchi railway and the main Bengal–Nagpur line, fringing the southern portion of Gumla sub-division, led eventually to the total deforestation of this hitherto untouched region (Damodaran 1995). The difficulties forest reserves posed for local people is indicated in Valentine Ball's (1995) travel journal in 1870, where he records the great distress that people living in Hazaribagh and Palamau suffered as a result of forest reserves.

The most perceivable impact of deforestation by the latter half of the nineteenth century was increasing food scarcity and the growing language of hunger. Famine studies to date have rarely included a study of environmental changes, though an examination of these changes is critical to understanding the capacity of certain communities to withstand drought and scarcity. A moral economy perspective attempts to ground famine theories more firmly in social and ecological contexts. Here, I consider the emphasis that the experiences of Indigenous people and their subsistence ethic bring to an understanding of famine and scarcity. In this argument, famine is seen as a process, a gradual development of impoverishment (Damodaran 1995).

The preceding decades of deforestation, demographic pressure and land use change had gradually pushed a relatively stable economy into crisis. Many of these changes had occurred as a result of exploitation by a new landlord class. The 1897 state famine report recorded that one-sixth of the population in Chotanagpur division had been reduced to serf status. The traditional economy could no longer cope with periods of scarcity. The careful husbandry and social systems designed to counter periodic subsistence crises had gradually been destroyed. As Detlef Schwerin notes, the land system which had been fundamentally changed through Hindu immigration from a joint ownership to a hierarchically structured individual ownership system hampered the unfolding of the economic creativity of the people. The peasantry was prevented from improving their holdings through the rapaciousness of petty landlords. In the northern districts of Chotanagpur, these processes had taken place much earlier than in the southern districts (cited in Damodaran 1995: 141).

Resistance as Resilience

Communities reacted to these changes with resistance. The history and dynamics of patterns of resistance and violence over the control and allocation of resources by the local communities of Chotanagpur and the predicament of ethnic identity and culture in the face of unrestrained globalising forces need to be explored more systematically. In attempting to understand the nature of resistance, one needs to examine the resilience and vitality of tribal culture in the face of exploitation and repression and in the face of people's crushing lack of access to their own resources. The response of communities was not slow in coming, and by the mid-nineteenth century there were a series of tribal rebellions.

Beginning with the unrest in Tamar in 1816 and the Munda rebellion in 1832, disaffection continued through the mutiny of 1857, and the last decades of the nineteenth century saw unrest in almost every district of Chotanagpur. W.J. Allen, who made an extensive tour of Singhbhum in 1861, noted "that the love of freedom was the general characteristic of the wild and hilly country of the savage Kols and Santhals" (cited in Purshottam Kumar, 1991: 87). The Birsa Munda uprising in the 1890s was the culmination of this period of rebellion. Birsa's rebellion originated against the forest laws of the British. British forest reservation laws had long proved irksome to the Mundas, and in the context of the degradation of their forest environment, exploitation by Hindu moneylenders, and a modernizing colonial state, they rose in protest. It can be argued that the despoliation of the forested landscape and the transformation of the local people's relationship with their environment in Chotanagpur in the nineteenth century was a powerful memory that was later revived in periods of cultural resistance. It was also through the mapping of the notion of the *diku* or "outsider" in these resistance movements that a new sense of community was renegotiated and a radical consciousness began to emerge. As Ranajit Guha notes: "for the Santhals under the Subah brothers and the Mundas under Birsa all stated their objectives to be power in one form or another" (Guha 1983: 10). The resistance movements of the latter half of the nineteenth century were critical to this growing consciousness. The effects of land alienation following from changes in colonial governance in the 1820s had an immediate and most visible effect in Tamar, where the exploitation of Hindu moneylenders, whose activities were bolstered by colonial courts, resulted in widespread protests which continued unabated until the 1830s. In the beginning, the communities sought to redress their grievances through colonial courts, wending their way long distances to Shergati in order to resolve land disputes which resulted from the seizure of Munda lands by Pathan moneylenders. When the courts failed to redress their grievances, protest seemed to be their only course of action.

The Birsa Munda rebellion against the forest regulations of the British was documented by the missionary John Baptist Hoffmann, a German Jesuit who became a strident critic of colonial policy towards tribes. Hoffmann was perturbed by the anger and the hostility of the Mundas, who were preparing under their leader Birsa to launch their attack on the government and the local missionaries by secretly plotting their rebellion in the hills and the forests. Birsa persuaded his followers that God had given him the task of liberating the tribals and instituting a new religion. Hoffman noted that the religious gatherings

> gave a harmless appearance to the numerous sardar meetings in which the intended rising was settled without arousing any serious suspicions in either government or missions. It facilitated the gathering of about 6,000 armed men

around Birsa in Chalkad in August 1895, after which it was announced that he would call fire from heaven to destroy the aliens … then a few young men who were still wavering between Christianity and the new religion came in and begged Hoffmann to leave immediately for Ranchi, because the very next morning the armed men with Birsa would start to massacre all the foreigners, adding that I as the nearest European to Chalkad was already designated as the first victim. Since I refused to move, they gave me up for lost and went away.

(cited in Damodaran 2006: 181)

The Birsa movement forced missionaries such as Hoffmann to recognize the strength of local grievances and the inability of the colonial state to deal with the grievances of the communities. For tribal converts like Birsa, then, the reimagination of community through the use of invented and traditional symbols became a more potent weapon than loyalist Christian discourse (KS Singh 1966).

In the twenty-first century, the protests flared up, but this time they are against mining. Felix Padel and Samarendra Das (2010) mention the current plans to expand bauxite mining and aluminium manufacture on the one hand and iron-mining and steel plant production on the other. There are also new plans to construct new mega-dams for supplying both industries. A British registered company named Vedanta Resources has an important profile in the industry. The company's annual report for 2005 noted that it engages in sustainable mining (Vedanta Resources, 2005). The reality is that mining is a highly destructive industry. As Roger Moody notes: "natural resource extraction cannot be reconciled with long term sustainability. For industry spokespeople to claim (as they often do) that there is such a thing as 'sustainable mining' is a transparent oxymoron" (Moody 2007). The company aimed to mine Niyamgiri Hill in Kalahandi district, which is a wildlife sanctuary and an elephant reserve. As a recent Centre for Science and Environment report notes, 75 percent of the area is covered with thick forests, with more than 300 species of vegetation including 50 species of medicinal plants. It has a number of perennial hill streams which serve the irrigation and drinking water needs of tribals living in the foothills. The hill is also considered sacred by the Dongria Kond tribals, who call it "Niyam Penu". Unfortunately, the area also has rich deposits of bauxite, almost 195 megatons of it. Another Indian company, Tata, also attempted to construct a new steel plant in the region. When the tribals protested, several of them were killed in police firings in Kalinganagar. This event highlighted the numerous iron ore and steel projects in the state and a highly controversial deal with the Korean company Posco to mine Orissa's iron and build a steel plant near Paradeep. As the Chief Minister of Orissa noted: "no one and I mean no one will be allowed to stand in the way of Orissa's development and the people's progress" (Damodaran 2017). However, Deogi Tina did stand in the way. She was a 35-year-old Ho woman who came from a village in Champa Koila, and her religion was the Ho religion, in which the hills and the mountains were the residing place of the deities in which she believed and which fed the streams and hills which were also sacred to her and her kinsfolk. If we are to examine how she died—she was shot from about five feet away—we can ascertain that Deogi Tina was executed in cold blood by the Orissa police in the presence of the District Magistrate, the District Collector, and the Superintendent of Police. Deogi was in fact assassinated by a state death squad. Subsequently, the bodies of twelve others killed in the protest were mutilated. So she joins the ranks of Ken Siro-Wiwa and other activists in Nigeria, Ecuador, Columbia, Peru and the Amazon who stand in the way of the extraction of minerals from their homelands by multinationals, which depend on the state as their middleman who will legitimate or force the allotment of lands they need to extract minerals.

The debate on the Anthropocene is wrapped up with debates on resilience and with security and the ways in which resilience is mediated on and how the power of community and place shapes resilience. We need to understand everyday forms of resilience through human agency, collective action and knowledge and to see resistance and resilience not as antonyms but as synonyms. Resistance needs to be seen as a means by which people change social processes and build alternatives. The challenging of neo-liberal perspectives and ways in which Resilience is recuperated through Resistance. We need to on bring the politics back to resilience thinking and to focus on agency and power in resilience science. We need to provide examples of good resilience and bad resilience that range from resilience in the aftermath of the New Orleans disaster, fishing communities in the Philippines to Mexico and Bolivia where communities less engaged with the market were more resilient to climate vulnerability. This paper has attempted to do this in the context of Eastern India.

Where do these debates lead us, and can one recuperate some emancipatory imaginings? One can launch bottom-up initiatives through active political engagement with local communities, and, as Mike Hulme (2011) tells us, by understanding alternative ways of being and thinking in these worlds. This does not exclude top-down solutions by the state and scientific elites to engage with ideas of regulation and stewardship in democratic and ethical ways and to seek for extraction what one can call our Montreal moment, where we successfully dealt with protecting the ozone layer through collaboration and negotiation. But it does recognize the limitations of a top-down approach.

Why does the history of colonial environmentalism, based mainly on the fear of climate change, matter to us today? The answer is quite simply that the Indian forest-reserve system, developed under British rule, became the basis for state land-use reservation on an enormous scale throughout the world. It is true that a variety of environmental movements developed after about 1870 in several temperate countries of the Global North within highly urbanized and prosperous societies. Here, the preservation of wilderness, in the form of national parks or game reserves, by the state or the preservation of erstwhile common land appealed to many mainly on the basis of aesthetics and species protection. Such reasons for this protection were sometimes transferred to the tropics, especially in post-war Africa, but the forest-reserve systems that came about there never acquired the global significance of those developed in India.

It is of course ironic that an exploitative colonial economic system should have provided the conditions for the emergence of an environmental awareness that did not develop in Europe or North America until the late nineteenth century. Of course, the methods of state-directed forest and water conservation were inadequate and easily distorted for commercial and elitist purposes. However, they provided a conservation infrastructure that is now, in many countries, available for sophisticated new kinds of community management.

Entirely novel kinds of globalizing ecological pressures are present today. The main threat to species-rich tropical forests is now posed by multinational timber companies based in China, Malaysia, Hong Kong, and Japan, for whom notions of sustainability have no meaning. These concerns can now buy access to the relatively untouched forests in Mozambique, the Amazon, and Melanesia and have been virtually impossible for poorer states to restrain. As a result, in areas largely outside forest reserves, millions of hectares of tropical forest are now disappearing annually at a time when the temperate forest areas of the rich countries, measured globally, are now increasing. The historical lesson to be learned, perhaps, is that for private capital and international trade interests global environmental well-being is generally an accounting irrelevance. Only local communities, with their histories of resistance, sometimes in effective collaboration with national governments and global environmental networks and institutions, can serve as guarantors of our fragile "island" Earth's well-being for future generations. The kind of climatic anxiety encapsulated in the intentions of the Kyoto Protocol (1997) and, more recently, of the Paris Agreement (2020) on global warming have been with us for centuries. Eco-apocalyptic arguments are not new. History shows that state intervention alone cannot control capital or corporate interests or deal with the deadly risks of climate change, forest loss, and species extinction. Environmentalism at the level of the locality and environmental resistance by communities against the inroads of capital can lead us into new emancipatory futures in the face of eco-apocalyptic arguments, and signal new possibilities.

Note

1 The IUCN is the International Union for Conservation of Nature (https://www.iucn.org/).

References

Annual Progress Report of the Forest Administration in Bengal (1898). Calcutta.
Ball, Valentine (1985) *Tribal and Peasant Life in India in the Nineteenth Century*. London: Thomas de la Rue & Co.
Bandyopadhyaya, Bibhutibhushan (2002) *Aranyak*, trans. Rimli Bhattacharya. Calcutta: Seagull Books.
Barton, Greg (2002) *Empire Forestry and the Origins of Environmentalism*. Cambridge: Cambridge University Press.
Brockway, Lucile H. (1979) *Science and Colonial Expansion: The Role of the British Royal Botanic Gardens*. New York: Academic Press.

Crutzen, Paul J. and Stoermer, Eugene F. (2000) The Anthropocene, *Global Change News-letter*, 41: 17–18. https://doi.org/10.1007/978-3-030-82202-6_2.

Damodaran, Vinita (1995) Famine in a forest tract: Ecological change and the causes of the 1897 famine in Chotanagpur, Northern India, *Environment and History*, 1(2): 129–158. https://www.jstor.org/stable/20722973.

Damodaran, Vinita (1997) Environment, ethnicity and history in Chotanagpur, India, 1850–1970, *Environment and History*, 3(3): 273–298. https://www.jstor.org/stable/20723049.

Damodaran, Vinita (2017) The locality in the Anthropocene: Perspectives on the environmental history of Eastern India'. In: Elliott, Alexander, Cullis, James, and Damodaran, Vinita (eds.), *Climate Change and the Humanities: Historical, Philosophical and Interdisciplinary Approaches to the Contemporary Environmental Crisis*. London: Palgrave Macmillan, pp. 93–116.

Damodaran, Vinita (2006) Colonial constructions of tribe in India: The case of Chotanagpur. In Lévai, Csaba(ed.), *Europe and the World in European Historiography*. Pisa: Pisa University Press, pp. 161–193.

Damodaran, Vinita and Padel, Felix (2018) Investment-induced displacement in Central India: A study in extractive capitalism, *Comparative Studies of South Asia, Africa and the Middle East*, 38(2): 396–411. https://doi.org/10.1215/1089201x-6982156.

Grove, Richard H. (1995) *Green Imperialism: Colonial Expansion, Tropical Island Edens and the Origins of Environmentalism, 1600–1860*. Cambridge: Cambridge University Press.

Grove, Richard, Damodaran, Vinita, and Sangwan, Satpal (eds.) (1998) *Nature and the Orient: The Environmental History of South and Southeast Asia*. Delhi: Oxford India Paperbacks.

Grove, Richard and Damodaran, Vinita (2006) Imperialism, intellectual networks and environmental change: Origins and evolution of global environmental history, Lok Sabha Committee on Scheduled Tribes 1676–2000: Part 1, *Economic and Political Weekly*, 41(41): 4345–4354. https://www.jstor.org/stable/4418810.

Guha, Ranajit (1983) *Elementary Aspects of Peasant Insurgency in Colonial India*. Delhi: Oxford University Press.

Hulme, Mike (2011) Meet the humanities, *Nature Climate Change*, Volume 1, pp. 177–179.

Kaplinsky, Raphael (2005) *Globalization, Poverty and Inequality: Between a Rock and a Hard Place*. Cambridge: Polity Press.

Kumar, Navtan (2009) *Mineral-rich Jharkhand fables the world*. Economic Times, June 20. https://economictimes.indiatimes.com/industry/indl-goods/svs/metals-mining/mineral-rich-jharkhand-fables-the-world/articleshow/4800450.cms.

Luxemburg, Rosa (1913) *The Accumulation of Capital*. https://www.marxists.org/archive/luxemburg/1913/accumulation-capital/ch26.htm.

Lewis, Simon L. and Maslin, Mark A. (2015) Defining the Anthropocene, *Nature*, 519: 171–180. https://www.nature.com/articles/nature14258.

Lok Sabha Report. (2005), *Atrocities on scheduled caste and scheduled tribes and patterns of social crimes towards them*. Parliament of India.

Merchant, Carolyn (1980) *The Death of Nature: Women, Ecology, and the Scientific Revolution*. New York: Harper & Row.

Moody, Roger (2007) *Rocks and Hard Places: The Globalization of Mining*. London: Zed Books.

Ortega Breton, H. and Hammond, P. (2016) 'Eco-Apocalypse: Environmentalism, Political Alienation, and Therapeutic Agency'. In: Ritzenhoff, K. and Krewani, A. (eds.), *The Apocalypse in Film: Dystopias, Disasters, and Other Visions about the End of the World*. Lanham, MD: Rowman & Littlefield Publishers, pp. 105–116.

Padel Felix and Das, Samarendra (2010) *Out of this earth, Out of This Earth: East India Adivasis and the Aluminium Cartel*, Delhi, Orient Blackswan.

Planetary Health Alliance n.d. https://www.planetaryhealthalliance.org/planetary-health.

Priyadarshi, Nitish (2008) Impact of mining and industries in Jharkhand, *South Asia Citizens Web*, November 16. http://www.sacw.net/article302.html#.

Rajan, Ravi (1998) 'Imperial environmentalism or environmental imperialism? European forestry, colonial foresters, and the agendas of forest management in British India 1800–1900'. In: Grove, Richard, Damodaran, Vinita, and Sangwan, Satpal (eds.), *Nature and the Orient: The Environmental History of South and Southeast Asia*, Oxford: Oxford India Paperbacks, pp. 324–372.

Richards, John F. (2006) *The Unending Frontier: An Environmental History of the Early Modern World*. Berkeley: University of California Press.

Schlich, W, Report (1885) *Forest Administration Report*. Calcutta.

Singh, K. S. (1966) The Dust-Storm and the Hanging Mist: A Study of Birsa Munda and his movement in Chhotanagpur. Calcutta: KLM.

Sorlin, Sverker (2014) 'Historians of the future: Emerging historiographies of the Anthropocene', Paper presented at the World Congress of Environmental History, Minho.

Tatpati, Meenal, Kothari, Ashish, and Mishra, Rashi (2018). The Niyamgiri story: Challenging the idea of growth without limits? In Singh, Neera, Kulkarni, Seema, and Pathak Broome, Neema (eds.), *Ecologies of Hope and Transformation: Post-Development Alternatives from India*. Pune: Kalpavriksh, pp. 76–113.

Tucker, Richard and Richards, John (1983) *Global Deforestation and the Nineteenth-Century World Economy*. Durham, NC: Duke University Press.

Vedanta Resources (2005) Annual Report. https://www.annualreports.com/HostedData/AnnualReportArchive/v/LSE_VED_2005.pdf.

Williams, Michael (2003) *Deforesting the Earth: From Prehistory to Global Crisis*. Chicago: University of Chicago Press.

Zalasiewicz, Jan (2010) *The Planet in a Pebble: A Journey into Earth's Deep History*. Oxford: Oxford University Press.

INDEX